Proceedings of the
Sixteenth Lake Louise Winter Institute

FUNDAMENTAL INTERACTIONS

Proceedings of the
Sixteenth Lake Louise Winter Institute

FUNDAMENTAL INTERACTIONS

Lake Louise, Alberta, Canada; 18–24 February 2001

Editors

A Astbury
B A Campbell
F C Khanna
M G Vincter

World Scientific
New Jersey • London • Singapore • Hong Kong

Published by

World Scientific Publishing Co. Pte. Ltd.

P O Box 128, Farrer Road, Singapore 912805

USA office: Suite 1B, 1060 Main Street, River Edge, NJ 07661

UK office: 57 Shelton Street, Covent Garden, London WC2H 9HE

British Library Cataloguing-in-Publication Data
A catalogue record for this book is available from the British Library.

FUNDAMENTAL INTERACTIONS
Proceedings of the Sixteenth Lake Louise Winter Institute

ISBN 981-02-4912-8

Printed in Singapore by FuIsland Offset Printing

PREFACE

The sixteenth annual Lake Louise Winter Institute, entitled "Fundamental Interactions", was held from February 18-24, 2001 at the Chateau Lake Louise located in the scenic Canadian Rocky Mountains. Pedagogical and review lectures were presented by invited experts. As well, a topical workshop was held in conjunction with the Institute, with contributed presentations by some of the participants. The sessions were scheduled in the mornings and in the evenings leaving the afternoons free to enjoy the many aspects of this beautiful part of western Canada.

The 2001 Lake Louise Winter Institute was devoted to precision measurements within the context of the Standard Model of Particle Physics and the possibilities for extensions to this model. The recent completion of the LEP program at CERN and results from other laboratories around the world have revealed to us the validity of this model to unprecedented precision. New results from neutrino experiments in Japan and North America have brought unexpected insight into the fundamental properties of these leptons. In addition, the state-of-the art in atomic trapping was clearly detailed to all participants.

We wish to express our most sincere gratitude to Lee Grimard for her efforts, organizational skills, and incredible patience (even in the most frustrating circumstances!) in bringing this Winter Institute to fruition, from the very first email to the publication of these proceedings. She was well aided by David Shaw and Norm Buchanan in the logistics and transportation of the participants to the Institute. We are indebted to them.

We wish to thank the Physics Department at the University of Alberta for the infrastructure support essential to this conference. Finally, we wish to acknowledge the generous financial support of the University of Alberta (through the university conference fund and the Dean of Science), the Institute of Particle Physics, and TRIUMF.

organizing committee: A. Astbury
B.A. Campbell
F.C. Khanna
M.G. Vincter

CONTENTS

ELECTROWEAK PHYSICS AT LEP2

RICHARD J. HEMINGWAY
IPP and Ottawa-Carleton Institute for Physics, Department of Physics,
Carleton University, Ottawa, Ontario, Canada K1S 5B6
E-mail: ryh@physics.carleton.ca

On 2 November 2000 the LEP machine was finally closed after 12 years of glorious running. With the 4 operating detectors, ALEPH, DELPHI, L3, and OPAL, an enormous wealth of new data at the highest centre-of-mass energies has been recorded. These lectures will focus on aspects of electroweak physics within the energy span of LEP2, namely 130-209 GeV. All current data are in very good agreement with the electroweak Standard Model.

1 Introduction

Without a shadow of doubt e^+e^- annihilation has been the most productive environment for new physics during the last 20-30 years. The early experiments at Orsay, Novosibirsk, Frascati, and SLAC have clearly demonstrated the simplicity and cleanliness of e^+e^- annihilation to fermion-pair final states. SPEAR and DORIS initiated charm spectroscopy in 1974-5, followed by DORIS and CESR with beauty spectroscopy in 1978. Experiments at PEP, PETRA, CESR, and TRISTAN have produced hundreds, if not thousands, of new results covering aspects of pure QED, weak interactions, strong interactions, 2-photon physics, spectroscopy of heavy quarks, particle lifetimes and decays, and precision tests of the Standard Model. In 1989 both SLC and LEP started operation on a higher energy scale to investigate the detailed properties of the Z boson. LEP was able to raise the beam energy in 1996 to allow production of W^+W^- boson pairs and, eventually, ZZ boson pairs. By the time LEP ceased operation, the e^+e^- centre-of-mass energy had reached 209 GeV, a factor of 3 above the previous generation of e^+e^- accelerators.

1.1 Setting the Scene, LEP1 to LEP2

LEP was formally approved in 1981 and the 4 experiments (Aleph, Delphi, L3, and OPAL) selected in 1982. Construction of the accelerator began in 1983 and the very first e^+e^- collisions were observed on 13 August 1989. The period 1989-1995, the LEP1 era, provided each experiment with approximately 5 million Z events. With the addition of superconducting RF cavities the LEP energy was gradually increased above W^+W^- threshold and heralded the LEP2 era (1996-2000) during which each experiment recorded approximately

10,000 W^+W^- pairs. The major physics goals of the LEP2 program were (a) to continue the precision tests of the validity of the Standard Model, (b) to make precise measurements of the W^+W^- final state (cross-section, W branching ratios, triple-gauge couplings, and m_W), (c) to continue the direct search for the Standard Model Higgs boson, and (d) to explore the increased phase-space for evidence of new physics, eg. SUSY. Only items (a) and (b) will be discussed in these lectures.

1.2 Major References

I invite you to access the following sources for basic details, newer results, and (as ever) the very latest results.

- CERN Yellow Reports: 1. *Physics at LEP2* [1], CERN 96-01, Volumes 1,2 edited by G. Altarelli, T. Sjostrand, F. Zwirner, and 2. *Reports of the Working Groups on Precision Calculations for LEP2 Physics* [2], CERN 2000-009, edited by S. Jadach, G. Passarino, R. Pittau. This latter reference is very important for details of the Monte Carlos and EW-libraries.

- Newer Results: see CERN-EP-2001-021 *A combination of preliminary EW measurements and constraints on the Standard Model*, and the references therein.

- Very Latest Results: see the website http://www.cern.ch/LEPEWWG/- for a continuously updated situation, reports, talks, figures, averages, and much more. Here you will find the LEP combinations which ultimately are accessed by the Particle Data Group. These combinations require sophisticated procedures and tests and involve many people from the 4 experiments. This is excellent work and we all owe them many thanks.

1.3 Basic Processes, Event Rates, and 2-f vs 4-f Diagrams

Figure 1 shows the Standard Model expectation for the cross-sections of the major processes at LEP2 over the centre-of-mass energy range 100-250 GeV [1]. Compared to the LEP1 era with $\sqrt{s} = m_Z$ giving a peak hadronic cross-section of approx 30 nb, all LEP2 cross-sections are several orders-of-magnitude smaller, with the more interesting ones in the range 1-20 pb. For example, at $\sqrt{s} = 196$ GeV with a luminosity of 100pb^{-1}, we would expect (with efficiency and purity both 100%) ~2,000 $q\bar{q}$, ~ 300 $\mu^+\mu^-$ or $\tau^+\tau^-$, $\sim 2,000$ W^+W^-, ~ 700 $\gamma\gamma$ with $\cos\theta_\gamma < 0.9$, and ~ 150 ZZ events. With the same luminosity at the Z we would record 3 million $q\bar{q}$!

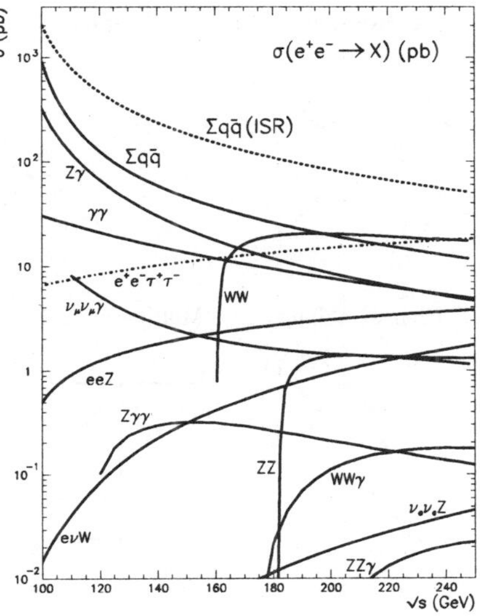

Figure 1: Some Standard Model cross-sections as a function of centre-of-mass energy.

One big difference between LEP1 and LEP2 physics is that the major processes of interest at LEP2 involve 4-fermion diagrams. This can be appreciated by looking at Figure 2 which shows the different classes of 4-fermion diagrams involving neutral currents [3]. The dominant 2-fermion diagram of LEP1 will, as we shall see later, contribute a background to the 4-fermion channels.

1.4 Luminosity, Centre-of-Mass Energy, and Beam Energy

As mentioned above, the LEP2 era started with the introduction of superconducting RF cavities in 1995. Each winter shutdown more and more RF capability was installed, culminating in 2000 when a record centre-of-mass energy of 209 GeV was achieved. Table 1 shows the integrated luminosity recorded by OPAL as a function of $\sqrt{s}$ for the period 1995-2000. These numbers are within a few % of those recorded by the other 3 experiments. In total, LEP delivered almost 1fb^{-1} over the 12 years of operation, with 189pb^{-1} recorded at LEP1 and 721pb^{-1} at LEP2. The luminosity-weighted centre-of-mass energy above W^+W^- threshold is $\sim$ 196 GeV. Each experiment had installed precision luminometers during the LEP1 era for the Z lineshape measurement and these provided typical luminosity uncertainities of $< 0.3\%$ at LEP2.

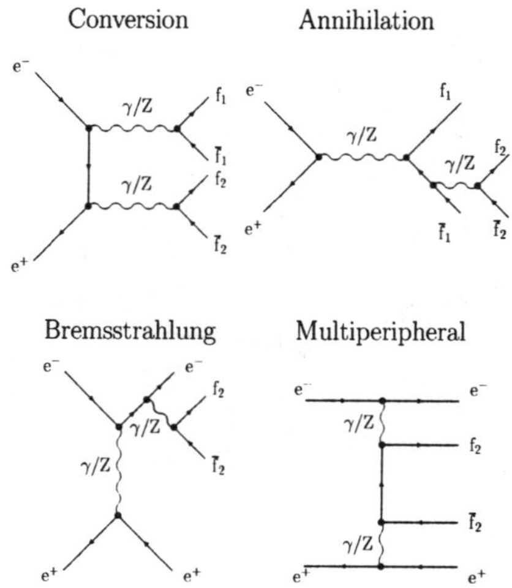

Figure 2: All possible classes of Neutral Current 4-fermion production diagrams.

Table 1: Summary of LEP2 integrated Luminosity during 1995-2000, as recorded by OPAL.

Year	1995	1996	1997	1998	1999	2000
$\sqrt{(s)}$ GeV	130 136	161 172	130 136 183	189	192 196 200 202	205 207
Luminosity pb^{-1}	3 3	10 10	3 3 57	187	30 78 79 38	80 140

Figure 3 shows the luminosity recorded by OPAL as a function of $\sqrt{s}$ over the entire LEP era. In the lower part of the figure can be seen the data recorded during LEP's final year. Also to be remarked is the 2-3pb^{-1} of Z calibration data that was recorded each year during the LEP2 period. This data was essential to maintain a well-calibrated detector for the lower-statistics LEP2 final states.

Table 2 provides a summary of the exact statistics of the major channels analysed by OPAL. You will notice that the ZZ channel (where we quote candidates-background) is not overwhelming in statistics due, primarily, to its small cross-section.

To be able to determine m_W with a precision of $\sim$ 25 MeV we need to control the LEP beam energy to 1 part in 10^4.[4] Unfortunately the LEP1 technique of resonant depolarisation runs out of steam near E_{beam}=65 GeV when

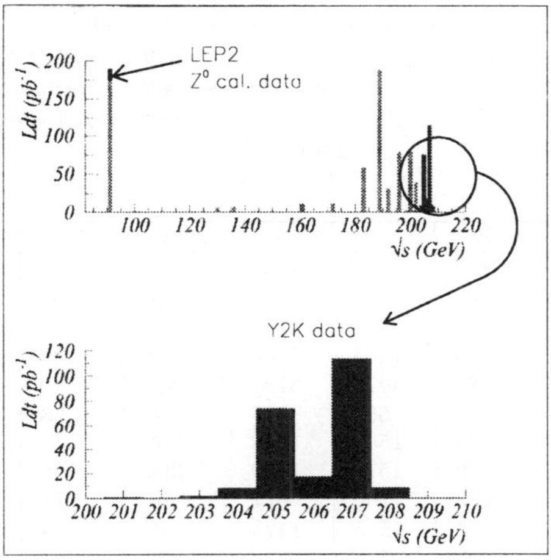

Figure 3: Total luminosity recorded by OPAL as a function of centre-of-mass energy.

the beam polarisation drops to zero.[5] However, it is hoped that a combination of the following techniques might just provide the needed precision.

- A combination of resonant depolarisation at E_{beam} < 65 GeV followed by magnetic extrapolation via NMR probes to higher energies has given $\Delta E_{beam} \sim 21$ MeV in 1999.
- A precision magnetic spectrometer in the LEP tunnel, with beam position monitors and accurately mapped Bfield, is hoped to provide $\Delta E_{beam} \sim$ 15 MeV.
- A machine physics relationship linking the synchrotron tune with applied RF voltage can provide an alternative measure of E_{beam}. This promises an uncertainty comparable to other methods.
- Use the 'radiative return' (see next section) events to calibrate E_{beam} with m_Z. This method, pioneered by Aleph [6], may be subject to large systematics and currently suffers from small statistics. Nevertheless, a combination of the 4 experiments may provide a competitive E_{beam} determination.

2 2-Fermion Physics

Although the Standard Model has been shown to provide an excellent description of 2-fermion physics at LEP1, it is important to continue these measure-

Table 2: Summary of actual event statistics, as recorded by OPAL. The 2-fermion final states require $\sqrt{s'}/\sqrt{s} > 0.85$, the Bhahba events require $|cos\theta| < 0.7$ and $\theta_{acol} < 10^0$ to enhance s-channel contribution, the $\gamma\gamma$ final state requires $cos\theta^* < 0.90$ (except for 183,189 GeV that have $cos\theta^* < 0.97$), and the ZZ final state is an estimate of the signal after background subtraction.

Year	Lumi pb^{-1}	$\sqrt{s}$ GeV	$q\bar{q}$	e^+e^-	$\mu^+\mu^-$	$\tau^+\tau^-$	$\gamma\gamma$	ZZ	W^+W^-
1995	5.3	133	380	210	56	24	66		
1996	10.0	161	370	285	45	38	102		28
1996	10.3	172	339	246	37	25	92		120
1997	57.3	183	1408	1260	174	123	620	4	877
1998	187.2	189	4072	3735	527	420	1740	50	3068
1999	29.6	192	662	577	81	63	215	19	431
1999	77.8	196	1555	1448	208	149	530	34	1277
1999	78.5	200	1519	1430	200	165	469	26	1127
1999	38.0	202	705	671	81	72	243	15	591
2000	80.0	205	1515	1423	217	154	486	31	1355
2000	140.0	207	2410	2338	353	267	767	53	2462
		TOTAL	14935	13623	1979	1500	5330	232	11336

ments at energies well above the Z-pole. The data at LEP2 allow new measurements at several discrete energy points over the range $130 < \sqrt{s} < 209$ GeV. The data from the 4 experiments have been combined: for the methodology and results see LEP2FF/00-03 [7] and http://www.cern.ch/LEPEWWG/lep2.

2.1 *Standard Model and Radiative Return*

One major complication of e^+e^- annihilation at LEP2 is the potential for Initial State Radiation (ISR) down to the Z-pole. If we wish to study full-energy annihilation events, then we must select those events with little or no such radiation. Figure 4 shows several distributions of $\sqrt{s'}$, the effective energy of the annihilation as estimated from the data. The Z-pole is clearly visible, particularly in the $q\bar{q}$ channel. Experiments typically choose $\sqrt{s'}/\sqrt{s} > 0.85$ for full-energy events and $\sqrt{s'}/\sqrt{s} > 0.10$ for inclusive measurements.

2.2 *Cross-sections and F-B Asymmetries for hadrons and leptons*

In general, cross-sections and angular distributions have been measured and averaged for $q\bar{q}$, e^+e^-, $\mu^+\mu^-$, and $\tau^+\tau^-$ final states. The e^+e^- Bhabha scattering has a dominant contribution from t-channel photon exchange and additional cuts (eg acolinearity and cms scattering angle) are made to enhance the s-channel. A compilation of the cross-section data from OPAL is shown

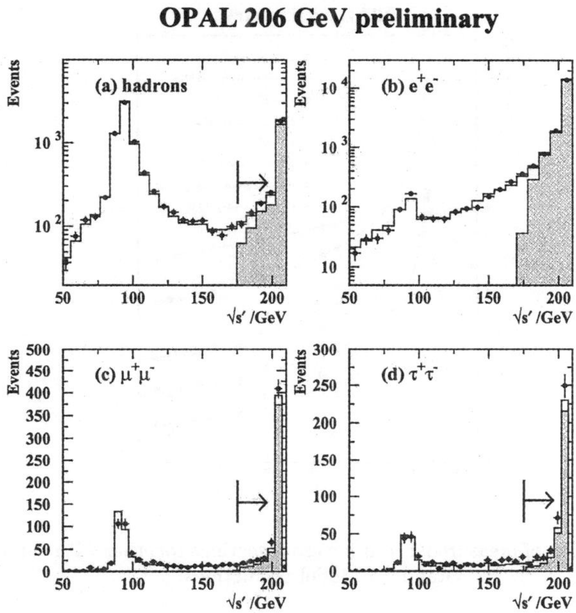

Figure 4: Distributions of reconstructed $\sqrt{s'}$ for selected events. The arrows show the position of the cut used to select non-radiative events. Points are data and histograms are Monte Carlo predictions with non-radiative events shaded.

in Figure 5 together with Standard Model expectations. One sees a very good agreement for all channels at all energies.

The combined LEP data on cross-sections for $q\bar{q}$, $\mu^+\mu^-$, and $\tau^+\tau^-$ are shown in Figure 6, together with the forward-backward asymmetries from final states where the outgoing charge is easily identified, $\mu^+\mu^-$ and $\tau^+\tau^-$. The Standard Model expectations are superimposed. The theoretical uncertainties are estimated to be $\sim 0.2\%$ on $q\bar{q}$ and $\sim 0.7\%$ on $\ell^+\ell^-$.

2.3 Cross-sections and F-B Asymmetries for heavy flavours

Heavy quark tagging allows cross-sections and asymmetries to be determined for both $c\bar{c}$ and $b\bar{b}$ final states. Figure 7 shows the ratio of the heavy flavour cross-sections to the total hadronic cross-section, R_b and R_c, and the values of the forward-backward asymmetries, A_{FB}^b and A_{FB}^c, together with Standard Model expectations. The data is in excellent agreement with the Standard

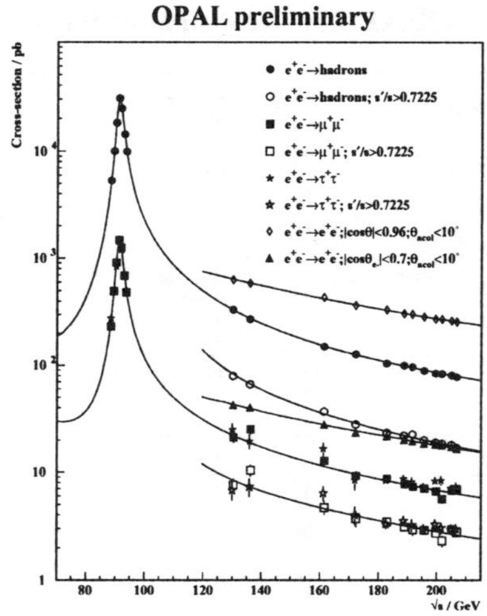

Figure 5: Compilation of measured 2-fermion cross-sections together with curves representing Standard Model predictions.

Model.

In conclusion, although the combined hadronic cross-sections are on average 2.5sd above Standard Model predictions [7], there is no significant evidence in the results for physics beyond the Standard Model.

3 Some Basic Checks

The excellent data on 2-fermion final states over such a large range of centre-of-mass energy has provided an opportunity to check the expected running of both the fine structure constant and the strong coupling constant, to check the size of the $\gamma - Z$ interference term, and to complete more neutrino counting tests.

3.1 Running of Fine Structure Constant

In processes involving virtual photon exchange, vacuum polarisation effects lead to a Q^2 dependence (running) of α_{em}. The leptonic contributions can be calculated but the effect of quark loops (non- perturbative QCD) must be estimated. There are two recent results from LEP which support the running of α_{em}. The L3 Collaboration has studied both small angle and large angle

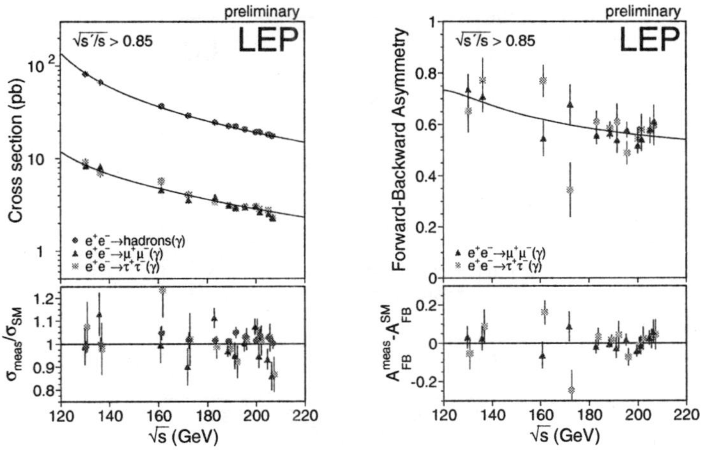

Figure 6: Combined LEP results on 2-fermion cross-sections and forward-backward asymmetries as a function of centre-of-mass energy. The lower plot sections show the ratio (or difference) of data to the ZFITTER prediction (curves).

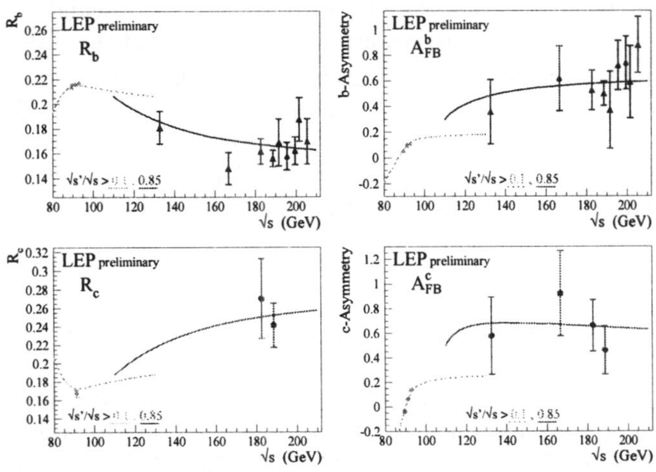

Figure 7: Combined LEP results on R_b and R_c, and A_FB^b and A_FB^c as a function of centre-of-mass energy. The curves represent the Standard Model prediction of ZFITTER.

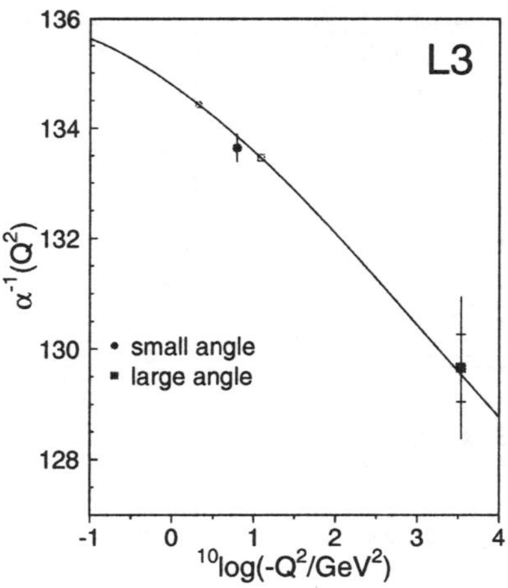

Figure 8: Measurements of the fine structure constant $\alpha_{\rm em}$ from small angle and large angle Bhabha scattering. The curve represents the Standard Model expectation.

Bhabha scattering[8]. In each case the data is divided into a small and large Q^2 sample and the change in $\alpha_{\rm em}$ required to represent the data over the relevant Q^2 span is determined. The results, shown in Figure 8, exhibit a 3.0sd and 2.9sd effect respectively, in agreement with the expected running of $\alpha_{\rm em}$.

The OPAL Collaboration has taken the non-radiative cross-sections and asymmetries for $q\bar{q}$, $\mu^+\mu^-$, and $\tau^+\tau^-$ over the range $130 - \sqrt{s} - 207$ GeV and used ZFITTER to determine the best value of $\alpha_{\rm em}$, keeping all other variables fixed.[9] They obtain $\alpha_{\rm em}^{-1} = 126.1^{+2.2}_{-2.1}$ at $\sqrt{s} = 190.7$ GeV which is 5.0sd below $\alpha_{\rm em}^{-1}(0)$. Figure 9 shows this result together with values from lower energy experiments.

3.2 Running of Strong Structure Constant

Using precision event shape distributions from the $q\bar{q}$ final state, the combined LEP data have been fitted to theory that uses second order calculations in α_s and NLLA. The event shapes are 1-T, M_H, C, B_T, B_W, and y_{23}^D and the derived values of α_s were presented at the LEP fest of October 2000: see http://lepqcd.web.cern.ch/LEPQCD/annihilations and are displayed in Figure 10. Excellent agreement with the expected running of α_s is evident over a wide range of $E_{\rm cm}$.

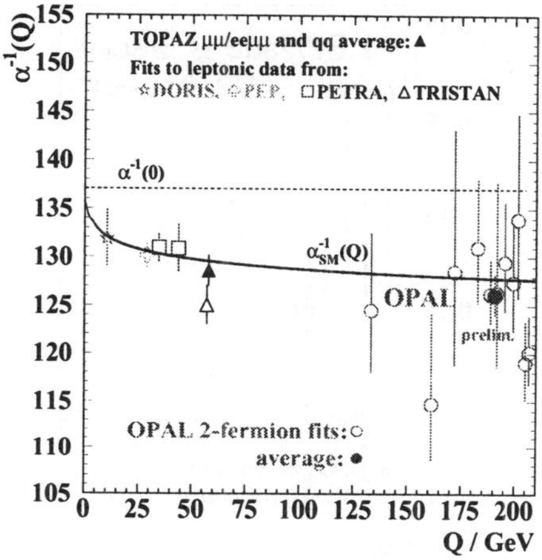

Figure 9: Measurements of the fine structure constant α_{em} as a function of Q. The curves represent the Standard Model expectation (solid = running, dotted = fixed).

3.3 Photon/Z Interference

Because the photon/Z interference term is very small in the region of the Z peak, the LEP1 lineshape fit uses its Standard Model value. Since the interference term becomes larger when one moves away from the Z pole, experimentally we can verify its Standard Model value. A previous test, eg by the VENUS Collaboration [10] at $\sqrt{s}$ = 58 GeV obtained a value for $\text{j}^{\text{tot}}_{\text{had}} = 0.196 \pm .083$ compared to the Standard Model value of 0.220. A new test by the L3 Collaboration[11] using the $q\bar{q}$ channel has determined $\text{j}^{\text{tot}}_{\text{had}} = 0.31 \pm 0.13 \pm 0.04$(theory) over the range $130 \leq \sqrt{s} \leq 189$ GeV to be compared with the Standard Model value of 0.21. Their data is shown in figure 11.

Figure 12 shows that the inclusion of this new LEP2 data would significantly reduce the uncertainty on m_Z if a lineshape fit were performed with $\text{j}^{\text{tot}}_{\text{had}}$ free.

3.4 Neutrino Counting

Although the LEP1 lineshape analysis estimates the number of light neutrino species indirectly via the measurement of the Z invisible width (giving $\text{N}_\nu = 2.984 \pm .008$), a more direct but statistically less powerful technique is to measure the cross-section for $e^+e^- \to \nu\bar{\nu}\gamma_{visible}$. These latter results have been summarised by Mnich [12] and provide $\text{N}_\nu = 3.00 \pm .08$ from all 4 experi-

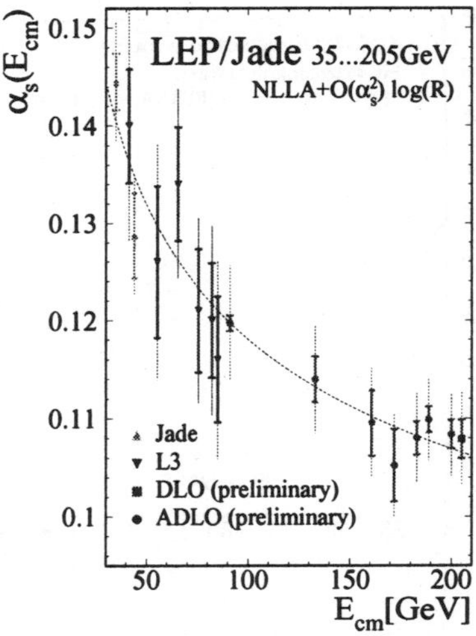

Figure 10: Running of the strong coupling constant α_s as derived from event shape data in e^+e^- annihilation.

ments at LEP1 and $N_\nu = 2.99 \pm .10$ from Delphi and L3 at LEP2. An example of the sensitivity of the $e^+e^- \to \nu\bar{\nu}\gamma_{visible}$ cross-section to the number of light neutrino species is shown in figure 13 by the L3 Collaboration [13]. The ratio of measured/Standard Model cross-section averaged over the 4 experiments is $0.965 \pm .028$, perfectly compatible with 3 neutrino species.

4 Constraints on New Physics

LEP has measured cross-sections and angular distributions for $e^+e^- \to q\bar{q}, e^+e^-$, $\mu^+\mu^-, \tau^+\tau^-, \gamma\gamma, W^+W^-$, and ZZ and, within errors, everything agrees with the Standard Model expectations. We will now use this data to set limits on a variety of models incorporating new physics. In general, the Standard Model Lagrangian is modified by the addition of a new contribution representing the new physics and incorporating model specific parameters. This new contribution will be multiplied by an effective coupling constant and divided by an effective scale. Typically, both the coupling constant and scale are unknown and, in general, an assumption is made for one of them when obtaining limits on the other. Thus, care should be taken when interpreting the results below.

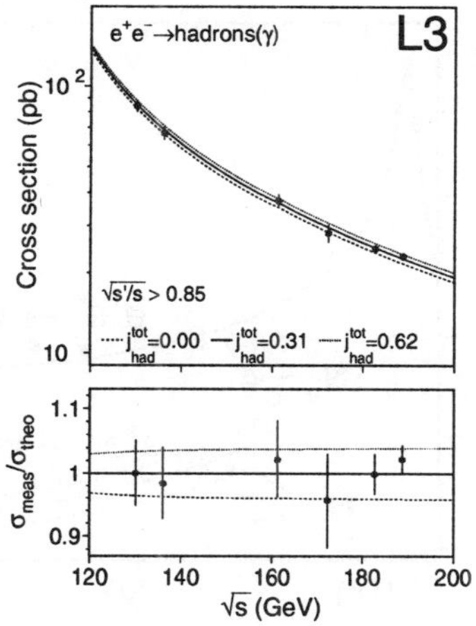

Figure 11: Hadronic cross-section as a function of centre-of-mass energy together with theory predictions for different values of j_{had}^{tot}.

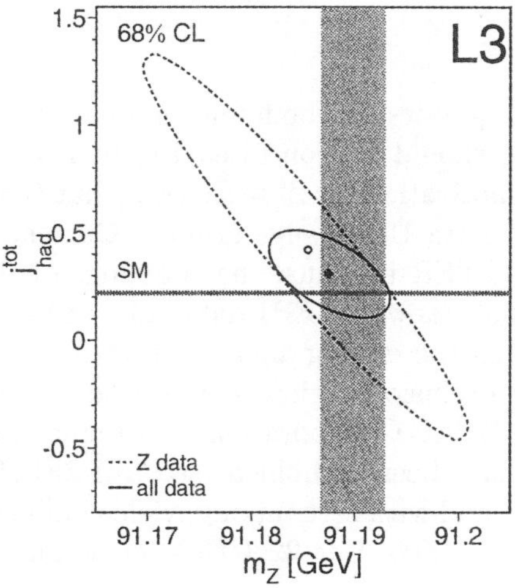

Figure 12: Allowed contours in the $m_Z - j_{had}^{tot}$ plane showing the improvement with new LEP2 data. The vertical band shows the m_Z range from a fit assuming the Standard Model value for γ/Z interference.

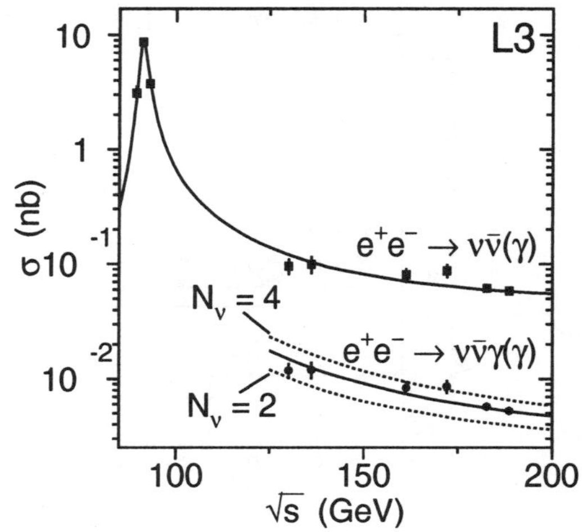

Figure 13: Cross-section measurements for $e^+e^- \to \nu\bar{\nu}\gamma$ compared to the Standard Model expectation for 2,3,4 neutrino species. The upper curve corresponds to the extrapolated full cross-section $e^+e^- \to \nu\bar{\nu}$.

4.1 Limits on Z′

An introduction to Z′ physics can be found in a paper by Altarelli, Mele, and Ruiz-Altaba [14] which should be consulted together with the mini-review in PDGT 2000[15]. One motivation for Z′ searches is that Grand Unified Theories, eg E(6), will contain extra U(1) gauge groups. Current analyses incorporate Z-Z′ mixing into ZFITTER in which the couplings of the bare state $Z^{0\prime}$ are model dependent. Analyses with LEP1 data only, eg by the Delphi Collaboration [16] , already limited the mixing angle to a few mrad. The LEP2 analyses generally extend the exclusion region in the Z′ mass-mixing angle plane as demonstrated by the OPAL Collaboration[17] in Figure 14. The data from the 4 LEP experiments have been combined (see LEP2FF/00-03 from the LEP-EWWG $f\bar{f}$ subgroup), and with zero mixing, yields limits on possible Z′ masses at 95% CL ranging from 400 GeV to 2260 GeV, depending on the specific model couplings.

4.2 4-f Contact Interactions

Eichten, Lane, and Peskin have demonstrated a method to detect possible quark or lepton substructure via the introduction of a Lagrangian representing 4-fermion contact interactions [18]. The coupling constant $(g^2/4\pi)$, which is

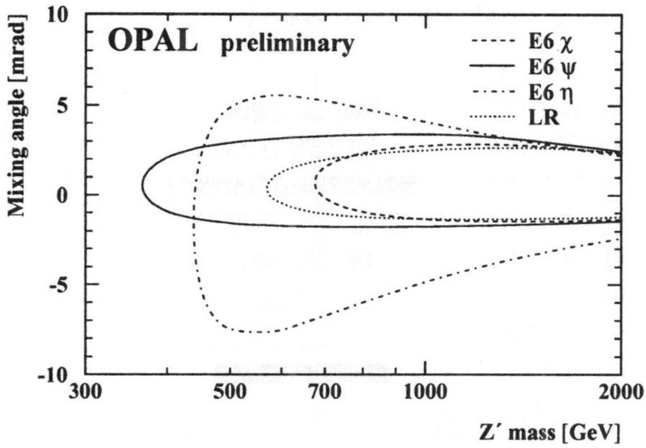

Figure 14: Exclusion contours at 95% CL in the Z′ mass-mixing angle plane for specific models. Only the inner area is allowed.

unknown, is generally set to unity by convention. Results on the scale on the contact interactions are given by Λ^- and Λ^+, where the sign characterises the chiral structure of the interaction. The LEP2 data have been fitted by the LEPEWWG (see LEP2FF/00-03) using (a) purely leptonic final states $\mu^+\mu^-$ and $\tau^+\tau^-$ and (b) heavy flavour $b\bar{b}$ and $c\bar{c}$ final states. The results are displayed in Figures 15 and 16 respectively. In the leptonic case the limits range from 8.0-23.9 TeV, and in the hadronic case from 1.3-14.0 TeV.

4.3 Exchange of R-parity violating sneutrinos

Since scalar neutrinos can decay to lepton-pairs via direct R-parity violating transitions, they can contribute to $\ell^+\ell^-$ final states via both s-channel and t-channel 2-fermion exchange diagrams. The analysis of LEP2 data thus allows limits to be derived on several products of couplings λ_{ijk} (where i,j,k represent generation indices) as a function of $\tilde{\nu}$ mass. A good example has been presented by the L3 Collaboration[19] at $\sqrt{s} = 189$ GeV and shown in Figure 17.

4.4 Exchange of Leptoquarks

In the reaction $e^+e^- \to f\bar{f}$ a contribution of t(or u)-channel leptoquark exchange is possible. Analyses can provide limits on scalar or vector leptoquark masses as a function of the relevant coupling constant. With a value of

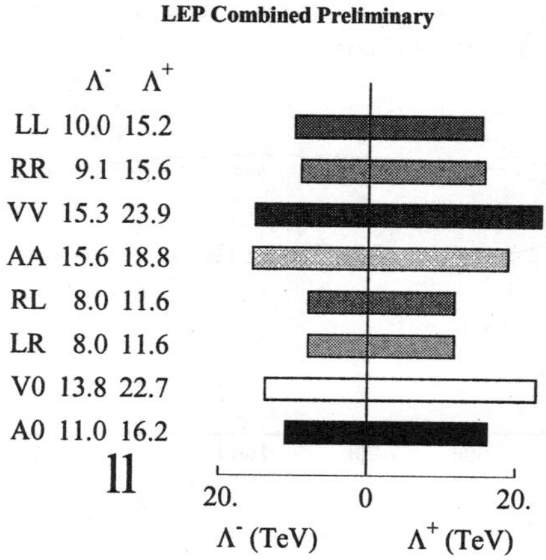

Figure 15: Limits on Λ for contact interactions from $\mu^+\mu^-$ and $\tau^+\tau^-$ final states.

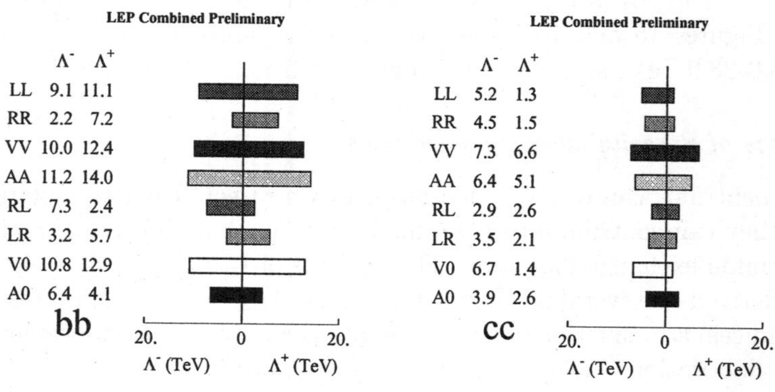

Figure 16: Limits on Λ for contact interactions from $e^+e^- \to b\bar{b}$ and $e^+e^- \to c\bar{c}$.

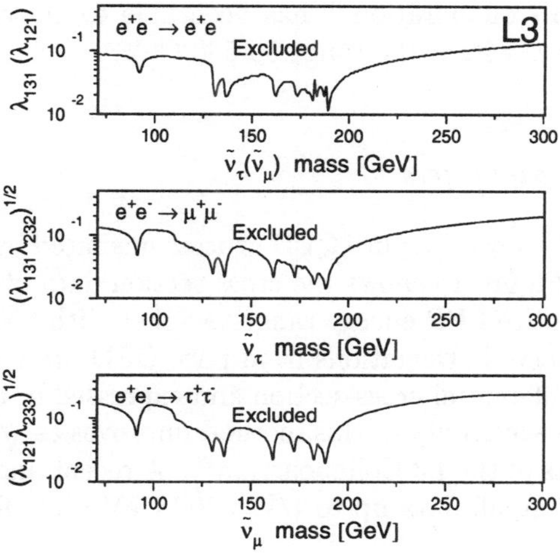

Figure 17: Upper limits at 95% CL on the coupling strengths λ_{ijk} of scalar leptons to leptons as a function of scalar neutrino mass.

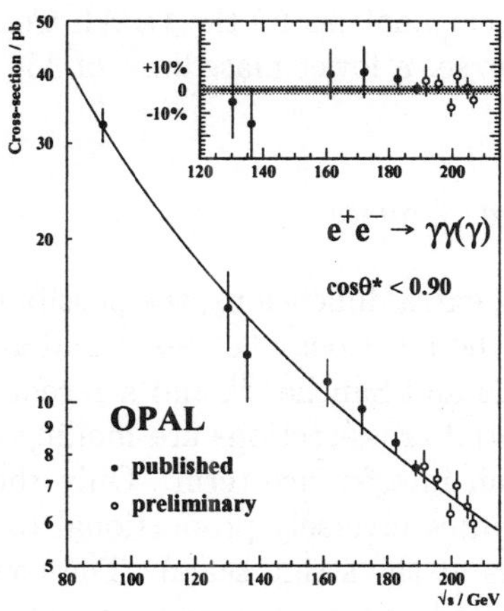

Figure 18: Measurements of the cross-section $e^+e^- \to \gamma\gamma$ as a function of centre-of-mass energy together with the Standard Model QED expectation. The upper box shows the data-Standard Model difference.

$g^2/4\pi = \alpha$ the L3 Collaboration [19] has been able to derive lower bounds for the leptoquark mass within the range 55-560 GeV.

4.5 QED cut-off parameters

The reaction $e^+e^- \to \gamma\gamma$ is a pure QED process mediated by electron exchange in the t-channel. Figure 18 shows the cross-section data of the OPAL Collaboration over the entire LEP energy range together with SM expectation. [9] No discrepancy is observed. Deviations from pure QED are typically determined from a fit of the differential cross-section and expressed in terms of cut-off parameters $\Lambda_\pm$. The sensitivity of this method improves as $\sqrt{s}$ increases, see for example the results of the L3 Collaboartion [20]. A recent analysis by the OPAL Collaboration [9] using all data up to $\sqrt{s} = 207$ GeV gives limits on $\Lambda_\pm$ in the 325-350 GeV range.

4.6 Exchange of Excited Electrons

The $e^+e^- \to \gamma\gamma$ reaction can also be influenced by the exchange of an excited electron (e^*) in the t-channel. The effect will depend on the e^* mass and on the coupling $g_{e^*e\gamma}$. A recent analysis by the OPAL Collaboration [9], with the restriction $g_{e^*e\gamma}=g_{ee\gamma}$, has set a lower mass limit of 354 GeV at 95% CL.

4.7 Models with Low-scale Gravity

In string models involving extra dimensions, the possibility of spin=2 graviton exchange could influence the reactions $e^+e^- \to f\bar{f}$ and also the final states $\gamma\gamma$, W^+W^-, and ZZ, see Mele and Sanchez [21] and a recent talk by P. Krieger at SUSY2K [22]. The differential cross-sections are modified by the addition of a pure graviton term and an interference term. Only the latter is relevant at LEP energies and contributes inversely proportional to the 4th power of M_s (the effective Planck mass at the string scale). The coupling constant is unknown (can only be calculated with a full knowledge of the underlying quantum gravitational theory) and is usually set to either λ = -1 or +1. 95% CL lower limits on M_s from an analysis of $\gamma\gamma$, ZZ, W^+W^- final states are 0.96 TeV and 0.92 TeV [21] and from $\mu^+\mu^-$, $\tau^+\tau^-$, $\gamma\gamma$, ZZ final states are 0.88 TeV and 0.89 TeV [23]. The latter results by the OPAL Collaboration are derived from the likelihood curves shown in Figure 19.

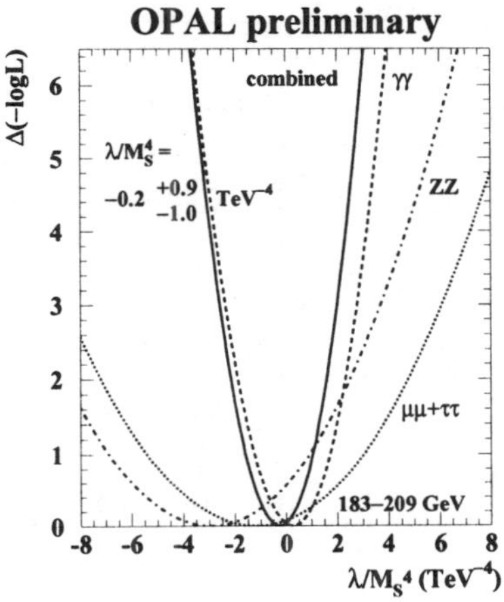

Figure 19: Log likelihood curves as a function of λ/M_s^4 for different final states together with the combined result (solid curve).

4.8 *Limits on FCNC single top production*

With a top quark mass of 174.3 ± 5.1 GeV, LEP2 cannot produce $t\bar{t}$ pairs. However, FCNC can produce a single top quark via a loop diagram leading to final states $t\bar{c}(\bar{u})$. Although Standard Model calculations of this process lead to impossibly small cross-sections [24], an investigation of this channel may provide surprises. Two current searches for such final states with $t \to bW$ decays have provided 95% CL upper limits of 0.48pb [25] and 0.35pb [26]. The Aleph Collaboration has expressed its result in the form of exclusion limits in the couplings corresponding to $t \to cZ$ and $t \to c\gamma$ [25]. This exclusion region is displayed in Figure 20 together with a previous exclusion from CDF. The κ_z limit is clearly improved and will continue to improve when all the LEP data are combined.

5 ZZ Physics

The reaction $e^+e^- \to$ ZZ, with a threshold near $\sqrt{s} = 183$ GeV, proceeds via the Standard Model diagrams NC02 involving the exchange of an electron in the t(u)-channel. The final state is of special interest because it constitutes an irreducible background to $e^+e^- \to H^0 Z^0$. Some of the final state characteristics that are used to identify Z-pair candidates are listed in Table 3 which

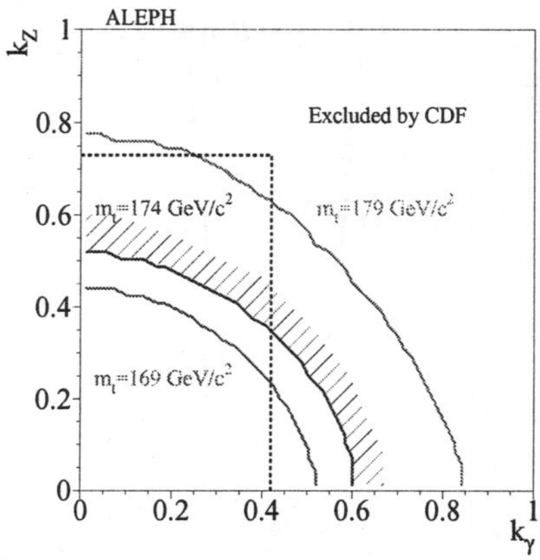

Figure 20: Exclusion curves at 95% CL in the κ_z-κ_γ plane for different values of m_t. The region excluded by CDF is shown by the dashed box.

is taken directly from a recent talk by G. Bella [27]. Remember that the Z branching ratios are 70% to $\mathrm{f\bar{f}}$, 10% to $\ell^+\ell^-$, and 20% to $\nu\bar{\nu}$. In general it is rather difficult to obtain 'clean' signals due to the large backgrounds, particularly those due to 2-fermion QCD and W^+W^--pairs. The channel $q\bar{q}\ell^+\ell^-$ has the highest efficiency*purity. A beautiful $q\bar{q}\mu^+\mu^-$ event at $\sqrt{s} = 205$ GeV recorded by the OPAL Collaboration is shown in Figure 21.

Table 3: Final States used to identify ZZ-pairs. ME represents missing energy, and leptons refers to electrons/muons/taus.

Final State	Fraction	Signature	Efficiency	Purity
$q\bar{q}q\bar{q}$	49%	4jets	30%	15 – 35%
$q\bar{q}\nu\bar{\nu}$	28%	2jets + ME	30%	60%
$q\bar{q}\ell^+\ell^-$	14%	2jets + 2leptons	50 – 80%	80 – 90%
$\ell^+\ell^-\nu\bar{\nu}$	4%	2leptons + ME	30%	45 – 55%
$\ell^+\ell^-\ell^+\ell^-$	1%	4leptons	40 – 60%	60 – 80%
$\nu\bar{\nu}\nu\bar{\nu}$	4%	invisible		

Analyses typically use many sequential cuts and/or multivariate techniques and impose mass constraints where possible (eg. $\mathrm{Z} \to \mathrm{f\bar{f}}, \ell^+\ell^-$). The event rates are small, see Table 2, and the channels are combined in a maximum likelihood fit that assumes Standard Model Z branching ratios to determine the reaction cross-section at each energy. Figure 22 shows the combined-LEP

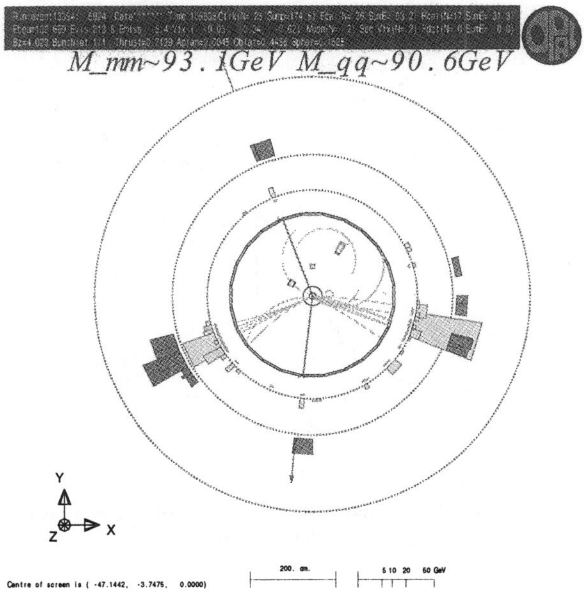

Figure 21: A classic example of $e^+e^- \to ZZ \to q\bar{q}\mu^+\mu^-$ as recorded by the OPAL Collaboration at $\sqrt{s} = 205\text{GeV}$.

cross-section together with theory expectation from YFSZZ and ZZTO, which have a 2% uncertainty. The agreement is excellent. In fact, the ratio of measured/theory cross-section, averaged over all $\sqrt{s}$ is 0.99 ± 0.06 [28].

All LEP Collaborations have analysed the ZZ cross-sections and angular distributions to search for anomalous neutral-current triple gauge couplings which are forbidden at tree-level within the Standard Model. No evidence has been found for non-zero couplings.

6 W^+W^- Physics

The main focus of the LEP2 program has been a detailed study of the $e^+e^- \to W^+W^-$ process to (a) confirm the presence of the non-abelian triple gauge couplings which are demanded by the Standard Model and to (b) make a precision measurement of m_W.

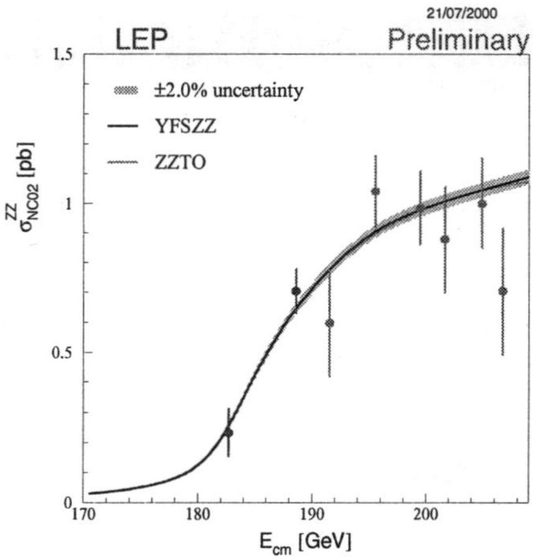

Figure 22: Measurements of the ZZ-pair cross-section (points) compared to the Standard Model prediction (curves).

6.1 Cross-section Determination

W^+W^--pair production takes place via the Standard Model diagrams CC03 which are (a) a t-channel ν_e exchange, (b),(c) s-channel γ, Z exchange involving the triple gauge couplings (TGCs) γWW and ZWW respectively. Diagram (a) is dominant near threshold and gives the forward W$^-$-peaking in the angular distribution. Some of the characteristics of the channels used to isolate W^+W^--pair candidates are given in Table 4, which is taken directly from a recent talk by G. Bella [29]. Remember that the W decays hadronically ($q\bar{q}\prime$) 67.5% and semi-leptonically ($\ell\bar{\nu}_\ell$) 32.5% of the time. The decay $W \to$ t$\bar{\text{b}}$ is not kinematically allowed. In comparison to ZZ production (see Table 3) the W^+W^- efficiencies and purities are significantly higher. There are still sizeable QCD backgrounds, particularly in the fully-hadronic channel, and these must be estimated carefully from Monte Carlo.

The candidate event samples are isolated after applying sequential data cuts and/or multivariate techniques. A beautiful $q\bar{q}\prime q\bar{q}\prime$ event at $\sqrt{s} = 208$ GeV recorded by the OPAL Collaboration is shown in Figure 23.

The W^+W^- signal is estimated by subtracting the Standard Model background and then applying a maximum likelihood fit to obtain the CC03 cross-section with/without a simultaneous extraction of the W branching ratios. If we assume Standard Model W branching ratios, the resulting all-LEP cross-

Table 4: Final States used to identify W^+W^--pairs. ME represents missing energy, and leptons refers to electrons/muons/taus.

Class	Final State	Fraction	Signature	Efficiency	Purity
Fully-hadronic	$q\bar{q}\prime q\bar{q}\prime$	45.6%	4jets	85%	80%
Semi-leptonic	$q\bar{q}\prime e\overline{\nu}_e$		2jets +	85%	95%
	$q\bar{q}\prime\mu\overline{\nu}_\mu$	43.9%	isolated lepton +	90%	98%
	$q\bar{q}\prime\tau\overline{\nu}_\tau$		ME	65%	85%
Fully-leptonic	$\ell\bar{nu}_\ell$	10.6%	2 acolinear jets + ME	45 – 80%	90%

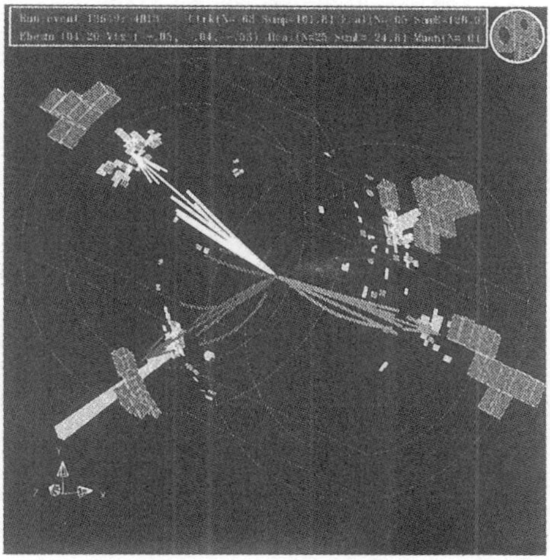

Figure 23: A classic example of $e^+e^- \to W^+W^- \to q\bar{q}\prime q\bar{q}\prime$ as recorded by the OPAL Collaboration at $\sqrt{s} = 208$GeV.

section is displayed in Figure 24 (see LEPEWWG/XSEC/2000-01[30]). The figure also shows the current theoretical predictions from RacoonWW/YFSWW3 and GENTLE that have inherent uncertainties of 0.4% and 0.7%, respectively. Clearly, the data are well-represented by the Standard Model predictions and provide a convincing case that all three CC diagrams are required, ie triple gauge couplings are needed. Figure 25 shows the latest LEP2 data in greater detail (see LEPEWWG/XSEC/2000-01).

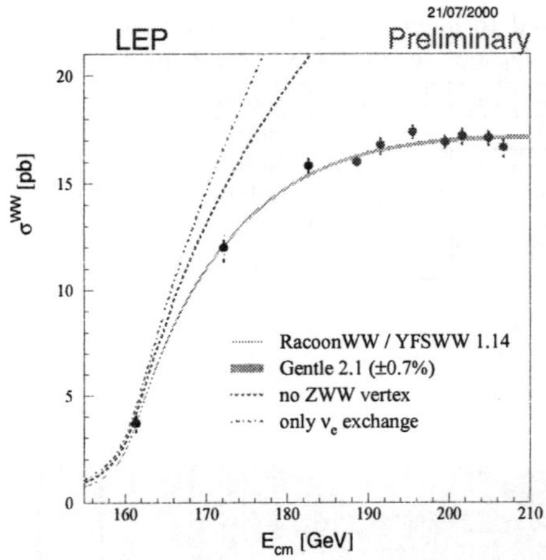

Figure 24: Measurements of the W^+W^--pair cross-section (points) compared to the Standard Model prediction (curves) with/without triple gauge couplings.

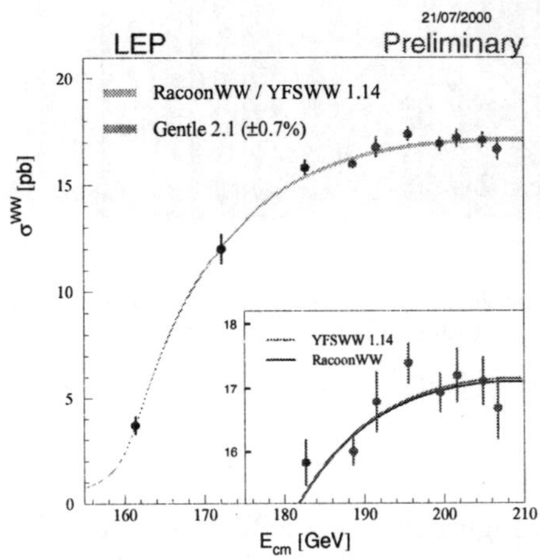

Figure 25: Measurements of the W^+W^--pair cross-section (points) compared to the Standard Model prediction (curves).

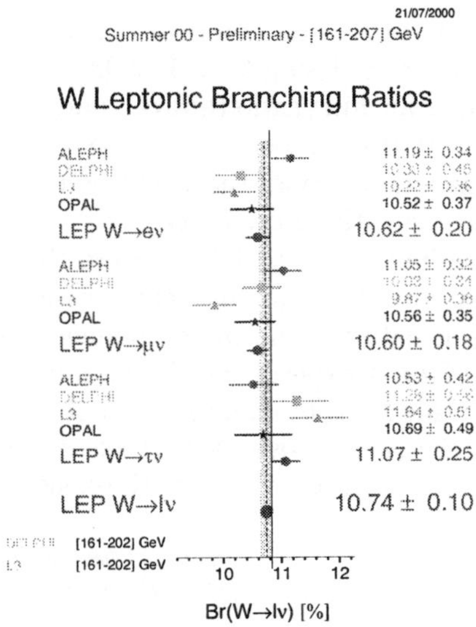

Figure 26: Measurements of the W leptonic branching ratios provided by the 4 LEP experiments together with the combined result. The solid vertical line represents the Standard Model expectation.

6.2 Branching Ratios, Lepton Universality, and CKM Matrix

If the maximum likelihood fit does not assume Standard Model W branching ratios, then these can be determined from the fit to the channel data. Figure 26 shows the W-leptonic branching ratios as determined by each of the 4 LEP experiments and their combined values. For each leptonic channel the individual measurements are in good agreement and the 3 combined values are consistent within 3.3% of equality thus confirming the Standard Model expectation of universal lepton couplings. Assuming lepton universality, the W to $\ell\bar{\nu}\nu_\ell$ branching ratio is determined to be $10.74 \pm 0.10\%$ per lepton decay channel, and the W to $q\bar{q}\prime$ branching ratio is determined to be $67.78 \pm 0.32\%$, in good agreement with Standard Model expectation (10.83% and 67.51% respectively). The precision is now better than that from the $p\bar{p}$ collider experiments [31].

The determined branching ratios can now be used to set new values on specific terms of the CKM matrix (see LEPEWWG/XSEC/2000-01 [30]). The

$q\bar{q}\prime$ branching ratio is related to a sum of 6 terms in the CKM matrix, of which V_{cs} is the least well determined. Using PDGT 2000 for the sum of the other 5 terms, the LEP hadronic branching ratio (above), and $\alpha_s(m_W^2) = 0.121 \pm .002$, a value of $V_{cs} = 0.989 \pm 0.016$ has been extracted. It should be noted that this value does not assume unitarity of the CKM matrix and that the error is dominated by the branching ratio uncertainty.

A more direct determination of V_{cs} follows from the measurement of R_c^W, the ratio of the charm to total hadronic branching ratio of the W. Since extra selections are needed to isolate the charm component, the resulting precision will be clearly inferior. The 4 experiments in combination have determined $R_c^W = 0.494 \pm 0.044$ and this leads to an extraction of $V_{cs} = 0.96 \pm 0.08$, significantly less precise than the value above. Nevertheless, the two determinations are in good agreement with each other, and with expectation.

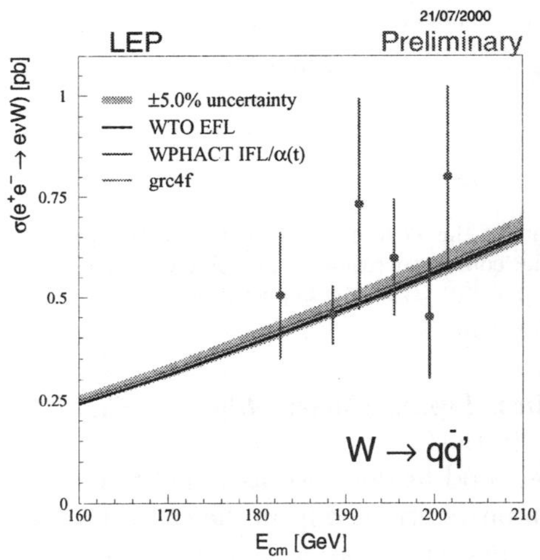

Figure 27: Measurements of the single-W cross-section (points) compared to Standard Model predictions (curves).

6.3 *TGC Determination and W-polarisation*

The W^+W^--pair cross-section measurements (see above) have already shown the necessity to include the two s-channel diagrams involving triple gauge couplings. These $W^+W^-\gamma$ and W^+W^-Z couplings are demanded by the gauge structure of the Standard Model. There are other reactions involving the

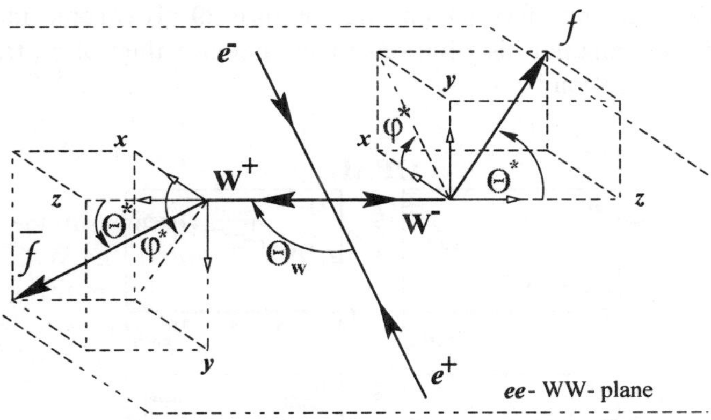

Figure 28: Diagram representing the angle definitions in the TGC analysis - from a talk by A. Macchiolo, EPS-HEP99, Tampere, Finland.

$W^+W^-\gamma$ coupling, in particular the single-W diagram leading to the $Wxle\overline{\nu_e}$ final state, and the single-γ diagram leading to the $\nu_e\gamma\overline{\nu_e}$ final state, both of which are accessible at LEP2. The form and strength of these vertex couplings are unambiguously predicted by the gauge structure of the Standard Model, see e.g. the paper of Bilenky et al. [32] and also the Yellow report CERN 96-01. Although the most general Lorentz-invariant Lagrangian describing TGCs has 14 terms, after insisting on gauge invariance and C+P conservation, only 3 terms remain which, in the Standard Model, are $\kappa_\gamma = 1$, $g_1^z = 1$, and $\lambda_\gamma = 0$, ie only 2 terms non-zero. The term 'anomalous' TGC refers to any TGC, including the 2 non-zero Standard Model TGCs, with values different from Standard Model expectation. The measured cross-sections for $e^+e^- \to W^+W^-$, and $We\overline{\nu_e}$, and the production/decay angular distributions for W^+W^--pair production are sensitive to the form and strength of all possible TGCs. Figure 27 shows the LEP cross-section for the single-W final state, for those cases with W hadronic decays. The data is seen to be in good agreement with the Standard Model expectation, albeit with a 5% theoretical uncertainty.

Figure 28, taken from a talk by A. Macchiolo at EPS-HEP99, shows the definition of the 5 angles needed to specify the complete production and decay distributions in the W^+W^- final state. We remark in passing that the entire angular phase-space is not always accessible due to ambiguities resulting from quark flavour identification, or missing neutrinos in the fully-leptonic channels, and problems with imperfect jet pairing. Figure 29 gives some idea of the sensitivity of these angular distributions to anomalous values of g_1^z, taken from the OPAL Collaboration [33].

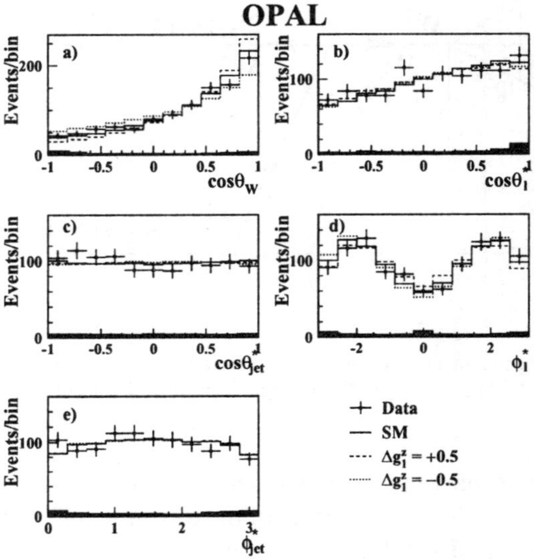

Figure 29: Sensitivity of angular distributions to anomalous values of g_1^z for data taken from semi-leptonic WW events at 189 GeV.

There are two methods in common use for an extraction of the TGCs. The first is model-dependent and determines the TGCs from maximum-likelihood fits to the cross-sections and angular distributions. The second is more general and evaluates the W^+W^- spin density matrix from which one can extract the helicity structure and polarisation properties. Figure 30 shows the log-likelihood curves from method 1 using the combined LEP data (see LEP-EWWG /TGC/2000-02 [34]) and Table 5 summarises the values and limits obtained for $\Delta g_1^z \equiv g_1^z - 1$, $\Delta\kappa_\gamma \equiv \kappa_\gamma - 1$, and λ_γ. In each case, the fits vary 1 parameter at a time, the others being kept fixed at their Standard Model values (ie =0). The results show no evidence for anomalous values of the couplings. More complex analyses looking for the other 14-3=11 parameters also give no evidence for non-zero values (this is good because they would violate C, P, or CP conservation). When all the LEP data is analysed, the precision on $\Delta g_1^z \equiv g_1^z - 1$, and λ_γ should be better than ± 0.02, and on $\Delta\kappa_\gamma \equiv \kappa_\gamma - 1$ better than ± 0.05 [35]. Note that these direct determinations of the TGCs con-

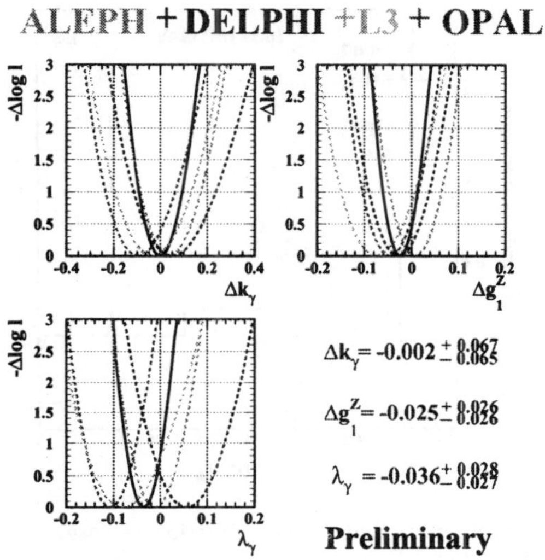

Figure 30: Log likelihood curves of the 4 LEP experiments and the combination (solid curve) for 3 triple gauge couplings.

trast with previous indirect estimations using EW radiative corrections at the Z pole [36].

Table 5: Table of values and limits for triple gauge couplings.

Parameter	Value	95% CL limits	Standard Model expectation
$\Delta g_1^z \equiv g_1^z - 1$	$-0.025^{+.026}_{-.026}$	[-0.074,+0.028]	0
$\Delta\kappa_\gamma \equiv \kappa_\gamma - 1$	$-0.002^{+.067}_{-.065}$	[-0.13,+0.13]	0
λ_γ	$-0.036^{+.028}_{-.027}$	[-0.089,+0.020]	0

The more general approach leading to helicity structure and polarisation has been formulated by Hagiwara, Peccei, and Zeppenfeld [37]. W's are expected in both transverse (helicity ±1) and longitudinal (helicity 0) polarisation states. One can employ the full 9*9 spin density matrix for W^+W^- or, more simply, just use the W-decay as a polarisation analyser to extract the fractional contributions of helicity -1, 0, and +1 (denoted by f_-, f_0, f_+ respectively). Hagiwara et alia [37] have predicted the transverse and longitudinal components as a function of the W production angle $\cos\theta_W$ at $\sqrt{s} = 190$ GeV. A recent study by the L3 Collaboration [38], whose results are displayed in Figure 31, shows that within

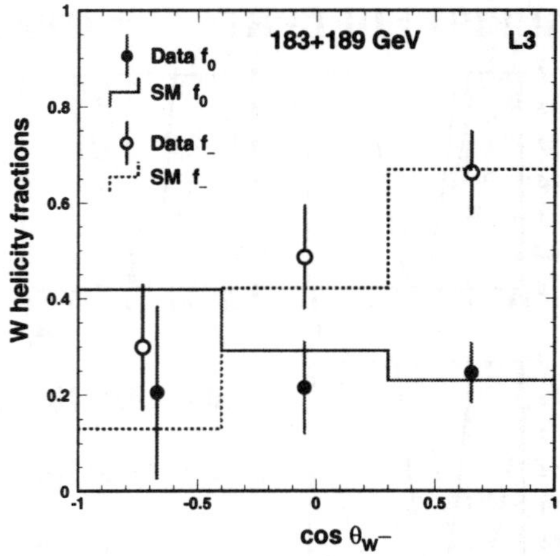

Figure 31: W helicity fractions f_0 and f_- as a function of cosθ_W. The Standard Model (KORALZ) predictions are represented by the histograms.

the present level of statistics the data and Standard Model expectations are in good agreement. Table 6 lists a summary of recent determinations [39,38,40] of f_0 together with Standard Model expectations.

Table 6: Table of values of longitudinal polarisation fraction, f_0.

$\sqrt{s}$ (GeV)	Collaboration	f_0 data	SM expectation
183	OPAL	$0.242 \pm 0.091 \pm 0.023$	0.272
183+189	L3	$0.261 \pm 0.051 \pm 0.016$	0.26
183+189	OPAL	$0.210 \pm 0.033 \pm 0.016$	0.257
183-202	L3	0.259 ± 0.035	0.248

6.4 Mass Determination

The determination of m_W is perhaps the most important, and certainly the most challenging, of the LEP2 precision measurements. See LEPEWWG/MASS/2000-01 [41] for more details. There are two standard methods [1] via the measurement of the cross-section ($e^+e^- \rightarrow W^+W^-$) variation near W^+W^- threshold, and [2] via the direct mass reconstruction from the W decay products. In the latter case, as seen in Table 4, there are three event categories: (a) the fully-hadronic channel $q\bar{q}'q\bar{q}'$, (b) the semi-leptonic channel $q\bar{q}'\ell\bar{\nu}_\ell$, and the fully-leptonic channel $\ell\bar{\nu}_\ell\ell\bar{\nu}_\ell$. As discussed by R. Strohmer[42], this last channel, which represents only 10.6% of all W^+W^- decays, has 2 undetected neutrinos and there are not enough constraints to perform a kinematic fit. Nevertheless, one can look at either the charged lepton energy spectrum or at a pseudo-mass estimator (which assumes the neutrinos are in the same plane as the leptons) to try and determine m_W. However, the resulting error on m_W ranges from 0.5 to 1.5 GeV and the channel is ignored for the time being.

The threshold method [1] (see CERN-PPE/97-154) was historically the first method used by the LEP Collaborations in 1996 when the beam energy was raised above W^+W^- threshold. Approximately 30 W^+W^- events were observed by each experiment at $\sqrt{s} = 161$ GeV and the resulting LEP-combined CC03 cross-section was determined to be 3.69 ± 0.45pb. Knowing the variation of cross-section with W mass (see Figure 32), this translates into $m_\mathrm{W} = 80.40^{+0.22}_{-0.21} \pm 0.03$ GeV, where the systematic error of 0.03 GeV comes from the LEP energy calibration. As we will see later, the statistical weight of this threshold measurement will be overshadowed by the more precise direct mass determinations.

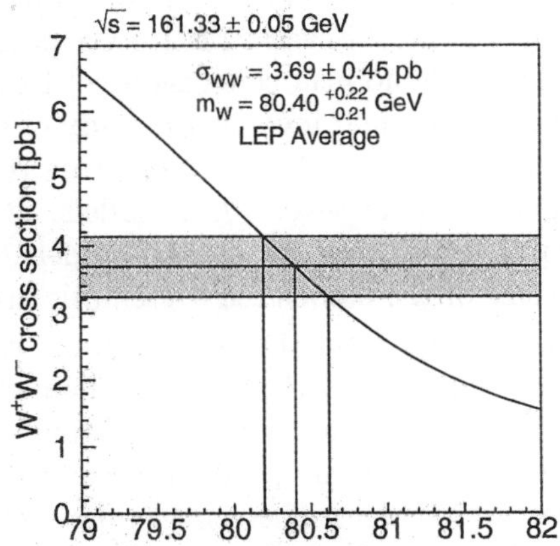

Figure 32: Extraction of m_W from the LEP measurement of the W-pair cross-section near threshold. The curve represents the Standard Model.

The direct mass reconstruction [2] utilises both the fully-hadronic and semi-leptonic W^+W^- decay channels. The fully-hadronic channel, which constitutes 45.6% of all W^+W^- decays, was discussed recently by S. Schmidt-Karst at Osaka[43]. The events are forced into 4 jets, energy/momentum constraints are applied, and the two Ws are constrained to have the same mass. Of the 3 possible 2-jet pair combinations, current methods to choose the correct pairing succeed in about 90% of the cases. The resulting mass distribution is fitted with a variety of methods and systematics are evaluated. Currently there are two serious complications associated with potential final state interactions (FSI).

The first of these is associated with Bose-Einstein Correlations (BEC), see a recent talk by A. Valassi at Osaka [44]. BEC produce an enhancement of identical bosons in regions of close proximity of phase-space, giving more $\pi^{\pm\pm}$ than π^{+-} at small values of Q. We already know that intra-W BEC exist, but if inter-W BEC were found to be present in the data then this could bias the determination of m_W. The Aleph Collaboration [45] has recently shown strong evidence for intra-W BEC (see Figure 33 for semi-leptonic WW events) but their current data do not strongly favour inter-W BEC (see Figure 34) in fully-hadronic WW events. It will take the combined-LEP data to make a more definite statement.

The second FSI complication is associated with Colour-Reconnection (CR), see a recent talk by P. DeJong at Osaka [46]. CR involves the exchange of coloured gluons during the non-perturbative hadronisation phase and could, in principle, give a bias to m_W. Several Monte Carlo models predict differing CR behaviour and there is no firm conclusion yet from the 4 experiments. However, a new, promising, method involving a study of particle flow was presented by the L3 Collaboration at Osaka [47] and this is shown in Figure 35 together with the expectations of several CR models. It is hoped that, with the full statistics from all 4 experiments, several models of CR could be ruled out, thus limiting the size of the associated systematic error.

Direct mass reconstruction from the semi-leptonic channel has been discussed recently by R. Strohmer at Osaka [42]. This channel represents 43.9% of all W^+W^- decays and comprises the 3 separate leptonic decays $e\overline{\nu}_e$, $\mu\overline{\nu}_\mu$, and $\tau\overline{\nu}_\tau$ (see Table 4). There is clearly no problem here with inter-W FSI, nor is there any difficulty with jet-choice. The semi-leptonic channel has good efficiency and high purity. Energy and momentum, and mass equality constraints are applied as for the fully-hadronic channel previously. However, the $\tau\overline{\nu}_\tau$ decay channel has additional missing neutrinos and the tau energy cannot be determined. One uses the mass from the hadronic decay in this case. Again, mass distributions are fitted with a variety of methods and systematic error contributions are evaluated.

Figure 36 shows some typical mass distributions from the 4 experiments. The shaded regions, which represents background contributions evaluated from MonteCarlo, are generally small and well-determined.

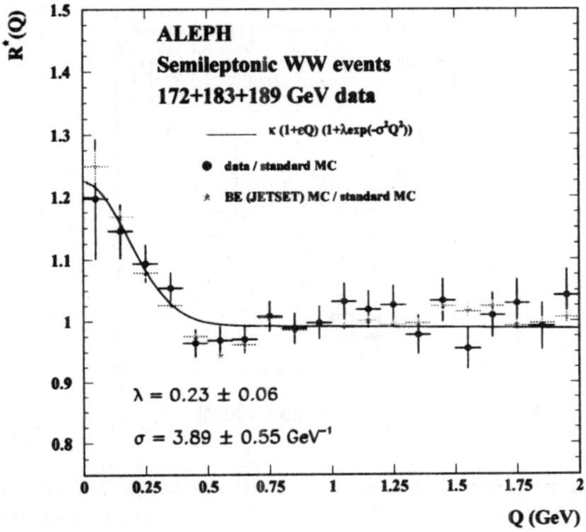

Figure 33: Bose-Einstein correlation function for semi-leptonic WW events. The solid curve represents a fit to the data.

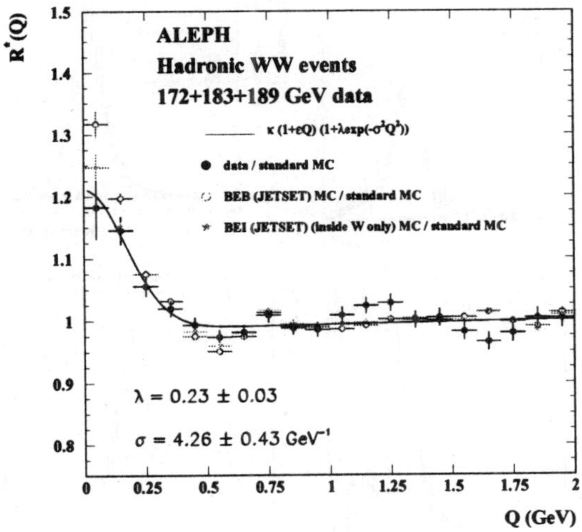

Figure 34: Bose-Einstein correlation function for fully-hadronic WW events. The solid curve represents a fit to the data. The open circles represent the prediction of a model that incorporates inter-W Bose-Einstein correlations.

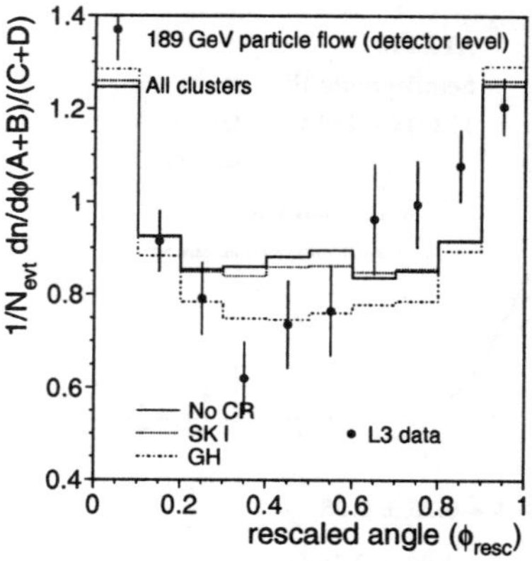

Figure 35: Ratio of particle flows between jets from the same W and between jets from different W's as a function of the rescaled angle. The histograms represent Monte Carlo predictions.

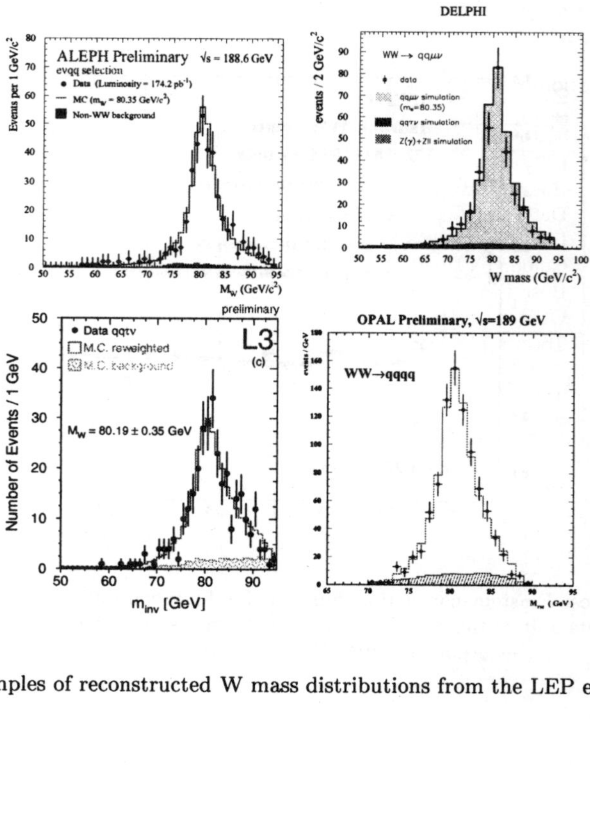

Figure 36: Examples of reconstructed W mass distributions from the LEP experiments.

6.5 Combination of Mass Values

The LEPEEWG (see LEPEEWG/MASS/2000-01) has combined the results of the direct reconstruction from the fully-leptonic and semi-leptonic channels after carefully taking into account the correlations in the systematic errors. The results are summarised in Table 7. One immediately notices the large values of systematic error associated with colour-reconnection and Bose-Einstein correlation in the fully-hadronic channel. We hope to reduce these by a substantial factor before the final analysis is complete. With the current values, the $q\bar{q}'q\bar{q}'$ statistical weight in the combination is but 27%. Another systematic that may reduce in the future is that labelled 'Hadronisation'. This is associated with the differing Monte Carlos currently employed and could benefit from improved 'tuning' of the models to data. One notices that the difference in m_W between the $q\bar{q}'q\bar{q}'$ and the $q\bar{q}'\ell\bar{\nu}_\ell$ channels is only $+5 \pm 51$ MeV, indicating that global FSI effects cannot be huge. This is somewhat reassuring. The current best measurement of m_{W} is the combination of all direct mass values together with the threshold mass value, and leads to a LEP result of $m_{\mathrm{W}} = 80.427 \pm 0.046$ GeV (see Table 8). It is hoped that a final LEP2 result will have a total error of ~ 30 MeV.

Table 7: Tabulation of systematic and statistical errors in the determination of the W-mass. Units are MeV.

Source of systematic error	$q\bar{q}'\ell\bar{\nu}_\ell$	$q\bar{q}'q\bar{q}'$	combined
ISR/FSR	8	10	8
Hadronisation	26	23	24
Detector effects	11	7	10
LEP beam energy	17	17	17
Colour reconnection	0	50	13
Bose-Einstein correlations	0	25	7
Others	5	5	4
Total systematic	35	64	36
Statistical error	38	34	30
Total error (MeV)	51	73	47
Mass m_{W} (GeV)	80.427	80.432	80.428

Table 8: Combination of LEP2 threshold and direct W-mass determinations.

Source	Mass value (GeV)
Threshold Method	$80.400 \pm 0.220 \pm 0.030$
Direct Mass Measurement	$80.428 \pm 0.030 \pm 0.036$
LEP2 combined	80.427 ± 0.046

6.6 Comparison of Direct and Indirect Mass Estimates

We can now compare the direct m_W measurement from LEP2 with (a) the direct measurements from the $p\bar{p}$ colliders and (b) the indirect measurements from the Standard Model fits to precision data of LEP1/SLD and neutrino scattering data (see Figure 37). Within the current level of statistics, there is no difference between the separate measurements. It certainly looks as though the Standard Model has no surprises, at least within the energy range over which we are testing it. It is an experimental triumph that the direct and indirect mass measurements agree so well.

Figure 38 shows the direct and indirect error ellipses in the 2-dimensional plane of $m_{\rm W}$ vs $m_{\rm t}$ as a function of the Standard Model Higgs mass. One notices two things: (a) the error ellipses overlap substantially and (b) the overlap area indicates a strong preference for a low mass Higgs.

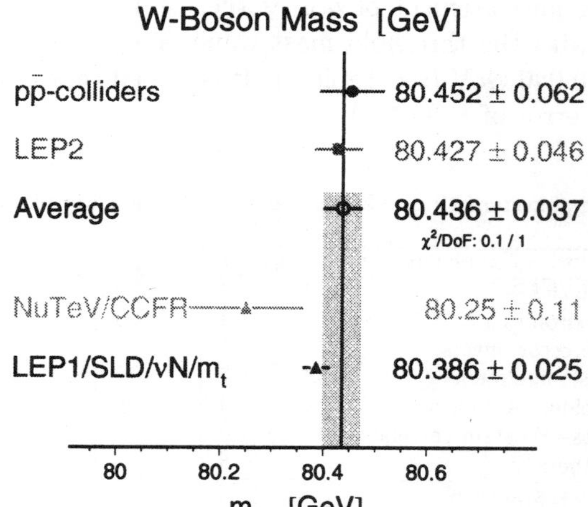

Figure 37: Comparison of W mass measurements. The direct measurements from LEP and $p\bar{p}$ colliders are averaged and compared to indirect determinations.

Finally, we show in Figure 39 the chisquare behaviour of the complete Standard Model fit as a function of the Standard Model Higgs mass in which all precision electroweak data has been used (LEP1 lineshape, SLD asymmetries, neutrino scattering, tau polarisations, direct measurements of $m_{\rm W}$ and $m_{\rm t}$, etc). For full details consult CERN-EP-2000-153 [48] for the combination procedures and CERN-EP-2001-021 [49] for the combination results themselves. In this figure the shaded region represents the area already excluded by di-

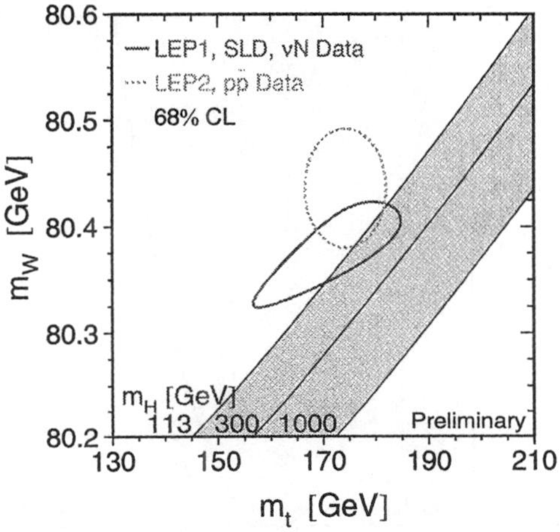

Figure 38: Comparison of the indirect measurements of m_W and m_t (solid contour) and the direct measurements (dashed contour). Also shown is the Standard Model relationship as a function of Higgs mass.

rect LEP Higgs searches (up to approximately 113 GeV at 95% CL [50]). The solid (dashed) curve provides a 95% CL upper limit for the Standard Model Higgs mass of 165 (206) GeV. This will provide a great incentive for dedicated searches at Tevatron2 and the LHC. Which laboratory will claim a discovery?

7 Conclusions

These two lectures have displayed the wealth and beauty of LEP2 electroweak physics. There is absolutely no evidence for any significant deviation from current Standard Model expectations. Long live the Standard Model of particle physics.

Acknowledgments

I would like to thank the organisers of the Lake Louise Winter Institute for the invitation to deliver these lectures and for their excellent hospitality in such a marvellous setting.

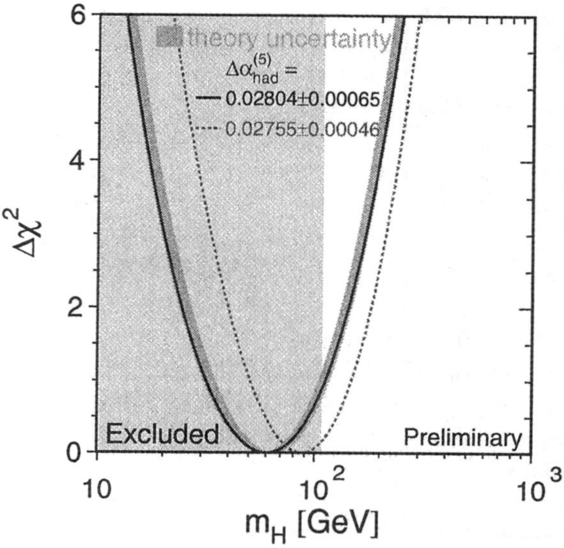

Figure 39: Chisquare versus Higgs mass. The theory uncertainty band represents missing higher order corrections. The excluded region is from direct Higgs searches. The dashed curve is the result obtained with a new determination of the fine structure constant.

References

1. *Physics at LEP2*, Editors: G. Altarelli, T. Sjostrand, and F. Zwirner, CERN 96-01.
2. *Reports of the Working Groups on Precision Calculations for LEP2 Physics*, Editors: S. Jadach, G. Passarino, R. Pittau, CERN 2000-009.
3. L3 Collaboration, M. Acciarri *et al*, *Phys. Lett.* **B413**, 191 (1997).
4. P. Renton, *Measurement of Beam Energy at LEP2*, talk at ICHEP 2000, Osaka, Japan. See http://ichep2000.hep.sci.osaka-u.ac.jp/
5. The LEP Energy Working Group, A. Blondel *et al*, *Eur. Phys. J.* **C11**, 573 (1999).
6. ALEPH Collaboration, R. Barate *et al*, *Phys. Lett.* **B464**, 339 (1999).
7. LEPEWWG $f\bar{f}$ Subgroup, *Combination of the LEP II $f\bar{f}$ Results*, LEP2FF/00-03. See http://lepewwg.web.cern.ch/LEPEWWG/
8. L3 Collaboration, M. Acciarri *et al*, *Phys. Lett.* **B476**, 40 (2000).

9. *Measurement of Standard Model Processes in e^+e^- Collisions at $\sqrt{s}$ ~203-209 GeV*, OPAL Physics Note PN469. See http://opal.web.cern.ch/Opal/
10. VENUS Collaboration, K. Yusa *et al*, *Phys. Lett.* **B447**, 167 (1999).
11. L3 Collaboration, *Determination of γ/Z Interference in e^+e^- Annihilation at LEP*, CERN-EP/2000-084.
12. J. Mnich, *Tests of the Standard Model*, CERN-EP/99-143.
13. L3 Collaboration, M. Acciarri *et al*, *Phys. Lett.* **B470**, 268 (1999).
14. G. Altarelli, B. Mele, and M. Ruiz-Altaba, *Z. Phys.* **C45**, 109 (1989).
15. Mini-review, *The Z′ Searches*, Reviews of Particle Physics, *Eur. Phys. J.* **C15**, 289 (2000).
16. DELPHI Collaboration, P. Abreu *et al*, *Z. Phys.* **C65**, 603 (1995).
17. OPAL Collaboration, *Limits on a Z′ Boson from e^+e^- to Fermion Pair Cross-sections and Asymmetries*, OPAL Physics Note PN372. See http://opal.web.cern.ch/Opal/
18. E.J. Eichen, K.D. Lane, and M.E. Peskin, *Phys. Rev. Lett.* **50**, 811 (1983).
19. L3 Collaboration, M. Acciarri *et al*, *Phys. Lett.* **B489**, 81 (2000).
20. L3 Collaboration, M. Acciarri *et al*, *Phys. Lett.* **B475**, 198 (2000).
21. S. Mele and E. Sanchez, *Study of Extra Space Dimensions in Vector Boson Pair Production at LEP*, CERN-EP/99-118.
22. *Searches for Extra Dimensions at LEP*, P. Krieger,talk given at SUSY2K, CERN, June 2000. See http://wwwth.cern.ch/susy2k/susy2ktalks/krieger.ps.gz
23. OPAL Collaboration, *Search for Low Scale Quantum Gravity in Extra Spatial Dimensions in Z^0Z^0 Production at LEP*, OPAL Physics Note PN440. See http://opal.web.cern.ch/Opal/
24. C. Huang, X. Wu, and S. Zhu, *Phys. Lett.* **B452**, 143 (1999).
25. ALEPH Collaboration, *Search for Single Top Production in e^+e^- Collisions at $\sqrt{s}$= 189-202 GeV*, CERN-EP/2000-102.
26. OPAL Collaboration, *Investigation of Single Top Quark Production at $\sqrt{s}$= 189 GeV*, OPAL Physics Note PN444. See http://opal.web.cern.ch/Opal/
27. G. Bella, *W and Z Couplings and Cross-sections*, talk at IVth Rencontre du Vietnam, Hanoi (July 2000), Vietnam. See http://opal.web.cern.ch/Opal/confrep/
28. S. Mele, *ZZ Cross-section Measurements*, talk at ICHEP 2000, Osaka, Japan. See http://ichep2000.hep.sci.osaka-u.ac.jp/
29. G. Bella, *W Properties and Decays at LEP2*, talk at Tau Workshop 2000, Victoria, Canada. See http://tau2000.phys.uvic.ca/

30. The LEP Collaborations ALEPH, DELPHI, L3, OPAL, and the LEP WW Working Group, LEPEWWG/XSEC/2000-01. See http://lepewwg.web.cern.ch/LEPEWWG/
31. A. Gurtu, *Precision Tests of the Electroweak Gauge Theory*, plenary talk at ICHEP 2000, Osaka, Japan. See http://ichep2000.hep.sci.osaka-u.ac.jp/
32. M. Bilenky *et al*, *Nucl. Phys.* **B409**, 22 (1993).
33. The OPAL Collaboration, *Measurement of W boson polarisations and CP-violating triple gauge couplings from W+W- production at LEP*, CERN-EP-2000-113, accepted by *Eur. Phys. J.* **C.**
34. The LEP Collaborations ALEPH, DELPHI, L3, OPAL, and the LEP TGC Working Group, LEPEWWG/TGC/2000-02. See http://lepewwg.web.cern.ch/LEPEWWG/
35. C. Sbarra, *LEP II Boson Cross-sections and Couplings*, talk at Moriond Electroweak 2000. See http://opal.web.cern.ch/Opal/confrep/
36. P. Molnar and M. Grunewald, *Phys. Lett.* **B461**, 149 (1999).
37. K. Hagiwara *et al*, *Nucl. Phys.* **B282**, 253 (1987).
38. L3 Collaboration, M. Acciarri *et al*, *Phys. Lett.* **B474**, 194 (2000).
39. OPAL Collaboration, G. Abbiendi *et al.* *EPJC* **8**, 191 (1999).
40. S. Jezequel, *Charged Triple Gauge Coupling at LEP and W polarisation*, talk at ICHEP 2000, Osaka, Japan. See http://ichep2000.hep.sci.osaka-u.ac.jp/
41. The LEP Collaborations ALEPH, DELPHI, L3, OPAL, and the LEP W Working Group, LEPEWWG/MASS/2000-01. See http://lepewwg.web.cern.ch/LEPEWWG/
42. R. Strohmer, *W Mass Measurements using Semileptonic and Fully Leptonic Events at LEPII*, talk at ICHEP 2000, Osaka, Japan. See http://ichep2000.hep.sci.osaka-u.ac.jp/
43. S. Schmidt-Karst, *W Mass Measurements using Fully Hadronic Events at LEP2*, talk at ICHEP 2000, Osaka, Japan. See http://ichep2000.hep.sci.osaka-u.ac.jp/
44. A. Valassi, *Bose-Einstein Correlations in W-pair events at LEP*, talk at ICHEP 2000, Osaka, Japan. See http://ichep2000.hep.sci.osaka-u.ac.jp/
45. ALEPH Collaboration, R. Barate *et al*, *Phys. Lett.* **B478**, 50 (2000).
46. P. De Jong, *Colour Reconnection in W Decays*, talk at ICHEP 2000, Osaka, Japan. See http://ichep2000.hep.sci.osaka-u.ac.jp/
47. *Search for Colour Reconnection Effects in* $e^+e^- \to W^+W^- \to$ *hadrons*, L3 Note 2560, submitted to ICHEP 2000, Osaka, Japan. See http://ichep2000.hep.sci.osaka-u.ac.jp/
48. *Combination procedure for the precise determination of* Z *boson param-*

eters from results of the LEP experiments, CERN-EP-2000-153.

49. *A Combination of Preliminary Electroweak Measurements and Constraints on the Standard Model*, CERN-EP-2001-021.
50. P. Igo-Kemenes, *Search for New Particles and New Phenomena: Results from e^+e^- Colliders*, plenary talk at ICHEP 2000, Osaka, Japan. See http://ichep2000.hep.sci.osaka-u.ac.jp/

THE NEUTRINO – A REVIEW

R.G. HAMISH ROBERTSON

Center for Experimental Nuclear Physics and Astrophysics,
University of Washington, Box 354290, Seattle, WA 98195
E-mail: rghr@u.washington.edu

The evidence for neutrino mass from experiments on atmospheric and solar neutrinos is now beyond significant doubt. It appears that active neutrinos are the major participants in these phenomena, although if the LSND experiment is confirmed, a light sterile neutrino would be unavoidable. Sufficient data now exist to allow reasonably quantitative statements about both the mixing and mass spectrum of the neutrinos. One sees a pattern very different from the quark sector.

1 Introduction

After seventy years of research, we still know remarkably little about the neutrino, but the last few years have seen some dramatic progress. One could say, in light-hearted but genuine tribute to this progress, that we know considerably less about the neutrino than we did 30 years ago. Three neutrinos are known experimentally by direct detection: electron, mu, and tau. That statement could not have been made before this year, when the DONUT collaboration reported[1] the first direct observation of the tau neutrino. While the existence of ν_τ was never seriously in doubt since the observation of the decay modes of the τ in the 1970s, we have learned not to be surprised at the oddness of neutrinos, and so it is reassuring to have discovered the tau neutrino produced as expected at Fermilab in charm decays. The Standard Model cast of particles is now complete with the notable exception of the Higgs.

The fermions can be arranged in left-handed doublets and right-handed singlets of weak isospin:

$$\begin{pmatrix} u \\ d' \end{pmatrix}_L \quad \begin{pmatrix} c \\ s' \end{pmatrix}_L \quad \begin{pmatrix} t \\ b' \end{pmatrix}_L$$

$$\begin{pmatrix} \nu_e \\ e \end{pmatrix}_L \quad \begin{pmatrix} \nu_\mu \\ \mu \end{pmatrix}_L \quad \begin{pmatrix} \nu_\tau \\ \tau \end{pmatrix}_L$$

$$(u)_R \quad (c)_R \quad (t)_R$$

$$(d')_R \quad (s')_R \quad (b')_R$$

$$(e)_R \quad (\mu)_R \quad (\tau)_R$$

There are no ν_R. This peculiarity was built into the Standard Model in recognition of the experimental fact that neutrino masses are much, much smaller than the corresponding charged-lepton masses (see Table 1). At the time, the masses of the neutrinos were consistent with zero, and that could be explained in the Standard Model by depriving neutrinos of right-handed fields. Masses must be Lorentz scalars, and one cannot make a scalar out of a purely left-handed object. In other words, a particle that is always left-handed must also move always at the speed of light (and have no mass), or one could pass it in a speeding Studebaker and see it falling behind spinning as a right-handed particle.

One can find right-handed fields even in the minimal Standard Model among the antineutrinos. A mass term in the Lagrangian can be constructed with a neutrino and an antineutrino if lepton number is not conserved, but the Higgs sector must then include either a pair of doublets or a triplet in order to couple the weak isospin. That is not strictly the Standard Model, but it does little violence to it as an extension. The resulting mass term provides a way to understand the smallness of neutrino mass without making it zero. If neutrinos are in fact identical to their antiparticles (Majorana neutrinos) it is plausible that they begin with a Dirac mass not very different from that of the corresponding charged leptons, but there are mass terms coupling neutrino and antineutrino. The right-handed components of those terms may be as heavy as the scale at which parity is broken. Upon diagonalizing to find mass eigenstates, two states, one very light (compared to the Dirac mass) and one very heavy, emerge. This is the "seesaw" mechanism. Appealing though the seesaw is, the smallness of neutrino mass may be completely unrelated to this, neutrinos could be Dirac particles, and our understanding of mass generation (spectacularly) deficient.

Table 1: Masses of the charged and neutral leptons.

Lepton	Mass	Unit
ν_e	<2.8	eV
e	510998.02	eV
ν_μ	< 0.19	MeV
μ	109.6583568	MeV
ν_τ	< 18.3	MeV
τ	1777.03	MeV

Standard Model neutrinos are not only massless, they are also as a consequence pure flavor states. Without mass to distinguish the neutrinos one from another, one is free to define a basis in which the neutrinos are flavor diagonal. As we shall see, neutrinos have mass and the mass eigenstates are not the same as the flavor eigenstates. Thus Table 1 is an anachronism that is shown here for the last time, it is to be hoped.

We have become accustomed to flavor mixing with the quarks. The quark mass eigenstates are not the same as the "current" or flavor eigenstates:

$$\begin{pmatrix} d' \\ s' \\ b' \end{pmatrix}_L = \begin{pmatrix} V_{ud} & V_{us} & V_{ub} \\ V_{cd} & V_{cs} & V_{cb} \\ V_{td} & V_{ts} & V_{tb} \end{pmatrix} \begin{pmatrix} d \\ s \\ b \end{pmatrix}_L$$

The CKM Matrix is nearly diagonal (displaying only the magnitudes of the central values of the experimentally determined matrix elements[2]):

$$\begin{pmatrix} d' \\ s' \\ b' \end{pmatrix}_L = \begin{pmatrix} 0.9750 & 0.224 & 0.004 \\ 0.224 & 0.9742 & 0.040 \\ 0.009 & 0.039 & 0.9992 \end{pmatrix} \begin{pmatrix} d \\ s \\ b \end{pmatrix}_L$$

One can define a similar matrix **U** to describe the mixing of neutrinos. (Naming it turns out to be more dangerous than defining it, and, recognizing the contributions of many others to the development of 3-flavor lepton mixing, e.g. [3] we will here simply call it **U**. It is also referred to as the MNS[4], or Maki-Nakagawa-Sakata-Pontecorvo [MNSP] matrix.),

$$\begin{pmatrix} \nu_e \\ \nu_\mu \\ \nu_\tau \end{pmatrix} = \begin{pmatrix} U_{e1} & U_{e2} & U_{e3} \\ U_{\mu 1} & U_{\mu 2} & U_{\mu 3} \\ U_{\tau 1} & U_{\tau 2} & U_{\tau 3} \end{pmatrix} \begin{pmatrix} \nu_1 \\ \nu_2 \\ \nu_3 \end{pmatrix}$$

In the Standard Model it is diagonal:

$$\begin{pmatrix} \nu_e \\ \nu_\mu \\ \nu_\tau \end{pmatrix} = \begin{pmatrix} 1 & & \\ & 1 & \\ & & 1 \end{pmatrix} \begin{pmatrix} \nu_1 \\ \nu_2 \\ \nu_3 \end{pmatrix}$$

but no fundamental symmetry enforces this. It is simply a statement that without neutrino mass there is no obstacle to choosing a representation that is diagonal because there is no physical observable to contradict that choice.

2 Neutrino Oscillations: Atmospheric Neutrinos

The experiments designed to search for proton decay, IMB[5] and Kamiokande[6], were obliged to deal quantitatively with atmospheric neutrino interactions as the major background to proton instability. It was known that the cosmic ray flux and resulting production rate of neutrinos in the upper atmosphere were uncertain at the level of about a factor of 2, but it gradually became apparent that the ratio of $\nu_\mu + \overline{\nu_\mu}$ to $\nu_e + \overline{\nu_e}$ was also not well predicted. The latter ratio, naively 2 (from pion and muon decay) and calculable to an accuracy of about 5%, was found to be low by a factor of typically 0.6. As it depends little on the cosmic-ray flux and normalization, the departure from the expected value was somewhat surprising, and neutrino-oscillation solutions were proposed. Towards the end of the operation of Kamiokande, evidence was accumulating for a zenith-angle dependence of the ratio (equivalent to a path-length dependence).

The Super-Kamiokande detector operated from April, 1996, to July 2001. With a fiducial mass of approximately 22,500 tons it acquired events at 10 times the rate of Kamiokande. Other detectors, MACRO and Soudan II, also accumulated atmospheric neutrino data. Evidence consistent with $\nu_\mu \rightarrow \nu_\tau$ neutrino oscillation has emerged.

- The zenith-angle dependence in the ν_μ flux departs greatly from the no-oscillation expectation and agrees closely with an oscillation description.
- The flavor ratio (ν_μ/ν_e) is 0.68 the expected magnitude.
- The zenith-angle dependence of the ν_e flux agrees closely with the no-oscillation expectation.
- There is no indication of matter effects at any energy, which supports the $\nu_\mu \rightarrow \nu_\tau$ interpretation (since matter effects are identical and cancelling for both species) and disfavors at the 99% CL the two-flavor sterile

scenario $\nu_\mu \to \nu_s$. However, significant admixtures of sterile neutrinos as a third component cannot be ruled out: up to 32% at 90% CL is permissible.

- Neutral-current production of π^0 is consistent with $\nu_\mu \to \nu_\tau$, although the cross section is uncertain to 25%.
- On a statistical basis, event topologies in Super-Kamiokande are consistent with τ production at about 2σ.
- Results from the K2K long-baseline experiment using the ν_μ beam from KEK support ν_μ disappearance at about the 2σ level.

The best-fit oscillation parameters are[15],

$$\begin{aligned} \Delta m^2 &= 2.4 \times 10^{-3}\ \mathrm{eV}^2 \\ \sin^2 2\theta &= 1.0 \end{aligned}$$

There is no evidence in the atmospheric data for the subdominant oscillation $\nu_\mu \to \nu_e$ channel at the same mass difference. Crucially important corroboration comes from the Chooz[7] and Palo Verde[8] reactor antineutrino oscillation searches, which have sensitivity to $\overline{\nu}_e$ disappearance in the same mass range.

In a combined analysis, Fig. 1, the Bari group[9] find an upper limit on the corresponding admixture between the electron neutrino and active non-electron flavors,

$$\begin{aligned} \tan^2\phi \equiv \tan^2\theta_{13} &\leq 0.03 && (90\%\ \mathrm{CL}) \\ \tan^2\psi \equiv \tan^2\theta_{23} &= 1.0(\times 2.3^{\pm 1}) && (90\%\ \mathrm{CL}) \\ \Delta m^2_{23} &= 3.0 \times 10^{-3}(\times 1.7^{\pm 1})\ \mathrm{eV}^2 && (90\%\ \mathrm{CL}). \end{aligned}$$

3 Neutrino Oscillations: Solar Neutrinos

Beginning with the Davis experiment in the late 1960's, every solar neutrino experiment has measured a flux lower than that predicted by solar models.

By 1997, data from 5 experiments (3 different types of experiment, Cl-Ar, Ga, and water Cherenkov) were providing information on different combinations of the various flux components that make up the solar spectrum. The current results are summarized in Table 2. Surprisingly, with only 3 independent types of measurement and 8 different neutrino sources in the sun, it is impossible to fit the data (well) without introducing neutrino oscillations

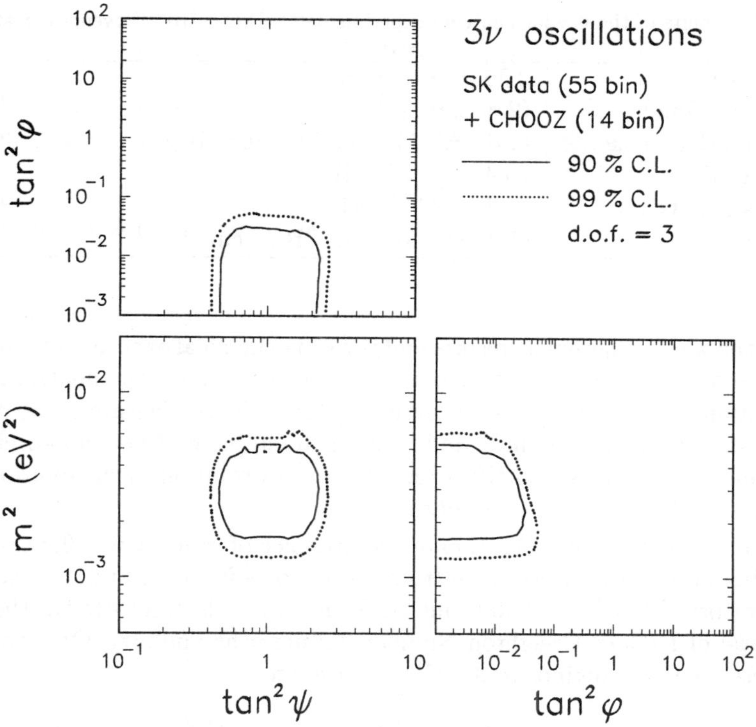

Figure 1: Combined analysis of atmospheric neutrino data and Chooz data to extract an upper limit on the magnitude of U_{e3} (Fogli *et al.*[9].)

or some other non-standard-model physics. Requiring only that the neutrino spectrum of each individual component is as measured in the laboratory (or, in the case of the *pp* flux, calculated), that only electron neutrinos come from the sun, and that the energy output agrees with the solar luminosity, one finds that any combination of the ^{7}Be and CNO flux is non-physical (i.e. negative) at more than 3σ. The model-independent analyses[18,19,20,21] made it clear for the first time that no astrophysical solution would resolve the solar neutrino problem, and the solution lay most likely with neutrino physics.

It could only be concluded that the shape of the ^{8}B spectrum was not as expected, containing more strength at high energies and less at low, and/or the neutrino flavor content is not pure electron, which alters the relationship be-

Table 2: Results of the 6 solar neutrino experiments (1 SNU = 10^{-36} events per atom per second).

Experiment	Result	Ref.
Cl-Ar	$2.56 \pm 0.16 \pm 0.16$ SNU	10
Kamiokande	$(2.80 \pm 0.19 \pm 0.33) \times 10^6$ ^{8}B ν_e cm^{-2} s^{-1}	11
SuperKamiokande	$(2.35 \pm 0.03 \pm 0.07) \times 10^6$ ^{8}B ν_e cm^{-2} s^{-1}	15
SAGE	$75.4^{+7.0}_{-6.8}{}^{+3.5}_{-3.0}$ SNU	13
Gallex+GNO	$74.1 \pm 5.4^{+4.0}_{-4.2}$ SNU	14
SNO	$(1.75 \pm 0.16) \times 10^6$ ^{8}B ν_e cm^{-2} s^{-1}	16

tween the water-Cherenkov results and the radiochemical experiments (because elastic scattering, unlike inverse beta decay, can occur via the neutral-current interaction with neutrinos of all active flavors). These features, not permitted in the Minimal Standard Model, are characteristic of neutrino-oscillation solutions[18]. In contrast to the standard-physics solution, such solutions can give an excellent account of all data.

A more direct demonstration of the presence of other active flavors in the solar flux besides the electron neutrino was to come from the Sudbury Neutrino Observatory. SNO[22] is a 1000-tonne D2O Cherenkov detector sited at the 6800-foot level of INCO's Creighton Number 9 mine near Sudbury, Ontario. SNO can detect solar neutrinos in 3 different processes,

$$\begin{aligned} &\nu_e + d \rightarrow p + p + e^- && \text{(CC)} \\ &\nu_x + d \rightarrow p + n + \nu_x && \text{(NC)} \\ &\nu_x + e^- \rightarrow \nu_x + e^- && \text{(ES)} \end{aligned}$$

By comparing the rates of CC reaction to the NC reaction, or CC to ES, one may deduce immediately if there are other flavors present, because the CC reaction detects only ν_e, the NC reaction is equally sensitive to all 3 active flavors, and the ES cross section for ν_μ and ν_τ is 0.154 that for ν_e. Those comparisons, in the energy regime where only ^{8}B neutrinos are present, are independent of solar physics. A determination of the *hep* neutrinos from the upper end of the spectrum is also needed to make the comparison rigorous; as expected, they play a minor role.

The construction of SNO began in 1990 and was complete in 1998. Water fill then commenced and after a period of calibration and adjustment, 'production' running began in November 1999. The first data from pure heavy water covering 241 live days between November 1999 and January 2001 were used[16] to derive a precise measurement of the ν_e flux above an electron kinetic energy of 6.75 MeV:

$$\Phi_{\rm SNO}^{\rm CC} = 1.75 \pm 0.07^{+0.12}_{-0.11} \pm 0.05 \ \times 10^6 \ {}^8{\rm B} \ \nu_e \ {\rm cm}^{-2}{\rm s}^{-1}$$

which may be compared with the flux measured[23] by Super-Kamiokande in elastic scattering,

$$\Phi_{\rm SK}^{\rm ES} = 2.32 \pm 0.03^{+0.08}_{-0.07} \ \times 10^6 \ {}^8{\rm B} \ \nu_e \ {\rm cm}^{-2}{\rm s}^{-1}$$

The results can be displayed graphically in a variety of ways. Figure 2 is a convenient form devised independently by Parke and by Hime[25].

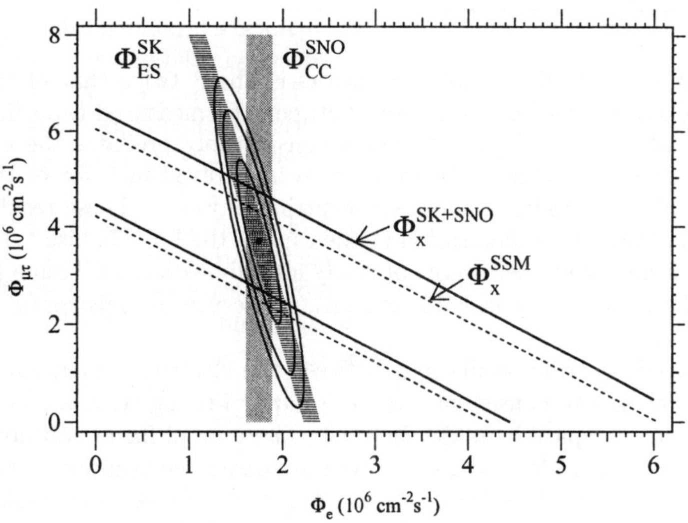

Figure 2: Extracting the flux of ν_e and $\nu_{\mu\tau}$ from a comparison of the data from SNO and Super-Kamiokande. The contours are *joint* confidence levels of 68%, 90%, and 99%.

The CC result is 3.3 σ below the ES result, indicating the presence in the solar neutrino flux of active flavors other than electron. Indeed, the fluxes measured (in units of 10^6 ^{8}B ν_e cm^{-2} s^{-1}) are,

$$\begin{aligned} \Phi_e &= 1.75 \pm 0.15 \\ \Phi_{\mu\tau} &= 3.69 \pm 1.13 \\ \Phi_{\rm total} &= 5.44 \pm 0.99 \\ \Phi_{\rm BPB01} &= 5.05^{+1.01}_{-0.81} \end{aligned}$$

Table 3: Two-flavor neutrino oscillation fits to the solar neutrino data, showing the 5 most probable solutions and the Goodness of Fit (Krastev and Smirnov[17]). The *hep* flux is constrained to the SSM value, ^{8}B is free. Shape information from SK[15] but not SNO[16] is used.

Solution	$\nu_e \rightarrow$	Δm^2 eV^2	$\tan^2\theta$	$\chi^2_{\rm min}$	G.O.F.
Large-angle	Active	5.0 x 10^{-5}	0.36	30.0	0.85
Vacuum	Active	1.4 x 10^{-10}	0.363 (2.7)	33.0	0.74
Low	Active	1.1 x 10^{-7}	0.69	34.2	0.69
Vacuum	Sterile	1.4 x 10^{-10}	0.35 (2.9)	36.4	0.59
Small-angle	Active	5.5 x 10^{-6}	0.0019	42.0	0.34

and one sees that the flux of non-electron flavors is about twice that of the electron flavor. There is excellent agreement between the measured total flux and the theoretically predicted value[24], which considerably predates the experimental one. Thus the solar neutrino problem is resolved in favor of the astrophysical calculations and demands new neutrino physics. (These results were known within the SNO collaboration at the time of the Lake Louise Winter Institute in February 2001, but were not ready for publication until June 18 of that year. To keep the text reasonably current, the results are incorporated in the written version.)

Matter-enhanced neutrino oscillations between the electron neutrino and the other active flavors are consistent with the data. Mixing with a sterile neutrino alone is not compatible at the 3σ level. There are 2-flavor solutions that describe the data well. Many analyses have been done; one example is that of Bandyopadhyay *et al.*[26]. The Large-Mixing-Angle (LMA) and the 'LOW' solutions survive, with LMA being considerably more probable (Fig. 3).

Krastev and Smirnov[17] find the best overall fit is for the LMA solution (see Table 3) but that other solutions are possible. Vacuum solutions tend to have significant shape distortions, which would be further disfavored if shape information from SNO were to be folded in to the analysis.

Day-night effects arise from matter regeneration in the varying path through the earth's core. No certain evidence for time variations beyond statistical fluctuations has shown up to date, and the most recent data from SK[27] (1496 days) yield,

$$2\frac{N-D}{N+D} = 0.020 \pm 0.020 \pm 0.013.$$

The absence of large day-night effects has already ruled out a large region of parameter space in the range $10^{-6} \leq \Delta m^2 \leq 10^{-5}$ eV^2 and $10^{-2} \leq \sin^2 2\theta$. As

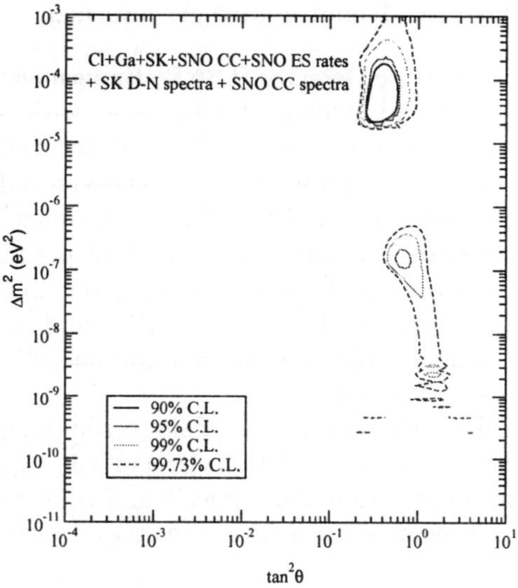

Figure 3: Two-neutrino mixing after SNO (from Bandyopadhyay *et al.*[26].)

Suzuki[27] shows (for further discussion, see also Bahcall *et al.*[28]), the small but general night enhancement matches better with the LMA solution than it does the SMA solution where effects are all negligible except when the sun is on the other side of the earth's core. The latter possibility seems disfavored at almost 3σ. The details of the LMA region are shown by Bahcall *et al.*[28]. The region of maximal day-night variation moves down in Δm^2 as the energy goes down, and provides a decisive marker for the LOW solution in Borexino[29]. However, if the LMA solution is the correct one, then the KamLAND experiment[30] will observe the disappearance of reactor $\overline{\nu}_e$ over a long baseline.

4 Neutrino Oscillations: Reactor and Accelerator

The Liquid Scintillator Neutrino Detector (LSND) Experiment[31] at Los Alamos National Laboratory has found evidence for $\overline{\nu}_\mu \rightarrow \overline{\nu}_e$ with $0.2 \leq \Delta m^2 \leq 2.0$ eV^2 and $0.04 \geq \sin^2 2\theta \geq 0.0015$. The statistical accuracy in the charge-conjugate channel $\nu_\mu \rightarrow \nu_e$ is not quite sufficient to provide a decisive confirmation, but the result is consistent. The KARMEN Experiment[32] at Rutherford-Appleton Laboratory reports no signal in a region of parameter space overlapping much of that explored by LSND. The small remaining region defines the parameters just given for LSND, plus a small island at Δm^2= 4.5 eV^2, $\sin^2 2\theta$= 2.5 x 10^{-3}. The Brookhaven E776 Experiment[33] excludes (in the $\nu_\mu \rightarrow \nu_e$ channel) that small island.

As has been described, the Chooz[7] and Palo Verde[8] long-baseline reactor antineutrino experiments show no oscillation effects for $\overline{\nu}_e$ at $\Delta m^2 \geq 10^{-3}$ eV^2.

Together with the results from atmospheric and solar neutrino measurements, these define the scenarios for neutrino mass and mixing that are most likely.

5 Mixing Matrix for Neutrinos

Without mass, a flavor state is just a matter of definition (representation), e.g.:

$$\nu_e = U_{e1}\nu_1 + U_{e2}\nu_2 + U_{e3}\nu_3$$

This relationship would stay the same at all times and places. But if the masses of ν_i are not zero, then

$$\nu_e = U_{e1}e^{-iE_1t}\nu_1 + U_{e2}e^{-iE_2t}\nu_2 + U_{e3}e^{-iE_3t}\nu_3 \ ,$$

where $E_i^2 = p^2 + m_i^2$ and the state evolves with time or distance.

The overall phase is unobservable

$$\nu_e = e^{-iE_1t}(U_{e1}\nu_1 + U_{e2}e^{-i(E_2-E_1)t}\nu_2 + ...)$$

and $m_i^2 << p^2$, so $E_j - E_i = \frac{(m_j^2-m_i^2)L}{2E}$ and observable effects depend only on mass-squared differences.

In the presence of matter the forward scattering amplitude for neutrinos is not zero, which leads to a refractive index:

$$n_l = 1 + \frac{2\pi N}{p^2} f_l(0)$$

which is different for l = e and $l = \mu$ or τ because W exchange can occur in addition to Z exchange for l = e.

$$n_l = 1 + \frac{2\pi N}{p^2}\left(-\sqrt{2}\frac{G_F p}{2\pi} + \mathrm{NC}\right)$$

The "matter oscillation length" L_0 is defined by

$$\sqrt{2}G_F N_e L_0 = 2\pi$$

and the refractive index differs only microscopically from 1:

$$n_l \simeq 1 - 1 \times 10^{-19}$$

Small though this difference is, the additional phase leads to important effects for neutrinos. Moreover, matter effects contribute a phase shift in the flavor basis, while mass shifts the phase in the mass basis. The composite propagation phase is non-diagonal, which can give rise to strong avoided-crossing resonance effects and large flavor changes even when the vacuum mixing is small.

The observation in solar and atmospheric neutrinos of flavor change is thus compelling evidence for both mass and mixing. As it turns out, because the solar data favor large mixing, our knowledge of the mixing is quite detailed, while knowledge of the mass spectrum is still fragmentary.

It is illuminating to write the **U** matrix as the product of 3 Euler rotations:

$$\begin{pmatrix} \nu_e \\ \nu_\mu \\ \nu_\tau \end{pmatrix} = \begin{pmatrix} U_{e1} & U_{e2} & U_{e3} \\ U_{\mu 1} & U_{\mu 2} & U_{\mu 3} \\ U_{\tau 1} & U_{\tau 2} & U_{\tau 3} \end{pmatrix} \begin{pmatrix} \nu_1 \\ \nu_2 \\ \nu_3 \end{pmatrix}$$

$$\begin{aligned} U &= \begin{pmatrix} 1 & 0 & 0 \\ 0 & \cos\theta_{23} & \sin\theta_{23} \\ 0 & -\sin\theta_{23} & \cos\theta_{23} \end{pmatrix} \times \\ &\times \begin{pmatrix} \cos\theta_{13} & 0 & e^{-i\delta_{CP}}\sin\theta_{13} \\ 0 & 1 & 0 \\ -e^{i\delta_{CP}}\sin\theta_{13} & 0 & \cos\theta_{13} \end{pmatrix} \times \\ &\times \begin{pmatrix} \cos\theta_{12} & \sin\theta_{12} & 0 \\ -\sin\theta_{12} & \cos\theta_{12} & 0 \\ 0 & 0 & 1 \end{pmatrix} \end{aligned}$$

Note, that writing **U** in this way builds in unitarity. However, unitarity for the 3x3 matrix is not a given and testing it is an important experimental objective for the future.

The product is the same form PDG[2] uses for the CKM matrix:

$$U =$$

$$\begin{pmatrix} c_{12}c_{13} & s_{12}c_{13} & s_{13}e^{-i\delta_{CP}} \\ -s_{12}c_{23} - c_{12}s_{23}s_{13}e^{i\delta_{CP}} & c_{12}c_{23} - s_{12}s_{23}s_{13}e^{i\delta_{CP}} & s_{23}c_{13} \\ s_{12}s_{23} - c_{12}c_{23}s_{13}e^{i\delta_{CP}} & -c_{12}s_{23} - s_{12}c_{23}s_{13}e^{i\delta_{CP}} & c_{23}c_{13} \end{pmatrix}$$

If neutrinos are Dirac, then only one CKM-like CP violating phase enters, but if neutrinos are Majorana, a separate phase for each submatrix is needed[36].

In the case where the solar and atmospheric mixing is maximal and there is no 13 mixing one arrives at the bimaximal mixing matrix[34,35]:

$$\begin{pmatrix} 1/\sqrt{2} & 1/\sqrt{2} & 0 \\ -1/2 & 1/2 & 1/\sqrt{2} \\ 1/2 & -1/2 & 1/\sqrt{2} \end{pmatrix}$$

But is U_{e3} really zero? And solar mixing is most likely large but not maximal, if the LMA or the LOW solution turns out to be the correct one. Inserting the appropriate mixing angles for the atmospheric ($\tan^2\theta_{23} = 1.0$) solar ($\tan^2\theta_{12} = 0.36$) and "13" ($\tan^2\theta_{13} = 0.0$) data, a "best bet" for **U** can be assembled:

$$\begin{pmatrix} \nu_e \\ \nu_\mu \\ \nu_\tau \end{pmatrix} = \begin{pmatrix} U_{e1} & U_{e2} & U_{e3} \\ U_{\mu 1} & U_{\mu 2} & U_{\mu 3} \\ U_{\tau 1} & U_{\tau 2} & U_{\tau 3} \end{pmatrix} \begin{pmatrix} \nu_1 \\ \nu_2 \\ \nu_3 \end{pmatrix}$$

Atmospheric

$$\begin{pmatrix} 1 & 0 & 0 \\ 0 & 1/\sqrt{2} & 1/\sqrt{2} \\ 0 & -1/\sqrt{2} & 1/\sqrt{2} \end{pmatrix} \times$$

Chooz

$$\times \begin{pmatrix} \sim 1 & 0 & e^{-i\delta_{CP}} \sin\theta_{13} \\ 0 & 1 & 0 \\ -e^{i\delta_{CP}} \sin\theta_{13} & 0 & \sim 1 \end{pmatrix} \times$$

LMA

$$\times \begin{pmatrix} 0.85 & 0.51 & 0 \\ -0.51 & 0.85 & 0 \\ 0 & 0 & 1 \end{pmatrix}$$

Armed with the matrix U, one can now calculate the amplitudes or intensities of the flavor components in the mass eigenstates, and vice versa. The flavor composition is indicated in Fig. 4.

An important consequence is the determination of the state that admixes with the electron neutrino in the solar neutrino problem. There is no mystery about whether ν_μ or ν_τ is involved, and how much; the atmospheric data fix the admixed component to be a 50:50 combination of the two. That in effect reduces the solar neutrino phenomenology to an equivalent two-neutrino situation, provided U_{e3} is negligible.

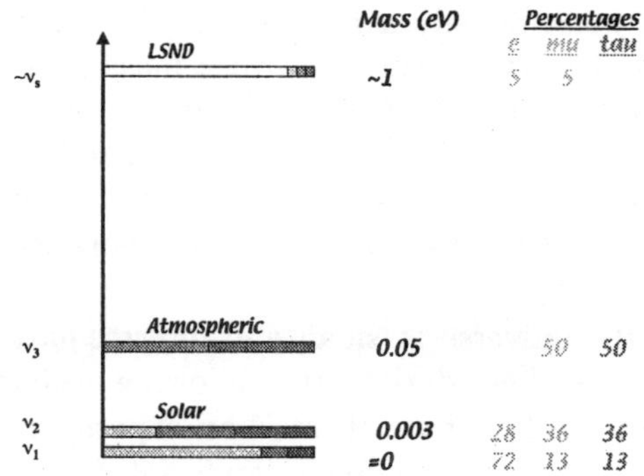

Figure 4: A viable mass and mixing scheme based on current knowledge of the mixing matrix derived from the atmospheric and solar data. Also shown but not quantitatively included in the intensities is an additional sterile state that would accommodate the LSND result without being seriously in conflict with other known information.

If the positive indications of neutrino oscillation from LSND, from atmospheric neutrinos, and from solar neutrinos are all correct, then a fourth, sterile, neutrino must exist, because the values of Δm^2 do not close cyclically for only 3 eigenstates. The 3+1 scenario shown in the figure indicates one way to accommodate the LSND result in the face of the experimental observation that both atmospheric and solar neutrino oscillations appear to involve active neutrinos. Another, 2+2, arrangement is also possible, in which case significant sterile admixtures in the atmospheric and solar solutions are required.

6 Neutrino Masses

Less is known about the masses at this juncture. Oscillation experiments measure only mass-squared differences, and so the spectrum indicated in Fig. 4 may be built on an offset, which can be substantial as it is limited only by the 2.8-eV laboratory limit from tritium beta decay[37]. The mass differences shrink as the offset increases, and the spectrum becomes very degenerate. Even the level order is uncertain: the order of ν_1 and ν_2 is fixed by the fact that matter-enhanced conversion is occurring in the sun, but the relative locations of the other two states, whether above or below, is not known. However, the significance of the oscillation results in linking together the mass values by experimentally known differences cannot be overemphasized. It is now possible to approach the mass question experimentally with "electron" neutrinos with no concern that one might be working on the wrong neutrino.

Neutrinos with a mass of only a few eV would contribute a significant fraction of the closure density. A species with a mass of $94h_0^2$ eV, where h_0 is the Hubble constant in units of 100 km s^{-1} Mpc^{-1}, would provide closure density. There is now convergence[2] on $h_0 = 0.71 \pm 0.07$, and the mass for closure would be 47 eV. If the mass eigenstates are at the limit given by tritium beta decay, and there are only 3 flavors, then neutrinos represent at most 18% of the closure density.

The structure of the universe at small and intermediate scales is incompatible with such amounts of hot dark matter, allowing probably no more than 10%. Nevertheless, for the first time there is an experimental upper bound, and good prospects that it will come down significantly.

There is also a lower bound that can be set by setting the mass of ν_2 to zero, in which case the atmospheric neutrino oscillation data yields a mass for ν_3 of 0.05 eV. Remarkably this value, small though it is, is comparable to the mass of all luminous stars.

Only two methods are presently known that have direct mass sensitivity that is at least roughly in the needed range, single beta decay (of tritium especially) and double beta decay.

Two new tritium experiments are in the wings: the Karlsruhe Tritium Neutrino experiment (KATRIN[38]) and an experiment[39] under research and development at the University of Texas at Austin. The objective of KATRIN is a mass sensitivity of 0.35 eV, and of the UTA experiment, 0.8 eV.

Many nuclei provide an opportunity to search for neutrinoless double beta decay; single beta decay is blocked by energy conservation. Because large high-resolution detectors can be made from Ge (enriched in ^{76}Ge), the current best limits come from the Heidelberg-Moscow collaboration[40,41] and the IGEX collaboration[42]:

$$< m_\nu^{Maj} >= \sum_m \lambda_m |U_{em}|^2 m_m <\sim 0.4 \text{ eV}$$

where m is an index summing over Majorana mass terms only and λ_m is the CP phase. For a comprehensive discussion of the connection between double beta decay and the oscillation and direct mass data, see Ref.[43]

7 Conclusions

The evidence for neutrino mass and mixing seems now to be beyond serious doubt. The mixing matrix in particular is becoming clear, and it looks very different from the CKM matrix. Understanding that phenomenon will guide our necessary extensions to the Standard Model.

This year has marked the end of the line for some old troupers of the Standard Model, the electron, mu, and tau neutrinos. Those neutrinos do not exist; in their place are the prosaically named ν_1, ν_2, ν_3, massive physical particles with an ecumenical view of flavor.

The experimental task is at once formidable and exciting. New oscillation experiments will tighten the limits on the allowed parameters. The masses may be just out of reach of present technology, or they may in fact materialize in the next generation of single and double beta decay experiments.

Acknowledgments

I gratefully acknowledge stimulating discussions with Bruce Campbell, Steve Elliott, Susan Gardner, and Stephen Parke. This work has been supported by the US DOE Division of Nuclear Physics.

References

1. B. Baller (DONUT Collaboration), Nucl. Phys. Proc. Suppl. **98**, 43 (2001).
2. Particle Data Group, Europ. Phys. J. **15**, 1 (2000).
3. B.W. Lee, S. Pakvasa, R. Shrock, and H. Sugawara, Phys. Rev. Lett. **38**, 937 (1977); B.W. Lee and R. Shrock, Phys. Rev. **D16**, 1444 (1977).
4. Z. Maki, M. Nakagawa, and S. Sakata, Prog. Theor. Phys. **28**, 870 (1962); B. Pontecorvo, Zh. Exp. Theor. Fiz. **53**, 1717 (1967).
5. R. Becker-Szendy *et al.*, Phys. Rev. **D46**, 3720 (1992).
6. Y. Fukuda *et al.*, Phys. Lett. **B335**, 237 (1994).

7. Y. Declais, Nucl. Phys. B (Proc. Suppl.) **70**, 148 (1999); M. Appolonio *et al.* *hep-ex*/9907037, Phys. Lett. **B466**, 415 (1999).
8. F. Boehm *et al.*, Phys. Rev. D **64**, 112001 (2001) *hep-ex*/0107009.
9. G. Fogli, E. Lisi, and A. Marrone, *hep-ph*/0105139.
10. B.T. Cleveland *et al.*, Astrophys. J. **496**, 505, (1998).
11. Y. Fukuda *et al.* Phys. Rev. Lett. **77**, 1683, (1996).
12. SuperKamiokande Collaboration, Y. Fukuda *et al.*, *hep-ex*/9805021, 1998; Y. Suzuki, these proceedings, 1999.
13. V. N. Gavrin [SAGE Collaboration], Nucl. Phys. Proc. Suppl. **91**, 36 (2001).
14. E. Bellotti, Nucl. Phys. Proc. Suppl. **91**, 44 (2001).
15. Y. Totsuka, in *Future Opportunities for Neutrino Physics*, Victoria, B.C., Nov. 8-11, 2001 (unpublished).
16. Q.R. Ahmad *et al.* (the SNO Collaboration) Phys. Rev. Lett. **87**, 071301 (2001).
17. P. Krastev and A. Yu. Smirnov, *hep-ph*/0108177.
18. N. Hata, S. Bludman, and P. Langacker, Phys. Rev. **D 49**, 3622 (1994).
19. S. Parke, Phys. Rev. Lett. **74**, 839 (1995).
20. K.M. Heeger and R.G.H. Robertson, Phys. Rev. Lett. **77**, 3720 (1996).
21. J.N. Bahcall, P.I. Krastev, and A.Yu. Smirnov, Phys. Rev. **D 58**, 096016 (1998).
22. J. Boger *et al.* the SNO Collaboration, Nucl. Instr. and Meth. **A449**, 172 (2000).
23. S. Fukuda *et al.*, Phys. Rev. Lett. **86**, 5651 (2001).
24. J.N. Bahcall, M.H. Pinsonneault, and S. Basu, *astro-ph*/0010346.
25. S. Parke, A. Hime, private communications, 2001.
26. A. Bandyopadhyay *et al.*, *hep-ph*/0106264.
27. Y. Suzuki, invited talk, Joint meeting of APS and JPS, Hawaii, October 17-21, 2001 (unpublished).
28. J.N. Bahcall, P.I. Krastev, and A.Yu. Smirnov, *hep-ph*/9905220.
29. G. Alimonti *et al.* Nucl. Phys. Proc. Suppl. **39**, 149 (1998).
30. J. Busenitz *et al.*, *Proposal for US Participation in KamLAND,* March 1999 (unpublished). May be downloaded from http://bfk0.lbl.gov/kamland/; C. Gonzalez-Garcia and C. Peña-Garay, *hep-ph*/0111432.
31. C. Athanassopoulos *et al.*, Phys. Rev. **C58**, 2489 (1998); see also K. Eitel, New Jour.Phys. **2**, 1 (2000).
32. K. Eitel, KARMEN Collaboration, Nucl.Phys.Proc.Suppl. **91**, 191 (2000); see also K. Eitel, New Jour.Phys. **2**, 1 (2000).
33. L. Borodovsky *et al.*, Phys. Rev. Lett. **78**, 274 (1992).

34. V.D. Barger, S. Pakvasa, T.J. Weiler and K. Whisnant, Phys. Lett. B **437**, 107 (1998) *hep-ph*/9806387.
35. G. Altarelli and F. Feruglio, Phys. Rept. **320**, 295 (1999).
36. K. Zuber, Phys. Rept. **305**, 295 (1998).
37. H. Barth *et al.* Nucl. Phys. **A654**, 988c (1999).
38. A. Osipowicz *et al.* (KATRIN Collaboration), Sub-eV sensitivity for the electron neutrino mass," *hep-ex*/0109033.
39. M. Fink, unpublished.
40. L. Baudis *et al.*, hep-ex/9902014.
41. P. Fisher, B. Kayser, and S. MacFarland, hep-ph/9906244.
42. D. Gonzalez *et al.*, Nucl. Phys. Proc. Suppl. **87**, 278 (2000).
43. Y. Farzan, O.L. Peres and A.Y. Smirnov, Nucl. Phys. B **612**, 59 (2001) *hep-ph*/0105105.

Beyond the Standard Model

G.G. Ross
Department of Physics, Theoretical Physics, University of Oxford,
1 Keble Road, Oxford OX1 3NP, U.K.

Abstract

We discuss ideas for physics going Beyond the Standard Model, including compositeness, supersymmetry and large new dimensions. The prospects for constructing the ultimate "Theory of Everything" are considered together with the implications for gauge coupling unification, fermion masses and mixing, and CP violation.

1 Introduction

In these lectures I will discuss the most promising attempts that have been made to go beyond the Standard Model of the strong and electroweak interactions.

In the next Section I will discuss the modern interpretation of the Standard Model as an effective field theory descending from some scale of underlying physics "Beyond the Standard Model". Many of the features of the Standard Model are immediately understandable in this context but the hierarchy problem, namely the problem of keeping the electroweak breaking scale low in the presence of a high scale of new physics, presents a serious difficulty.

In Section 3 I will discuss the "bottom up" approach to go beyond the Standard Model. Such attempts can be classified by the way they solve the hierarchy problem. One possibility is that the states of the Standard Model are not all fundamental and that there is a scale of compositeness which should be accessible to the next generation of experiments. A second possibility is that there is a new symmetry, supersymmetry, which protects the electroweak breaking scale from corrections involving the high scale of new physics thus solving the hierarchy problem. In this case there must be new supersymmetric partners to the Standard Model states whose mass cannot be much more than a TeV and thus should also be accessible to the next generation of experiments. The third possibility is much more radical, requiring that there is *no* scale of new physics far beyond a TeV. In this case one must explain the origin of the weakness of the gravitational interaction which is conventionally explained by the appearance of the Planck scale some $10^{16}TeV$. In order to do so it is necessary to suppose that gravity propagates in more than four space-time dimensions. The new dimensions are hidden from us because they are compactified at some low scale. In practice this means something less than of order a millimeter!

In Section 4 I will adopt the "top down" approach in which we attempt to determine what lies beyond the Standard Model by constructing a "Theory of Everything" which gives rise to the Standard Model at low scales. This ambitious project is largely motivated by the desire to extend the unification which comes with the Standard Model, possibly even including the gravitational interactions. The most promising top-down ideas include Grand Unification in which the strong, weak, and electromagnetic interactions are fully unified and Superstring Unification in which this unification extends to the gravitational interactions.

In Section 5 I discuss in more detail the new ideas surrounding the possibility of large new dimensions. Of particular interest is a question how they fit into top-down models and whether there are testable predictions capable of discriminating theories involving large new dimensions from their more conventional cousins.

Finally in Sections 6-9 I discuss the implication of these ideas for the parameters of the Standard Model including quark masses, mixing angles, and CP violation, family symmetries, charged lepton masses and neutrino masses.

2 The Standard Model

The Standard Model is a renormalisable quantum field theory based on the the underlying local gauge symmetry, $SU(3) \times SU(2) \times U(1)$. Associated with this gauge symmetry are the gauge bosons, the eight gluons, the charged weak bosons, $W^{\pm}$, and the neutral weak boson, Z, together with the photon. The matter states fill out three generations of quarks and leptons, (u, d, e, ν_e), (c, s, μ, ν_μ) and (t, b, τ, ν_τ). Each generation has identical properties under the Standard Model gauge group. Under $SU(2)$ the representations are chiral with the left-handed belonging to $SU(2)$ doublets and the right- handed states $SU(2)$ singlets. The three generations however have different mass, corresponding to different Yukawa couplings in the Standard Model Lagrangian. In addition, in order spontaneously to break the symmetry, the Standard Model has an electroweak doublet of Higgs scalars fields. After spontaneous symmetry breaking all but one neutral component of the scalars is "eaten" providing the longitudinal components of the $W^{\pm}$ and Z. In order to specify the Standard Model completely it is necessary to define 19 fundamental parameters which include the three gauge couplings, together with the masses, mixing angles and CP violating phase of the quarks, plus the masses of the charged leptons. Essentially all strong and electroweak phenomena are explained by the Standard Model in terms of these fundamental parameters. The tests are at a high level of precision[1] as shown in Figure 1 and have tested the theory at the quantum level, being sensitive to radiative corrections. All the states of the Standard Model have been found with the exception of the Higgs scalar. However indirect tests looking for its effects via radiative corrections indicate that it is there and cannot be much above the energies presently being probed. This is shown in Figure 2[1].

The only area where the Standard Model is known to fail is in the prediction of the neutrino masses and the lepton mixing angles. In the Standard Model these are absent, but a relatively minor extension of the Standard Model to include right-handed neutrinos is able to describe this data too.

2.1 Questions raised by the Standard Model

Despite its success many physicists think the Standard Model is incomplete because it leaves so many questions unanswered. These include:

1. Why renormalisable?

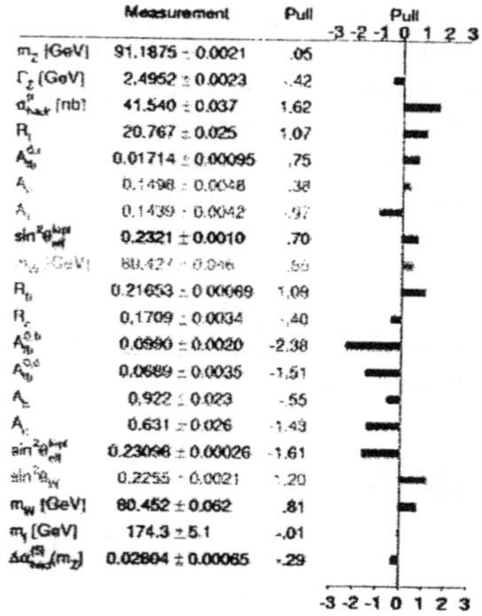

Figure 1: Precision tests of the Standard model showing the deviation of the measurements from the Standard Model predictions.

2. Why a local gauge theory?
3. Why chiral fermions?
4. Why the Higgs?
5. Why $SU(3) \times SU(2) \times U(1)$?
6. Why the fermion multiplet structure : why is $Q_q = nQ_l/3$; why left-handed doublets...?
7. Why 3 generations?
8. Why 19 parameters?
9. What about gravity?

2.2 The Standard model as an Effective Field Theory

In these lectures I will try to review the progress that has been made in answering these questions. Surprisingly much of this structure follows immediately if one supposes that the Standard Model is an *effective field theory* descending from a fundamental theory at some high scale, M[2]. This follows because the effective Lagrangian describing the light fields, ϕ_{light}, obtained on integrating

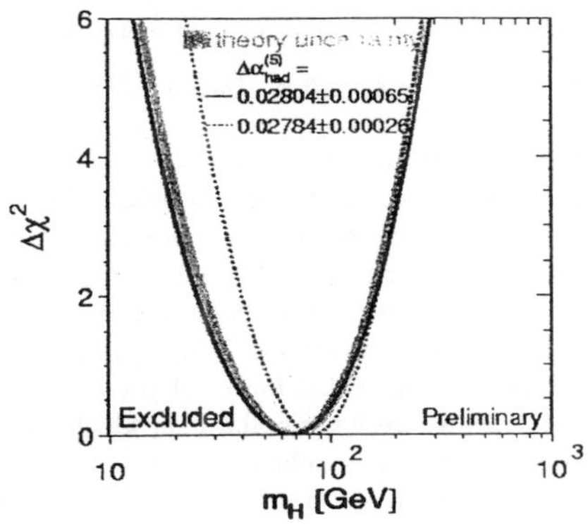

Figure 2: The χ^2 distribution showing the need for a light Higgs from its radiative effects.

out the heavy fields (with mass of $O(M)$) may be written in a Taylor series ordered in inverse powers of M

$$\mathcal{L}(\phi_{light}, \phi_{heavy}) \rightarrow \mathcal{L}^{eff}(\phi_{light}) = \Sigma_n a_n O_n^{d_n} (\frac{1}{M})^{d_n-4} \tag{1}$$

where $O_n^{d_n}$ is a local combination of light fields labelled by its engineering dimension. The best known example of this is the Fermi theory of weak interactions. Weak interactions are described by a term such as $G_F(\overline{\mu}\gamma_\mu\nu)(\overline{\nu}\gamma^\mu e)$ which applies at energy scales much less than the W boson mass and comes from integrating out the heavy W boson. Here $G_F \propto 1/M_W^2$. It may be seen from eq.1 that low dimension terms dominate for energies $<< M$. This means the dominant terms of the effective theory describing boson and fermion fields define a *renormalisable* theory with the terms

$$\mathcal{L}^{eff} \sim \phi^4,\ \overline{\psi}\psi\phi,\ M^2\phi^2,\ M\overline{\psi}\psi,\ \phi\partial_\mu\phi A^\mu,\ \overline{\psi}\gamma_\mu\psi A^\mu,\ M^2 A^\mu A_\mu ... \tag{2}$$

However if this effective Lagrangian is to have *any* light fields the large mass terms of eq.2 must be absent. The implications of this condition for the possible types of field are

- **Light fermions.** If $\mathcal{L}$ has a chiral symmetry light fermions are possible :

$$\psi_L \rightarrow e^{i\alpha}\psi_L,\ \psi_R \rightarrow \psi_R\ \ forbids\ \ M\overline{\psi}\psi \equiv M(\overline{\psi}_L\psi_R + \overline{\psi}_R\psi_L) \tag{3}$$

where the subscripts L and R refer to the helicity components. It is significant that all the known light fermions are chiral in the sense that the left-handed components all transform as doublets under the group while the right-handed components are singlets. Thus they cannot acquire mass while the electroweak group is unbroken, consistent with the effective field theory interpretation.

- **Light vector bosons.** If $\mathcal{L}$ has a local gauge symmetry light bosons are possible :

$$A_\mu \to A_\mu + \partial_\mu \Lambda \ \ forbids \ \ M^2 A_\mu A_\mu \tag{4}$$

In this case the leading dimension terms have $d_n = 4$ and define a *renormalisable local gauge field theory* of the gauge bosons interacting with scalars or fermions. Again this is consistent with the Standard Model because all observed vector fields are associated with an underlying local gauge symmetry. One can even say more about the nature of the gauge boson coupling to fermions. The general coupling has the form $A^\mu(a\overline{\psi}_L\gamma_\mu\psi_L + b\overline{\psi}_R\gamma_\mu\psi_R)$. We have just argued that a light fermion should have no mass term $M\overline{\psi}\psi$ in this effective theory. Thus if the fermion is massive it must get its mass from spontaneous symmetry breaking from a term $\overline{\psi}_L\psi_R\Phi$ where Φ is a scalar field which acquires a vacuum expectation value (vev). If the gauge symmetry is unbroken by this vev and the corresponding gauge boson is massless then, necessarily, the transformation properties of ψ_L and ψ_R under this gauge symmetry must be the same to allow the mass term. This implies $a = b$ in the coupling of the massless gauge boson to the massive fermion *i.e.* massless gauge bosons couple in a vectorlike manner. On the other hand if the gauge symmetry is spontaneously broken by the vev of Φ then there is no requirement that $a = b$ and hence massive gauge bosons may have parity violating couplings [1].

- **Light Scalars.** If $\mathcal{L}$ has a spontaneously broken global symmetry there are light Goldstone bosons :

$$\phi \to \phi + a \ \ forbids \ \ M^2\phi^2 \tag{5}$$

However ϕ cannot carry gauge quantum numbers as is required for the Higgs scalar of the Standard Model, otherwise a $M'^2\phi^2$ term is allowed with $M' \sim gM$ where g is the gauge coupling. It is at this point that the Standard Model appears inconsistent with the interpretation of it as an effective field theory coming from an underlying physics at high scale. The Standard Model requires light Higgs but this argument says that all scalars including the Standard Model Higgs should have mass close to the scale of new physics. This is the "hierarchy problem".

2.3 Candidate solutions to the hierarchy problem

There are three main ideas to solve the hierarchy problem :

Compositeness. In this case some of the states of the Standard Model are composite and there is a new strong interaction responsible for generating the composite state. In order to avoid the hierarchy problem the states associated with this new strong interaction must be quite light, of order one TeV.

Supersymmetry. The only possible symmetry capable of keeping a Higgs boson light (as is needed in the Standard Model) is supersymmetry (plus a chiral symmetry). Supersymmetry assigns scalars to supermultiplets with fermion partners and thus their properties are related :

$$\phi\,(J=0) \ \to \psi\,(J=\frac{1}{2}) \ \ requires \ \ m_\psi = m_\phi$$

[1] This is likely but not inevitable - there are left-right symmetric models in which the massive boson couples in a parity conserving way

Thus if the fermion mass m_ψ is forbidden by a chiral symmetry as discussed above the scalar mass will also vanish. In this case there must be supersymmetric partners to the Standard Model states. As we shall see, in order to prevent radiative corrections reintroducing the hierarchy problem, the states must be quite light, less than or of order 1 TeV.

Large new dimensions. A more radical explanation to the hierarchy problem is that there are *no* large mass scales associated with new physics including no fundamental Planck scale. As we shall discuss this is possible if gravity propagates in more than four space-time dimensions. In this case the four dimensional Planck scale, $M_P \sim 10^{19} GeV$, responsible for setting the strength of the gravitational interactions, is not a fundamental quantity but is related to the underlying higher dimensional Planck scale. Provided this is low, less than one TeV, the hierarchy problem does not arise. In order to implement this solution to the hierarchy problem it is necessary that Kaluza Klein states associated with the large new dimensions should be present with a mass of order one TeV.

The conclusion of all this is that an effective field theory describing scalar, fermion and vector fields light on a scale M associated with an underlying fundamental theory must be a renormalisable gauge field theory - just the structure of the Standard Model- in which the massless gauge bosons have vectorlike couplings (as do the gluons and the photon) while the massive gauge bosons may have parity violating couplings. Further, all light fermions should be chiral, just as is observed. The striking agreement between these features of an effective field theory and what we observe strongly suggests to me that the Standard Model is indeed an effective field theory descending from an underlying theory which applies at a high energy scale. However this beautiful agreement breaks down when we consider the Standard Model scalar sector responsible for the spontaneous symmetry breaking of the electroweak symmetry. To accommodate the light scalars needed for spontaneous symmetry breaking one of the three possibilities just discussed must be realised. I turn now to a more complete discussion of these possibilities.

3 Beyond the Standard Model - bottom up

3.1 Supersymmetry - The supersymmetric spectrum

If the Standard Model is to be incorporated in a field theory with N=1 supersymmetry, it is first necessary to assign the states of the Standard Model to N=1 supersymmetric representations[3]. Let us start with the SU(3) x SU(2) x U(1) vector bosons. The only super-multiplet available which contains vectors is the vector supermultiplet containing states of helicity 1 and $\frac{1}{2}$. Thus we must assign the vector bosons to vector supermultiplets transforming in the same way under SU(3) x SU(2) x U(1) as the Standard Model gauge bosons; this is shown in Table 1. It is clear that this means we have to introduce gauginos, spin $\frac{1}{2}$ fermionic partners of the gauge bosons, which transform under the gauge group in exactly the same way as their vector partners.

Turning now to the assignment of the quarks and leptons to supermultiplets it appears that there is the option of assigning them to vector supermultiplets, with vector partners, or to chiral supermultiplets, with scalar partners. However the former possibility is not available, unless we increase the size of the gauge group, for in the Standard Model the transformation properties

$$\begin{pmatrix} g_\mu^{a=1..8} \\ \widetilde{g}^{a=1..8} \end{pmatrix} + \begin{pmatrix} W_\mu^{i=1..3} \\ \widetilde{W}^{i=1..3} \end{pmatrix} + \begin{pmatrix} B_\mu \\ \widetilde{B} \end{pmatrix}$$

$$\begin{pmatrix} q_L \\ \widetilde{q}_L \end{pmatrix} + \begin{pmatrix} u_L^c \\ \widetilde{u}_L^c \end{pmatrix} + \begin{pmatrix} d_L^c \\ \widetilde{d}_L^c \end{pmatrix} + \begin{pmatrix} l_L \\ \widetilde{l}_L \end{pmatrix} + \begin{pmatrix} E_L^c \\ \widetilde{E}_L^c \end{pmatrix}$$

$$\begin{pmatrix} \widetilde{H}_1 \\ H_1 \end{pmatrix} + \begin{pmatrix} \widetilde{H}_2 \\ H_2 \end{pmatrix}$$

Table 1: The MSSM multiplet content. Standard Model states have no tilde. The first line shows the gauge bosons and their fermionic partners the gauginos. The second line shows the quarks and leptons together with their scalar partners the squarks and sleptons. Finally the Higgs sector is shown with two Higgs doublets together with their fermion partners the Higgsinos

Figure 3: Graphs generating contributions to the Higgs mass. In the Standard Model only the first graph is present. In its supersymmetric generalisation the second graph also contributes and cancels the contribution of the first graph in the limit supersymmetry is exact.

of the quarks and leptons is different from that of the gauge bosons of the Standard Model (cf Table 1). Thus we cannot identify any of the gauginos of Table 1 with the quarks or leptons and the only possibility is to assign the latter to *chiral* supermultiplets with the transformation properties of the quarks and leptons. Thus we must introduce scalar partners to the quarks and leptons transforming in the same way as their fermionic partners as shown in Table 1.

Finally we must assign the Higgs bosons of the Standard Model to a supermultiplet. Having spin 0, they can only be assigned to a chiral supermultiplet and thus they must have fermionic partners called higgsinos. However the introduction of an SU(2) doublet of fermions of definite helicity introduces an anomaly in the SU(2) x U(1) gauge theory, breaking the $U(1)$ gauge symmetry. This is unacceptable and to avoid it we are forced to introduce *two* Higgs chiral supermultiplets with *opposite* hypercharge (so that the higgsinos have cancelling contributions to the anomaly) as shown in Table 1. It turns out that both Higgs multiplets are needed in supersymmetry to give quarks and leptons a mass so this is indeed the minimal choice (i.e. attempts to identify the H_2 supermultiplet with a lepton supermultiplet without introducing an H_1 would have failed).

To summarise, the supersymmetric version of the Standard Model requires more than twice the number of particles; gauginos, partners for each of the gauge bosons, and squarks and sleptons, scalar partners for each of the quarks and leptons and two Higgs doublets with fermionic "Higgsino" partners.

Of course we know that the supersymmetry can not be exact because otherwise, for example, there would be a degeneracy between the supersymmetric partners of quarks, the squarks, with

SUSY state	Upper bound	Most "natural"
LSP	50GeV	40 GeV
NLSP	90GeV	
$C^{\pm}$	90GeV	55GeV
$\tilde{g}$	370GeV	270 GeV
$\tilde{t}$	300GeV	220 GeV

Table 2: Naturalness bounds on the mass of SUSY states, the lightest and next-to-lightest (LSP and NLSP), the chargino, $C^{\pm}$, the gluino, $\tilde{g}$, and the lightest squark the stop $\tilde{t}$. The upper bound corresponds to a fine tuning by a factor 10 [5]. .

the quarks. Such states have not been observed and so supersymmetry must be broken in such a way as to give large mass to the squarks. However, we also know that these new superpartners cannot be too heavy. In particular the radiative corrections of Figure 3 will generate a mass squared for the Higgs-boson. Ignoring the fermion mass, evaluation of the graph in Figure 3a, with pointlike coupling, h, of the elementary Higgs to the quarks, gives a contribution to the Higgs mass

$$m_H^2 \approx \frac{h^2}{16\pi^2}\Lambda^2 \tag{6}$$

where Λ^2 is the cut-off of the virtual momentum, k^2, corresponding to the scale at which new states corresponding to physics beyond the Standard Model should be included. The second graph gives a contribution

$$m_H^2 \approx -\frac{h^2}{16\pi^2}(\Lambda^2 - m_S^2) \tag{7}$$

Summing the two contributions gives eq.(6) with the cut-off replaced by the supersymmetry breaking scalar mass squared. Therefore the condition that the Higgs scalar should be of the order of the electroweak scale places a condition on the supersymmetry breaking scale. This is the only evidence we have for low energy supersymmetry. Quantifying this we arrive at the values of Table 2 showing the naturalness bounds (i.e. assuming no large "unnatural" cancellations occur in the calculation) on the mass of SUSY states [4, 5] together with the most "natural" value [5]. Note that different measures[6] of naturalness give varying results and so one should not take specific values to seriously, but the overall message that supersymmetric states should be quite light is clear.

Once one has extended the Standard Model to include low energy supersymmetry, one is able to build a fully unified theory at the very high scale without introducing any fine tuning problems. One hopes that such a scheme will be capable of explaining the origin of the symmetries, the multiplet structure and the couplings of the Standard Model leading to the final "Theory of Everything"

3.2 Compositeness - do we really need supersymmetry?

The hierarchy problem seems to imply that all scalars should be heavy ... but what about the π, K, η etc? The discussion presented above assumed that the scalar field was elementary but

QCD	QT$_C$D
$\frac{g_3^2}{4\pi}(1GeV)=1$ $f_\pi = 100MeV$ $M_W = 30MeV$	$\frac{g_{TC}^2}{4\pi}(1TeV)=1$ $f_{\pi_{TC}} = 100MeV$ $M_W = 80GeV$
Hadrons$\sim 1GeV$	Techni-hadrons$\sim 1TeV$
Pseudo-Goldstone bosons, $\pi \sim 100MeV$	Pseudo-Goldstone bosons, $2GeV$ upwards

Table 3: Technicolour properties

in QCD the π, K, η are not present in the original Lagrangian but appear as bound states of quarks. The distinction between elementary and composite scalars becomes clear at radiative order. Consider the contribution, eq.(6), to the Standard Model Higgs boson coming from Figure 3a. If we take Λ very large, the Grand Unification scale or the Planck scale, we encounter the hierarchy problem because the suppression factor $\frac{h^2}{16\pi^2}$ cannot prevent the Higgs being driven to an unacceptably high mass. For composite scalars the calculation changes due to the non-pointlike nature of the coupling of bound states such as the π, K, η to the quarks. In this case $h \to he^{-k^2/\Lambda_{QCD}^2}$ because, above the scale Λ_{QCD}, the strong interaction becomes weak and can no longer support a bound state. As a result evaluation of the one loop graph gives

$$m_H^2 \approx \frac{h^2}{16\pi^2}\int_0^\infty e^{-2k^2/\Lambda_{QCD}^2}dk^2 \approx \frac{h^2}{16\pi^2}\Lambda_{QCD}^2. \tag{8}$$

We see that composite states are not sensitive to arbitrarily high scales and thus the hierarchy problem is avoided provided the scale of compositeness is low.

But can the Higgs scalar of the Standard Model be composite? To answer this note that with *no* Higgs scalar the $W^\pm$ and Z would still be massive! This is because QCD generates a quark condensate $<\overline{u}u>=<\overline{d}d>=O(\Lambda_{QCD}^3)$ which breaks $SU(2)_L \times SU(2)_R$ to $SU(2)_{L+R}$. Since the left-handed $SU(2)$ is local gauge group factor of the Standard model we see that this condensate breaks the electroweak group. In this case the pions, $\pi^{\pm 0}$, the would-be Goldstone bosons associates with the spontaneous breaking of the global symmetry $SU(2)_L \times SU(2)_R$ are eaten providing the longitudinal components of the $W^\pm$ and Z and generating their mass with mass $M_W = g_2 f_\pi/2$. Since the pion coupling strength is small, $f_\pi \approx 100MeV$, this mass is far too small. In any case we know that pions exist i.e. they haven't been eaten! However this analysis does suggest that the Standard Model could break through a new composite force which is strong at a much higher scale. The original idea was technicolour and introduced new $SU(2)$ quark doublets and singlets which also carried a new "technicolour" quantum number corresponding to an $SU(N)$ technicolour gauge theory. The basic technicolour family is

$$\begin{pmatrix} U_L \\ D_L \end{pmatrix}_i, \; U_{R,i}, \; D_{R,i}, \quad i = 1..N \tag{9}$$

Due to the technicolour interaction a condensate forms $<\overline{U}U>=<\overline{D}D>=O(\Lambda_{TC}^3)$ and generates electroweak symmetry breaking with $M_W = g_2 f_{\pi_{TC}}/2$. Provided one scales up the properties of technicolour as shown in Table 3 one obtains acceptable electroweak breaking.

As we shall discuss, precision tests disfavour additional electroweak doublets so the original idea of technicolour is untenable. However it is possible to modify the idea without introducing new fermions and thus avoiding this problem. The idea is that the top quark has new strong interactions and that its condensate breaks the electroweak group. From Table 3 we see that if the new interaction of the top quark should be strong at a scale $\Lambda \approx 1TeV$ a top quark condensate may be expected to form with $< \bar{t}t >\sim \Lambda^3$ generating the correct mass for the weak gauge bosons. In this case the Higgs will be a $\bar{t}t$ bound state.

3.3 Large New Dimensions

The third explanation for the low cut-off scale follows from the fact that in theories with new space dimensions, there may be no fundamental scale above a TeV[7]! In such "compactified" theories all but four space-time dimensions are small in the sense that phenomena associated with them only propagate over distance scales much smaller than are currently probed. In such theories the weakness of the gravitational interaction does not require a fundamental Planck scale of $O(10^{18}GeV)$. The reason is readily seen if we compute the Newtonian potential between two point masses in the underlying (4+δ) dimensional theory. It is given by

$$V_{gravity}(r) = \frac{m_1 m_2}{M_{Planck,4+\delta}^{\delta+2}} \cdot \frac{1}{r^{\delta+1}}$$

where $M_{Planck,4+\delta}$ is the fundamental mass scale in the (4+δ) dimensional theory (the string tension in a string theory). This form applies at distances much smaller than the size, R, of the δ additional compact dimensions. At distances larger than their size we have

$$V_{gravity}(r)|_{r>>R} = \frac{m_1 m_2}{M_{Planck,4+\delta}^{\delta+2}} \cdot \frac{1}{R^\delta} \cdot \frac{1}{r}$$

This shows that the Planck scale in our 4 dimensional theory is made larger by the spreading of flux in the new dimensions. It is given by

$$M_{Planck,4}^2 = M_{Planck,4+\delta}^{\delta+2}\, R^\delta \tag{10}$$

For a large new dimension (measured in the fundamental unit $M_{Planck,4+\delta}$ we have $(M_{Planck,4+\delta}\, R)$ $>> 1$ and so we can see that the large size of the Planck mass (and the associated feebleness of the gravitational force) may be due to the existence of a new large dimension. Given that the fundamental scale is now $M_{Planck,4+\delta}$ the cutoff for the field theory radiative corrections of the Standard Model should be identified with this scale because above it the states of the theory propagate in more than four dimensions and the radiative corrections must be computed in this dimension. Thus the hierarchy problem is solved if $M_{Planck,4+\delta} \leq O(1TeV)$.

3.3.1 Kaluza-Klein excitations

What are the first manifestations of a new compact dimension? The original Kaluza-Klein formulation involved toroidal compactification in which the new dimension can be mapped onto a circle with a (factorisable) space-time metric given by

$$ds^2 = -dx^0 dx_0 + dx_i dx^i + dx_\alpha dx^\alpha, \quad i = 1, 2, 3, \quad \alpha = 4, ..., 4+\delta \tag{11}$$

The first indication in our four dimensional world of a new large dimension is the appearance of a tower of Kaluza-Klein states, excitations of the states propagating in the additional dimension. This is simply illustrated for the case of one additional dimension compactified on a circle. Since the extra dimension is compact a general field $\phi(x,y)$ may be expanded in terms of harmonics of the new dimension with coordinate y giving

$$\phi(x,y) = \sum_{n=-\infty}^{\infty} \phi^n(x)e^{iny/R} \tag{12}$$

such that $\phi(x,y) = \phi(x, y+2\pi R)$. A massless mode in 5D satisfies $(\partial_x^2 + \partial_y^2)\phi(x,y) = 0$. Using eq(12) one immediately finds

$$\partial_x^2\phi(x,y) = m^2\phi(x,y),\ m^2 = \left(\frac{n}{R}\right)^2 \tag{13}$$

In this simple case the four dimensional tower of Kaluza-Klein modes, $\phi^n(x)$, are equally spaced with mass n/R.

For the factorisable metric case the implications are dramatic for the graviton Kaluza-Klein modes must be very light. Using this bound and eq(10) one obtains from eq(13) the mass gap between Kaluza-Klein modes of the graviton

$$\begin{aligned} \Delta m &= \left(\frac{M_{Planck,4+\delta}}{1TeV}\right)^{\frac{\delta+2}{2}} 10^{\frac{12\delta-31}{\delta}} eV \\ &= 10^{-3}eV,\ 20KeV,\ 7MeV,\ 100MeV;\ \ \delta = 2,4,6,8 \end{aligned} \tag{14}$$

Although each mode couples with gravitational strength only, and thus is very difficult to observe, at high energies the inclusive cross section for producing the kinematically accessible tower of Kaluza Klein states is only suppressed by inverse powers of $M_{Planck,4+\delta}$ their effects should be visible in LHC experiments.

Recently there has been renewed interest in the case the metric is not factorisable[8]. For the case of a single extra dimension the line element has the form

$$ds^2 = e^{-\rho(x_4)}(-dx^0dx_0 + dx_idx^i) + dx_4dx^4, \quad i = 1,2,3, \tag{15}$$

where $x_4 = R\theta/2\pi$, $0 < \theta \leq 2\pi$. An explicit example was recently provided by Randall and Sundrum. They considered the case of two parallel 3-branes sitting on the fixed points of an S^1/Z_2 orbifold. The 5D spacetime is essentially a slice of AdS_5 and the tensions of the two 3-branes are chosen so that the 4D spacetime appears flat. This last requirement forces the one of the two branes to have negative tension. The solution to Einstein's equations has the form of equation(15) with $\rho(r) = kR$ where k is the five dimensional curvature. The exponential "warp" factor, $\varpi = e^{-kR}$, in the metric then generates an hierarchy of the mass scales between the two branes. All the fundamental mass scales are of order $M_{Planck,4}$. The graviton is localised to the positive tension brane and matter is on the negative tension brane. Masses on the negative tension brane are of order $\varpi M_{Planck,4}$. Due to the exponential dependence of the warp factor on the size of the new dimension one may readily generate the desired mass hierarchy with $\varpi = \mathcal{O}(10^{-15})$ even though the size, R, of the orbifold is only some 40 times k^{-1} which is of order the Planck length.

I will return to a discussion of the phenomenology of large new dimensions in Section 5.

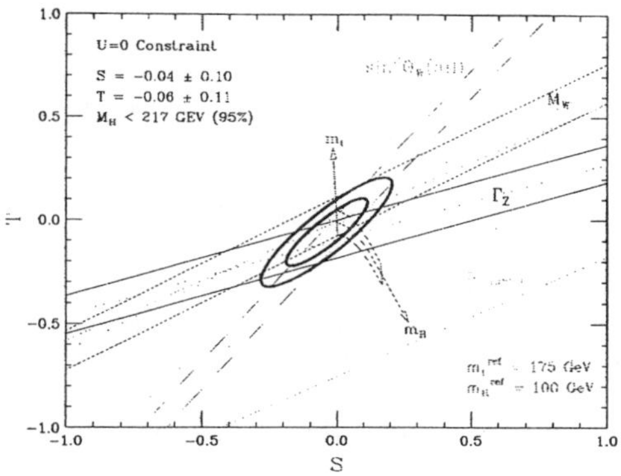

Figure 4: Fit of the precision electroweak data to S and T .

3.4 Precision Tests : vacuum polarisation

We have seen that the Standard Model can be viewed as an effective field theory involving new physics far above the electroweak scale. The beauty of this is that it immediately answers the first three of the fundamental questions in Section 2.1 that the structure of the Standard Model raises. However in order to answer the fourth question it is necessary to have new physics no higher than the TeV scale; compositeness, supersymmetry or large new dimensions being the best candidates for this new physics. Such new physics will have an effect at lower scales through radiative corrections and so the precision tests of the Standard Model can already be used to test these ideas and severely constrain the possibilities. The most obvious way to do these tests is to compute all the quantities appearing in Figure 1 including the effects of the new physics. It turns out that the most significant radiative effects are the vacuum polarisation type corrections to gauge boson propagators and a convenient way to summarise the precision tests is to consider the variables S and T [9] (or ϵ_1, ϵ_3 [10]) which are weak isospin conserving and weak isospin violating parameters related to the vacuum polarisation of interest.

The data are plotted in Figure 4 [11]. The normalisation of S and T is such that a typical non-Standard-Model contribution occurs at $O(1)$. The Standard Model 98% confidence level contours are centered about the origin and the sensitivity to the Higgs mass is indicated. One may see that there is little room for new contributions. How does this compare with the expectation in the various schemes outlined above?

3.4.1 Supersymmetry

Supersymmetry neatly avoids conflict with the S and T measurements because its contribution decouples in the sense that $S, T \sim M_Z^2/4\pi M^2$ and so is very small in the limit that the SUSY mass scale M becomes large. The same is true of the contribution of the graviton KK modes.

3.4.2 Technicolour and topcolour

Theories which are non-decoupled are very strongly constrained by this data. These include technicolour, theories with a fourth generation, composite models of quarks and leptons. For example in technicolour the change δS in S is given by [9]

$$\delta S = +0.3 N_D \frac{N_{TC}}{3}$$

where N_D is the number of new technidoublets and N_{TC} is the number of technicolours. Even for $N_D = 1$, $N_{TC} = 3$ this is inconsistent with the precision measurement, $S = -.04 \pm 0.1$.

As mentioned above there have been suggestions how to modify the technicolour idea so as not to conflict with the precision tests. Topcolour is an example, breaking the electroweak symmetry through a top quark condensate [13]. In this case the third generation transforms under one $SU(3)_1$ group which becomes strongly coupled close to a TeV, while the remaining light generations transform under another $SU(3)_2$ which remains weakly coupled at this scale. The $SU(3)_1 \times SU(3)_2$ group is broken to its diagonal subgroup $SU(3)_{QCD}$ at a high scale. Thus topcolour does not introduce additional doublets (the top quark still comes in just three $(SU(3)_1)$ colours) and hence does not change S significantly. However it turns out that there is a conflict between the value of the condensate needed to give the $W^\pm$ and Z their mass and that needed to give the top quark its mass, the latter being too large if one fits the former. A way out of this is to use a top condensation see-saw [14, 15], introducing an $SU(2)$ singlet topcolour triplet fields, χ_L, χ_R with charge 2/3 as for the top quark. Now topcolour drives condensates for $< \overline{\chi}_L t_R >$ and $< \overline{t}_L \chi_R >, < \overline{\chi}_L t_R >$ and $< \overline{t}_L \chi_R >$ as well. In the case the $< \overline{\chi}_L \chi_R >$ condensate dominates the top quark mass is given by a see-saw form

$$M_t \simeq \frac{< \overline{\chi}_L t_R >< \overline{t}_L \chi_R >}{< \overline{\chi}_L t_R >}$$

Using this one can allow for $< t_L \chi_R >$ to be large enough to give the correct $W^\pm$ and Z masses while keeping the top quark mass at its measured value. Note that in this model the Standard Model Higgs doublet is the composite state $(\overline{\chi}_R t_L, \overline{\chi}_R b_L)$. A consequence of this type of composite model is the existence of several composite scalars. If the new strong interactions involve only the χ_R and the $(t, b)_L$ doublet, then the low energy theory is the Standard Model with an additional heavy Higgs boson. If all components of the t, b and χ feel the new strong interactions, the composite Higgs sector includes three weak doublets and three singlets. Therefore an experimental determination of the nature of the Higgs sector is an important discriminator of this type of theory. Note that, in contrast to other mechanisms of dynamical electroweak symmetry breaking, this mechanism is of the decoupling type and reduces to the Standard Model as the scale of the new strong interaction is increased. Of course in this limit there is a fine tuning price to pay, just as in the case of Supersymmetry.

3.4.3 Large new dimensions

As for supersymmetry, the effect of graviton KK modes on S, and T also decouples as the higher dimension mass scale increases. It is of interest to see why *any* significant contributions occur because at first sight one might expect negligibly small corrections proportional to gravitational

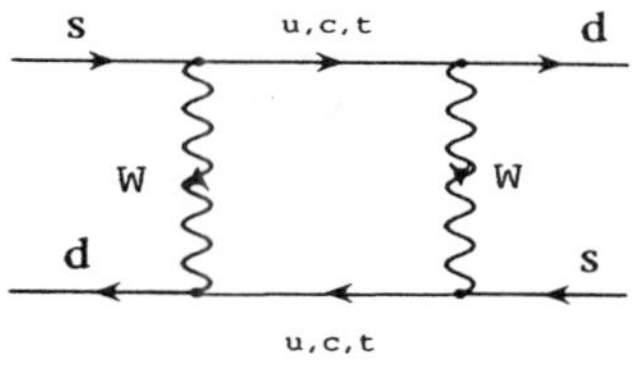

Figure 5: Box diagram generating $\Delta S = 2$, $\Delta Q = 0$ processes in the Standard Model.

strength couplings. Consider for example the KK graviton corrections to the Standard Model processes $e^+e^- \to f\bar{f}$, $q\bar{q} \to f\bar{f}$[19]. In this case the Born amplitude due to graviton excitation exchange is given by

$$A \approx \frac{1}{M^2_{Planck,4}} \sum_n T_{\mu\nu} F^{\mu\nu\rho\sigma} T_{\rho\sigma} \tag{16}$$

where $T_{\mu\nu}$ is the energy momentum tensor and $F^{\mu\nu\rho\sigma}$ is the KK propagator. The sum is over the tower of KK modes. Since the spacing is given by $1/R$ and the cutoff should be $M_{Planck,4+\delta}$ there will be $M_{Planck,4+\delta}$modes contributing for 1 additional dimension or $(M_{Planck,4+\delta}R)^\delta$ modes for δ additional dimensions. From eq(10) we have

$$\frac{M^2_{Planck,4}}{M^2_{Planck,4+\delta}} = M^\delta_{Planck,4+\delta}\, R^\delta \tag{17}$$

and so finally $A \propto 1/M^2_{Planck,4+\delta}$, suppressed by the higher dimension mass scale rather than the 4D Planck scale. For $M_{Planck,4}$ of $O(1TeV)$ these effects may be visible but clearly they are of the decoupling form. Note that the amplitude has a characteristic angular dependence corresponding to tensor exchange, that the total cross section is unaltered in leading order (to be compared with other possible new physics contributions such as heavy vector or scalar exchange), and that the amplitude will have a characteristic $L - R$ asymmetry. Thus is this signal is observed it will be relatively easy to determine its nature. We shall discuss direct probes for large new dimensions in Section 5.

3.5 Precision tests : Flavour changing neutral currents

Due to the GIM mechanism the Standard Model has no flavour changing neutral current at tree level but these appear through radiative corrections, for example via the graphs of Figure 5. The fact that this structure fits observation on rare kaon decays provides a stringent test of the Standard Model beyond tree level. In fact it was by requiring that $\Delta S = 2$, $\Delta Q = 0$ processes be within experimental limits that Gaillard and Lee were able to place a limit on the charm quark mass of $O(1GeV)$ long before it was found in the laboratory[16]. Of course any theory going beyond the Standard Model should not disturb this agreement and this provides another strong constraint on such models.

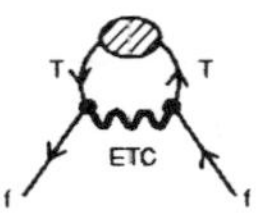

Figure 6: Diagram generating light quark mass by coupling to TC quark via ETC gauge boson messenger.

3.5.1 Technicolour and extended technicolour

The attempts to describe fermion masses in technicolour theories rely on the introduction of further "extended technicolour" gauge interactions[17] which couple normal quarks to light quarks. At radiative order these interactions generate fermion masses via the graphs of Figure 6 of the form

$$m_q \propto \frac{<\overline{Q}Q>}{M_{ETC}^2}$$

In order to give a mass of the correct size the mass, M_{ETC}, of the extended technicolour boson must not be too large, $M_{ETC} < 1TeV$. However closure of the algebra of the underlying gauge symmetry needed to generate the extended technicolour gauge bosons requires that they should also have coupling to light quarks alone, introducing flavour changing processes at tree level via the Born graphs involving ETC gauge boson exchange. In order adequately to suppress these it is necessary for $M_{ETC} > 1TeV$ in conflict with the need to generate light quark masses. Ways out of this contradiction have been considered. For example, it may be that technicolour and extended technicolour do not behave in the same way as QCD and that there are significant strong interaction corrections to the calculation of the radiative generation of fermion masses as in Figure 6. This analysis illustrates the very strong constraints flavour changing neutral current processes place on new physcs. Let us discuss how supersymmetry copes with these constraints.

3.5.2 Supersymmetry

In the case of supersymmetry there are also new contributions to rare processes which provide strong constraints on the masses of the new supersymmetric states. An example is shown in Figure 7. Evaluation of these graphs leads to the following constraints [20]

$$\begin{aligned} \tfrac{m_{\tilde{d}}^2 - m_{\tilde{s}}^2}{m_{\tilde{d}}^2} &\lesssim 6.10^{-3}\left(\tfrac{m_{\tilde{d}}}{Tev}\right)\left(\tfrac{\sin\tilde{\theta}_c}{\sin\theta_c}\right)^2 \\ \tfrac{m_{\tilde{e}}^2 - m_{\tilde{\mu}}^2}{m_{\tilde{e}}^2} &< 10^{-1}\left(\tfrac{m_{\tilde{e}}}{Tev}\right)^2\left(\tfrac{\sin\tilde{\theta}_e}{\sin\theta_c}\right)^2 \end{aligned} \qquad (18)$$

where the tilde refers to the relative mixing angle between the Standard Model state and its SUSY partner.

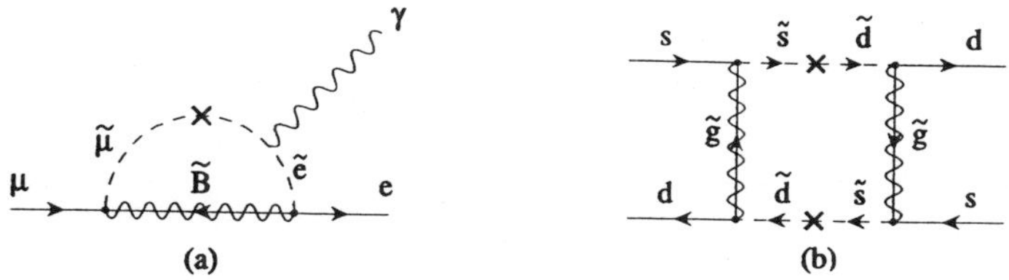

Figure 7: Diagrams generating FCNC processes in supersymmetry.

x	$\sqrt{\lvert \mathrm{Re}\,(\delta^d_{12})^2_{LL}\rvert}$	$\sqrt{\lvert \mathrm{Re}\,(\delta^d_{12})^2_{LR}\rvert}$	$\sqrt{\lvert \mathrm{Re}\,(\delta^d_{12})_{LL}\,(\delta^d_{12})_{RR}\rvert}$
0.3	1.9×10^{-2}	7.9×10^{-3}	2.5×10^{-3}
1.0	4.0×10^{-2}	4.4×10^{-3}	2.8×10^{-3}
4.0	9.3×10^{-2}	5.3×10^{-3}	4.0×10^{-3}
x	$\sqrt{\lvert \mathrm{Re}\,(\delta^d_{13})^2_{LL}\rvert}$	$\sqrt{\lvert \mathrm{Re}\,(\delta^d_{13})^2_{LR}\rvert}$	$\sqrt{\lvert \mathrm{Re}\,(\delta^d_{13})_{LL}\,(\delta^d_{13})_{RR}\rvert}$
0.3	4.6×10^{-2}	5.6×10^{-2}	1.6×10^{-2}
1.0	9.8×10^{-2}	3.3×10^{-2}	1.8×10^{-2}
4.0	2.3×10^{-1}	3.6×10^{-2}	2.5×10^{-2}
x	$\sqrt{\lvert \mathrm{Re}\,(\delta^u_{12})^2_{LL}\rvert}$	$\sqrt{\lvert \mathrm{Re}\,(\delta^u_{12})^2_{LR}\rvert}$	$\sqrt{\lvert \mathrm{Re}\,(\delta^u_{12})_{LL}\,(\delta^u_{12})_{RR}\rvert}$
0.3	4.7×10^{-2}	6.3×10^{-2}	1.6×10^{-2}
1.0	1.0×10^{-1}	3.1×10^{-2}	1.7×10^{-2}
4.0	2.4×10^{-1}	3.5×10^{-2}	2.5×10^{-2}

Table 4: Limits on $\mathrm{Re}\,(\delta_{ij})_{AB}\,(\delta_{ij})_{CD}$, with $A, B, C, D = (L, R)$, for an average squark mass $m_{\tilde{q}} = 500\mathrm{GeV}$ and for different values of $x = m^2_{\tilde{g}}/m^2_{\tilde{q}}$. For different values of $m_{\tilde{q}}$, the limits can be obtained multiplying the ones in the table by $m_{\tilde{q}}(\mathrm{GeV})/500$.

x	$\|(\delta^l_{12})_{LL}\|$	$\|(\delta^l_{12})_{LR}\|$
0.3	4.1×10^{-3}	1.4×10^{-6}
1.0	7.7×10^{-3}	1.7×10^{-6}
5.0	3.2×10^{-2}	3.8×10^{-6}
x	$\|(\delta^l_{13})_{LL}\|$	$\|(\delta^l_{13})_{LR}\|$
0.3	15	8.9×10^{-2}
1.0	29	1.1×10^{-1}
5.0	1.2×10^{2}	2.4×10^{-1}
x	$\|(\delta^l_{23})_{LL}\|$	$\|(\delta^l_{23})_{LR}\|$
0.3	2.8	1.7×10^{-2}
1.0	5.3	2.0×10^{-2}
5.0	22	4.4×10^{-2}

Table 5: Limits on the $\left|\delta^d_{ij}\right|$ from $l_j \to l_i\gamma$ decays for an average slepton mass $m_{\tilde{l}} = 100\text{GeV}$ and for different values of $x = m^2_{\tilde{\gamma}}/m^2_{\tilde{l}}$. For different values of $m_{\tilde{l}}$, the limits can be obtained multiplying the ones in the table by $(m_{\tilde{l}}(\text{GeV})/100)^2$.

SUGRA	Gauge Mediation	Anomaly mediation
$m_{\tilde{u}_i}=m_{\tilde{d}}=m_{\tilde{E}}=m_0\vert_{M_P}$ $M_{\tilde{\lambda}_i}=M_{1/2}\vert_{M_P}$	$m^2_{\tilde{f}}(M_{GM})=\frac{C^f_i}{\sqrt{n}}M_i^2(M_{GM})>>m_{3/2}$ $M_{\tilde{\lambda}r}(M_{GM})=\frac{\alpha_r(t)}{4\pi}M_0>>m_{3/2}$	$m_{\tilde{i}}=-\frac{1}{4}\frac{\partial\gamma}{\partial g}\beta_g m^2_{3/2}+m_0^2$ $M_{\tilde{\lambda}g}=\frac{\beta_g}{g}m_{3/2}$
$A^{u_i}=A^{d_i}=A^{l_i}=A_0\vert_{M_P}$	$A^{u_i}=A^{d_i}=A^{l_i}=0\vert_{M_{GM}}$	$A^y=-\frac{\beta_y}{y}m_{3/2}$

Table 6: The soft SUSY breaking parameters in various breaking schemes. Here M_P is the Planck mass, M_{GM} is the mass of the messanger in gauge mediation and n is the number of messengers. The quantities γ and β are the ususal renormalisation group functions $\gamma(g,y)=\frac{dlnZ}{dln\mu}$, $\beta_g(g,y)=\frac{dg}{dln\mu}$ and $\beta_y(g,y)=\frac{dy}{dln\mu}$ where y is a Yukawa coupling and we have only included terms involving it in the expression for A_y.

In more detail the constraints distinguish between left and right handed squarks. In Table 4 and 5 we list these constraints for $\delta=\Delta_{AB}/\widetilde{m}^2$ where $A,B=L,R$ taken from reference [18]. One of the questions anyone studying supersymmetry breaking must answer is why there should be such a close degeneracy between squark states or between slepton states?

Soft SUSY breaking parameters The general parameterisation of soft supersymmetry breaking mass terms involves many (~ 110) new parameters

$$
\begin{aligned}
L_{SB} &= -\frac{1}{2}\sum_{\widetilde{A},\widetilde{q},\widetilde{u},\widetilde{d},\widetilde{l},\widetilde{E}} M_{\widetilde{A}}\lambda_{\widetilde{A}}\lambda_{\widetilde{A}} - m_{\widetilde{q}}|\widetilde{q}|^2 - m^2_{\widetilde{u}}|\widetilde{u}|^2 - m^2_{\widetilde{d}}|\widetilde{d}|^2 - m^2_{\widetilde{l}}|\widetilde{l}|^2 - m^2_{\widetilde{E}}|\widetilde{E}| \\
&\quad -m_0(\sum_{u,q} h^u A^u \widetilde{q}\widetilde{u}^c H_2 + \sum_{d,q} h^d A^d \widetilde{q}\widetilde{d}^c H_1 + \sum_{l,E} h^l A^l \widetilde{l}\widetilde{E}^c H_1) \\
&\quad -m_0(B\mu H_1 H_2) + h.c.
\end{aligned}
\tag{19}
$$

It is clearly impossible to work with so many parameters. For this reason most attempts to study SUSY phenomenology have used plausible models of supersymmetry breaking to simplify the analysis. The models considered broadly fall into three classes, SUGRA models in which gravitational strength couplings communicate SUSY breaking from a hidden to the visible sector[22], Gauge mediated models in which family blind gauge interactions communicate SUSY breaking from a hidden to the visible sector [23] and Anomaly mediated models in which the supersymmetry breaking is linked to the existence of the superconformal anomaly[24]. The implications for the soft terms of eq(19) are given in Table 6.

The near degeneracy of squark and slepton masses for the case of gauge mediation and anomaly mediation follows because they are generated by gauge (flavour blind) interactions. In the case of SUGRA it is argued that the gravitational interactions are also flavour blind although it is known in the context of string theories that this is not always the case. Even if degeneracy is achieved at a high scale, radiative corrections will split this degeneracy in the particle masses. Luckily the dominant corrections are due to flavour blind gauge interactions so this does not reintroduce problems with flavour changing neutral currents. However such corrections are important when considering the phenomenology of supersymmetry. Figure 8[25] shows the SUSY spectra for a variety of supersymmetry breaking models.

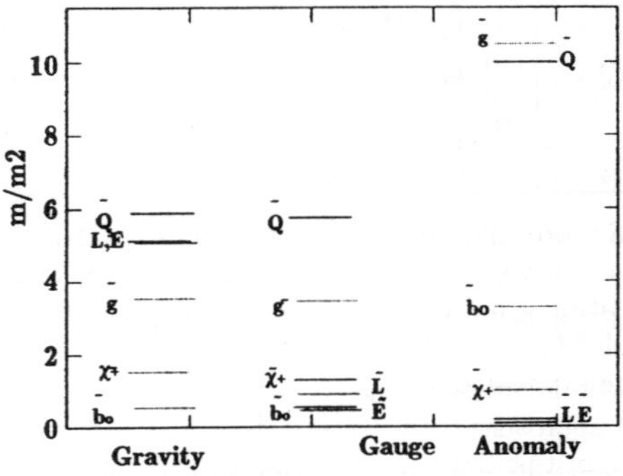

Figure 8: SUSY spectra for two SUGRA models, gauge mediation and anomaly mediation, including radiative corrections needed to determine the masses at laboratory energy scales.

3.6 Score sheet : Bottom up approach

In Table 7, I summarise the results of the comparison of the candidate models to solve the hierarchy problem with the precision tests of the Standard Model and beyond.

In this Table I have included the analysis of Professor Marciano of the indications of a departure from the Standard model expectation for $(g-2)_\mu$. As he points out, only in SUSY is there a natural way to get a large corrections. I have also indicated that SUSY is needed to accommodate a light Higgs, near the value preferred by the indirect measurements of Figure 2. The reason is that in SUSY Higgs, like other scalars, are protected from getting very large masses. Moreover the quartic Higgs coupling, undetermined in the Standard Model, are determined by the gauge couplings in te MSSM. As a result at tree level there are simple mass relations

$$\begin{aligned} m^2_{h,H} &= \frac{1}{2}[m^2_A + m^2_Z \mp \sqrt{(m^2_A+m^2_Z)^2 - 4m^2_A m^2_Z \cos^2 2\beta} \\ m^2_{H^\pm} &= m^2_W + m^2_A \end{aligned} \tag{20}$$

where m^2_A is a free parameter controlled by supersymmetry breaking. One sees from eq(20) that the lightest Higgs state is lighter than m_Z at tree level. It turns out that there are large radiative corrections due to the large top Yukawa coupling [21]. For a top mass less than $180GeV$ the lightest Higgs in the MSSM must be lighter than $140GeV$. Clearly this is a very important prediction of the MSSM.

	Technicolour	Large extra dimensions	SUSY
M_W/M	✓	✓	✓
Precision tests : decoupling	×	✓?	✓
$(g-2)_\mu$	×?	×?	✓
Light Higgs	×	×(✓$_{SUSY}$)	✓

Table 7: Score sheet for models solving the hierarchy problem

4 Beyond the Standard Model - top down

In the discussion of the Standard Model in Section 2.1 we listed a series of questions left unanswered. The bottom up approach suggested extensions of Standard Model capable of answering the first four questions. In this Section I will discuss the "top down" approach which may be able to shed light on the remaining questions. In particular why should we have the Standard Model gauge group, why the complicated fermion spectrum with left-handed quarks and leptons being assigned to doublets and the right-handed states assigned to singlets, why are the charges of the quarks 1/3 integer and why are there three generations? Finally what determines the parameters of the Standard Model?

These questions strongly indicate the need for further unification. The unification of the strong, weak and electromagnetic interactions is incomplete in the Standard Model because each corresponds to its own gauge group factor and each has its own gauge coupling constant. Grand Unification seeks to provide complete unification by embedding the Standard Model gauge group in a simple non-Abelian gauge group such as $SU(5)$. In this the three Standard Model couplings are related to the single coupling of $SU(5)$. As we shall discuss this can be made consistent provided unification occurs at a very high scale and radiative corrections are included when determining the strong weak and electromagnetic couplings as measured in the laboratory. Due to the appearance of this high scale the hierarchy problem is very severe. For this reason Grand Unification only makes sense if it is combined with low-energy supersymmetry. As we shall discuss, the detailed predictions coming from this synthesis provide us with indirect evidence for low-energy supersymmetry.

Complete unification of the fundamental forces requires the inclusion of gravity. The only candidate quantum theory of gravity that we have is string theory. Unification with gravity also seemingly implies unification at a very high scale because the fundamental mass scale associated with a gravitational interaction is the Planck scale although recent developments studying theories with extra dimensions suggest we may be able to circumvent this. In Section 4.2 I will discuss string unification and the experimental hints that such uification is realised in nature.

4.1 Grand Unification

The prototype Grand Unified Theory is based on the group $SU(5)$ a rank 5 group with just enough neutral generators to accommodate those of the rank 5 Standard Model[28]. In $SU(5)$ the states of a single family are assigned to just 2 representations, a $\overline{5}$ and a 10. Thus the representaion content of the Standard Model is simplified. The assignment to $\overline{5}$ and 10 are shown in eqn(21)

$$
\begin{aligned}
SU(5) &\supset SU(3)\otimes SU(2)\otimes U(1) \\
\psi_{\overline{5}} &= \begin{bmatrix} d_1^c \\ d_2^c \\ d_3^c \\ e^- \\ \nu_e \end{bmatrix}_L \\
\chi^{10} &= \frac{1}{\sqrt{2}} \begin{bmatrix} 0 & u_3^c & -u_2^c & -u^1 & -d^1 \\ -u_3^c & 0 & u_1^c & -u^2 & -d^2 \\ u_2^c & -u_1^c & 0 & -u^3 & -d^3 \\ u^1 & u^2 & u^3 & 0 & e^+ \\ d^1 & d^2 & d^3 & e^+ & 0 \end{bmatrix}_L
\end{aligned} \tag{21}
$$

where c denotes the charge conjugate and the numerical indices are $SU(3)$ indices. It may be seen that the quarks and leptons belong to the same multiplet. Clearly $SU(5)$ simplifies the Standard Model spectrum. Enlarging the GUT gauge group can simplify it even more. Going to $SO(10)$ one finds that a complete family fits into a single 16 dimensional representation which has the $SU(5)$ decomposition $16 \supset 10+\overline{5}+1$. The additional state is an $SU(3)\otimes SU(2)\otimes U(1)$ singlet and accommodates the right-handed neutrino needed to give neutrino masses.

Of course $SU(5)$ must be broken and it is important to consider the pattern of the symmetry breaking leading to the Standard Model. In the case of $SU(5)$, this is given by

$$
SU(5) \underset{<\Sigma_{24}^A>}{\longrightarrow} SU(3)\otimes SU(2)\otimes U(1) \underset{<H_5>}{\longrightarrow} SU(3)\otimes U(1) \tag{22}
$$

where one can see that the single 24-dimensional adjoint representation, Σ, leads to the breaking of $SU(5)$ immediately to the Standard Model $SU(3)\otimes SU(2)\otimes U(1)$ and subsequently the electroweak breaking is triggered by a doublet contained in a 5-dimensional Higgs representation, H_5. The most interesting prediction from $SU(5)$ follows because there is just a single gauge coupling. Thus the weak mixing angle which is given in terms of the ratio of the two coupling constants associated with $SU(2)\otimes U(1)$ is determined. The prediction is

$$
\sin\theta_W{}^2 = \frac{T_r(T_{3L}^2)}{T_r(Q^2)} = 3/8 \tag{23}
$$

but this applies at the unification scale, M_X. As we shall discuss including radiative corrections to the couplings to run them down to low scales one finds that the 3/8 initial value leads to quite acceptable values at laboratory energies for the weak mixing angle but only in a supersymmetric GUT.

The Yukawa couplings of the theory are also restricted by $SU(5)$ in a way that gives predictions for fermion masses. This may be seen from the equation

$$
L_Y = h\overline{\psi}_{5R}^{p}\chi_{p,qL}^{10}H_5^q \Rightarrow m_e = m_d = h < h_5^5 > \tag{24}
$$

The mass of electron is equal to the mass of down quark at the unification scale. The same is true for the heavier generations. As we will discuss in Section 4.1.4, even after including the radiative corrections needed to determine the masses at the laboratory energy scale, these predictions are not realistic apart from that of the third generation, so some modification is needed. It is possible to generate such a modifications for example through the coupling

$$L'_Y = h'\overline{\psi}^p_R H^{sq}_{45p}\chi_{sqL} \Rightarrow m_l = 3m_{Q_d} \tag{25}$$

where H_{45} is a 45 dimensional representation of Higgs fields. The Yukawa coupling generates a relation between quark masses which says that the strange quark would be 1/3 of the muon mass, a much more acceptable relation. We shall see how this can also lead to a good prediction for the down quark mass.

4.1.1 Unification of gauge couplings

The Grand Unified prediction for the gauge couplings, such as occurs in $SU(5)$, applies at the scale, M_X, at which the symmetry is unbroken. To compare with experiment these predictions must be radiatively corrected to take account of the different running of the strong, weak and electromagnetic couplings below M_X. In the Standard Model this was analyzed by Georgi, Quinn and Weinberg [29] using the RG equations for $\alpha_i = g_i^2/(4\pi), i = 2,3$ and $\alpha_1 = \frac{5}{3}g_i^2/(4\pi)$ where g_i are the $SU(3), SU(2), U(1)$ couplings

$$\alpha_i^{-1}(q^2) = \alpha_i^{-1}(\mu^2) + \beta_{0i} ln(q^2/\mu^2) + .. \tag{26}$$

where the one-loop β function is simply evaluated knowing the light matter content

$$2\pi\beta_{0i} = \begin{pmatrix} 0 \\ \frac{-22}{3} \\ -11C^2 \end{pmatrix} + N_f \begin{pmatrix} \frac{4}{3} \\ \frac{4}{3} \\ \frac{4}{3}C^2 \end{pmatrix} + N_H \begin{pmatrix} \frac{1}{10} \\ \frac{1}{6} \\ 0 \end{pmatrix} \tag{27}$$

and N_f, N_H denote the number of families and Higgs doublets respectively. In SU(5) $C^2 = 1$ and $\alpha_i(M_X^2) = \alpha_G$; the same applies to many of the phenomenologically viable GUTs and superstring unification schemes. Given these equations the low-energy couplings should evolve in energy to meet at the unification value α_G. However, the recent precision measurements of $\sin\theta_W$ are inconsistent with these predictions; the couplings fail to meet by more than six standard deviations [30].

The situation is quite different in the MSSM due to the effects of the new supersymmetric states on the evolution of the couplings. Including this gives [31]

$$2\pi\beta_{0i} = \begin{pmatrix} 0 \\ -6 \\ -9C^2 \end{pmatrix} + N_f \begin{pmatrix} 2 \\ 2 \\ 2C^2 \end{pmatrix} + N_H \begin{pmatrix} \frac{3}{10} \\ \frac{1}{2} \\ 0 \end{pmatrix} \tag{28}$$

In fig 9 the evolution of the couplings (including two-loop effects) is shown and it may be seen that the couplings do meet in a point. This happens if the mass of the new supersymmetric states (assumed degenerate here) are low $M_{SUSY} \approx 10^{2.5\pm1}$. On the basis of these results it is tempting to argue that there is evidence both for new forms of (relatively) light supersymmetric matter

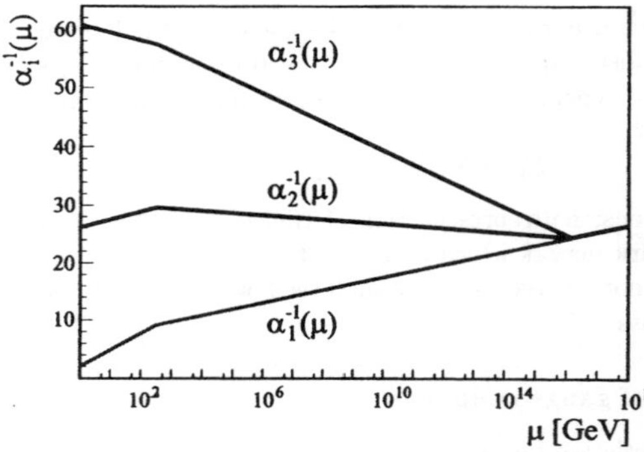

Figure 9: Evolution of running couplings. The break at low scales corresponds to the transition from non-supersymmetric to supersymmetric β functions.

and an underlying unified theory[2]. This conclusion is so dramatic that it needs to be evaluated with considerable caution. The most obvious reservation is that it relies on the extrapolation of the Standard Model, albeit extended to its supersymmetric version, some twelve orders of magnitude beyond the energy scale at which it has been tested. No theory has proved to be so robust in the past so it is understandable if one views this extrapolation with some caution. Even given the framework of Grand Unification there is considerable uncertainty for the relative values of the couplings may differ in different unification schemes. Within the (large) class of theories which give the $SU(5)$ predictions there are corrections coming from virtual states with mass $\geq O(M_X)$ which have not been included in the analysis and which can affect the predictions substantially. These predictions are even more sensitive to additional light states beyond those needed for the minimal supersymmetric Standard Model (Indeed it is possible to bring the predictions of non-supersymmetric $SU(5)$ into agreement with experiment if additional light Higgs doublets are added). Furthermore if the GUT should not break immediately to the Standard Model there will be several intermediate scales of breaking and again the results will change substantially.

Despite these caveats it is remarkable that the simplest possible extension of the Standard Model to include supersymmetry coupled with the simplest assumption about Grand Unification yields predictions in detailed agreement with experiment provided the new supersymmetric states

[2]It is perhaps appropriate to comment about the meaning of this fit, for the three gauge couplings are described in terms of three effective parameters M_X, α_X and the effective supersymmetry mass scale, meaning there will always be a fit and apparently no test of unification! However this is not quite fair for the resulting values of the parameters must be reasonable if the scheme is to make sense. Thus M_X should be less than the Planck scale but large enough to inhibit proton decay in Grand Unified theories. Also we would like α_X to be within the perturbative domain although it may be sensible to contemplate non-perturbative unification. Finally the supersymmetry mass scale must be large enough to explain why no supersymmetric states have been found and small enough to avoid the hierarchy problem (as we will see the latter gives a very strong constraint). As we will discuss these conditions are indeed satisfied.

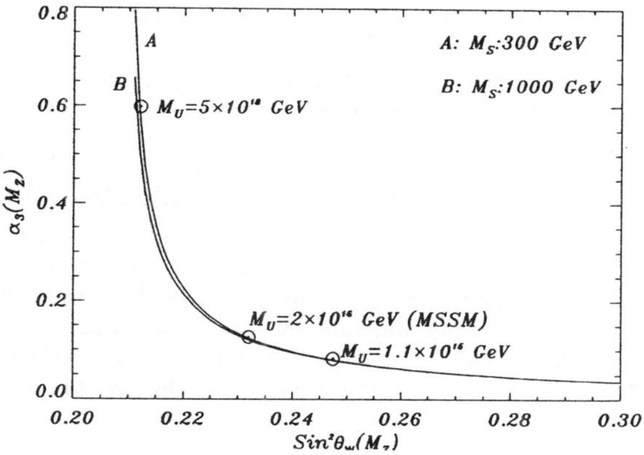

Figure 10: A plot of the prediction for the gauge couplings in the MSSM. The curves A and B have different values for the SUSY threshold, M_S, as shown. The dependence on the unification scale M_U is also shown. The relative area between the curves to the overall area for a reasonable range of α_3 and $\sin^2(\theta_W)$ gives a measure of the accuracy of the prediction.

are very light. Indeed the level of agreement is remarkable.

The precision of the prediction is better illustrated in Figure 10[32] where one may see that the predicted region in the $\sin\theta_W - \alpha_s$ plane occupies only a very small part of the a priori allowed region. A better way to illustrate this may be obtained by using the assumed unification to predict the value of the strong coupling in terms of the weak and electromagnetic couplings. A measure of the significance of this result is given by the relative area of the theoretical prediction compared to the overall area of reasonable predictions. This is quite impressive already as it appears to be better than one part in one hundred. Using the measured value of the strong coupling one may use this plot to determine the prediction for $\sin^2\theta_W$. One finds

$$\sin^2\theta_W = 0.2314 \pm 0.002 - 0.25(\alpha_s - 0.119))$$

to be compared with the experimental value $\sin^2\theta_W = 0.2312 \pm 0.0002$.

4.1.2 Neutrino masses in GUTs - the see-saw mechanism

The implications of the neutrino oscillation experiments are very exciting, for non-zero neutrino mass means a departure from the Standard Model and neutrino oscillations indicate violation of lepton family number, again lying beyond the Standard Model. The first question that needs to be answered, is why are the neutrino masses so small. Grand Unification provides the most promising explanation, namely that neutrino masses are small due to the "see-saw" mechanism [34] in which the light neutrino are suppressed by a very large scale associated with the onset of new (Grend unified?) physics. The "see-saw" mechanism follows naturally in the case that right-handed neutrinos exist.

Suppose that there is no weak isospin 1 Higgs field and hence there are no mass terms of the $\nu_L\nu_L$ type. In this case there are two possible neutrino masses

$$m_{Dirac}\overline{\nu_L}\nu_R + M_{Majorana}\nu_R\nu_R \equiv \left(\overline{\nu_L} \quad \nu_R \right) \begin{pmatrix} 0 & m_D \\ m_D & M_M \end{pmatrix} \begin{pmatrix} \overline{\nu_L} \\ \nu_R \end{pmatrix} \tag{29}$$

The Dirac mass is similar to an up quark mass and one's naive expectation is that they should have similar magnitude. On the other hand, the Majorana mass term is invariant under the Standard Model gauge group and does not require a stage of electroweak breaking to generate it. For this reason, one expects the Majorana mass to be much larger than the electroweak breaking scale, perhaps as large as the scale of the new physics beyond the Standard Model; for example, the Grand Unified scale or even the Planck scale. Diagonalising the mass matrix gives the eigenvalues

$$\begin{aligned} m_{Heavy} &\simeq M_M \\ m_{Light} &\simeq \frac{m_D^2}{M_M} \end{aligned} \tag{30}$$

The see-saw mechanism generates an effective Majorana mass for the light neutrino (predominantly ν_L) by mixing with the heavy state (predominantly ν_R) of mass M_M. It is driven by an effective Higgs $\Phi^{I_W=1}$ made up of $H^{I_W=1/2}\ H^{I_W=1/2}\ /\ M_M$ (hence the two factors of m_D in eq.(30)). Eq.(30) shows that a large scale for the Majorana mass gives a very light neutrino. For example with $M_M = 10^{16}$ GeV and m_D taken to be the top quark mass gives

$$m_{Light} \simeq 3.10^{-3}\ \text{eV}$$

This estimate shows that it is quite natural to have neutrinos in a mass range appropriate to give, for example, solar neutrino oscillations. However, in many cases, larger masses capable of explaining the other oscillation phenomena are possible because the Majorana mass for the right handed neutrinos is often smaller than the Grand Unified mass. A Majorana mass for the right-handed neutrino requires a Higgs carrying right-handed isospin 1 (in analogy with the left-handed case when it needed left-handed isospin 1). If this field is not present (for example in level one string theory this is always the case) one may get a *double see-saw* because the Majorana mass for the right-handed neutrino is also generated by an effective Higgs, made up of $H^{I_{W,R}=1/2}H^{I_{W,R}=1/2}/M'$, where M' denotes a scale of physics beyond the Grand Unification scale. Taking this to be the Planck scale (probably the largest reasonable possibility) and $< H^{I_{W,R}=1/2} >$ to be the Grand Unified scale (it breaks any Grand Unified group) one finds

$$M_M \simeq \frac{(10^{16})^2}{10^{19}}\ \text{GeV}$$

giving

$$m_{Light} \simeq 1\ \text{eV}$$

Thus, one sees that the see saw mechanism naturally gives neutrino masses in the range relevant to neutrino oscillation measurements. Moreover, as the neutrino mass is proportional to the

Dirac mass squared, taking the Dirac mass of each family of neutrinos to be of the order of the equivalent up quark mass, one obtains a large hierarchy between different families of neutrino. This is what is required if one is to explain several oscillation phenomena, for it allows the existence of several mass differences.

4.1.3 SUSY-GUT Nucleon Decay

One of the features of $SU(5)$ is its prediction for proton decay. Given that Super Kamiokande will extend the lifetime measurement of the proton by a factor of 10-20 it seems appropriate at this time to review the implication of Grand Unification for proton decay [35]. The original diagram suggested in SU(5) generating proton decay involves vector boson exchange leading to the amplitude $\sim \frac{1}{M_x^2}qqq\ell$ and the decay rate $\Gamma_{p\to\pi^0 e^+} \propto \frac{m_p^5}{M_X^4}$. Since the amplitude is due to the exchange of a new vector boson X_μ the amplitude is dimension 6 and is suppressed by two powers of the X-boson mass. Allowing for couplings of order one, the current bound on the proton lifetime leads to the bound on the X- boson mass, $M_X > 1.5.10^{15}$GeV.In supersymmetric Grand Unification there are additional sources of nucleon decay. In this it is the exchange of a coloured Higgsino triplet that leads to proton decay and as a result the amplitude is only suppressed by one power of the heavy fermion mass (a dimension 5 operator). The bound on this mass scale is some 8 . 10^{23}GeV if the couplings involved are of order one. Thus if this scale is associated with the Grand Unified scale, then there must clearly be some supression due to small couplings. For example, in a supersymmetric Grand Unified theory, the graph involves Yukawa couplings to the light quarks which are so small that the proton decay rate prediction is acceptable. In this case the decay mode for this second class of diagram is largely into strange quarks because of the larger Yukawa coupling to heavy quarks, leading to the dominant decay modes [36]:

$$\begin{array}{llll} p\rightarrow & K^{\dagger}\overline{\nu}_{\mu}, & \pi^{+}\nu, & K^{0}\mu^{+} \\ & 1 & :0.5 & :0.007 \\ n\rightarrow & K^{0}\overline{\nu}_{\mu}, & \pi^{0}\overline{\nu}_{\mu} & \\ & 1.8 & :0.24 & \end{array}$$

The overall rate is model dependent. In a realistic SO(10) model it is very close to the present limit. In other models such as flipped SU(5) [38] the dimension 5 operators are suppressed and the dimension 6 operators are dominant, leading to the prediction that the proton should decay into $\pi^0 e^+$as in the original non-supersymmetric SU(5). Again the rate may be expected to be close to the upper bound.

Since the dimension 5 operators with $O(1)$ couplings require a scale of $O(10^{24}Gev)$ to keep the rate within experimental bounds one may be worried that corrections at the Planck scale will lead to unacceptably fast processes; for example black hole evaporation or wormhole production [35]. Since there is no reason to expect a small coupling in these processes they may give an amplitude which is too large. Similarly in a superstring theory couplings that are allowed by the symmetries are typically of order one and again in this case one would obtain unacceptably fast proton decay from the exchange of heavy string states. As we discuss below we do not even have to have Grand Unification while preserving the successful predictions of Grand Unification. In this case there is no absolute necessity for the proton to be unstable at all and indeed one can construct models in which the low energy symmetry forbids proton decay within measurable limits [39]. Given the caveat discussed above concerning the generation of large dimension 5 operator contributions in string theories, one might think that this latter possibility is the most likely; i.e. the theory should have a symmetry which kills the dimension 5 operators once and for all. In this case one is left with the dimension 6 operators as the dominant source of nucleon decay.

4.1.4 SUSY Unification

As I have stressed supersymmetry is necessary to avoid the hierarchy problem in a Grand Unified theory. The resulting prediction for gauge coupling unification provides a very strong hint that low energy supersymmetry is realised in nature. There are further unification predictions which support this idea.

Fermion mass unification As may be seen from eq.(24) fermion masses may be related too. For the third generation the relation $m_b = m_\tau$ is not viable but, as in the case of gauge couplings, this prediction applies at the GUT scale and must be radiatively corrected. Gauge corrections make the mass increase at lower scales while Yukawa interactions decrease the mass. For the b, τ system the gauge couplings dominate and, since only the b is coloured, QCD interactions will make the b quark heavier at low scales. The detailed prediction for m_b and m_τ depends on the soft supersymmetry breaking parameters but, remarkably, for a choice of soft mass parameters consistent with the other unification constraints, excellent agreement is found with their measured values[40]. Unfortunately it is not possible to set the first and second generations of down quarks and leptons equal. However a simple modification involving the additional couplings of eq.(25) does give excellent predictions for the light generations too. We shall discuss this in Section 8.

Soft mass unification A further unification prediction is what I call "soft mass unification". In the spirit of looking for simplicity at high energies, this is the assumption that, at the unification scale, the scalar masses and the gaugino masses are separately degenerate with values $m_0, m_{\frac{1}{2}}$ (a similar universality assumption may be made for the other soft parameters). At low energy radiative corrections splits these degeneracies. For example the squarks and gluinos will become heavier than the winos and leptons simply because the gauge interactions causes the masses to rise and QCD is the strongest gauge interaction. Lacking information about the mass spectrum of the supersymmetric particles this of course does not lead us to any testable predictions. However, when one applies the same ideas to the Higgs boson something very interesting happens. One of the main arguments for a low-energy supersymmetry was the need to explain the hierarchy of masses and in particular the difference between the Planck scale and the electroweak breaking scale. It is therefore reasonable to ask how the supersymmetric unification we have constructed explains this difference. Remarkably the theory contains a mechanism for symmetry breaking which automatically selects the correct breaking pattern, namely $SU(3) \otimes SU(2) \otimes U(1) \rightarrow SU(3) \otimes U(1)$. To see how this comes about note that the coefficient of the quartic Higgs coupling term is given in terms of gauge couplings by supersymmetry whereas it is an unknown parameter in the Standard Model. The masses of the two Higgs fields are given by soft mass unification at the unification scale

$$m_{H_1}^2(M_X) = m_{H_2}^2(M_X) = m_0^2 + \mu^2 \tag{31}$$

where μ is the Higgsino mass term corresponding to the term $\mu H_1 H_2$ in the superpotential. At lower energies the Higgs mass terms are given via the RG equations in terms of the same mass parameters discussed above. The Higgs masses evolve differently[41]

$$\frac{dm_{H_i}^2}{dlog(Q)} = \frac{1}{8\pi^2}(-3M_2^2 g_2^2 - \frac{3}{5}M_1^2 g_1^2 + 3h^2\delta_{i2}(m_{\tilde{t}}^2 + m_{\tilde{t}^c}^2 + m_2^2 + A^2)),\ i = 1,2 \tag{32}$$

where the M_i are the gaugino masses and the tilde denotes the supersymmetric state. Thus the difference between m_1, m_2 is due to the term involving the top Yukawa coupling, h. This term will drive m_2^2 negative if it is large enough, triggering electroweak breaking. Now we can see why it is $SU(2)$ and not $SU(3)$ that is broken by this radiative breaking mechanism for, although the RG equations for both the Higgs scalars and the top (and bottom) squarks have destabilising terms due to the top Yukawa coupling only the squarks have large stabilising terms due to QCD interactions. Thus it is the Higgs $(mass)^2$ that is driven negative, and it is $SU(2)$ that is broken by the resultant Higgs vev.

¿From eq 32 we see that the initial value of h may be chosen to give the correct value of M_Z for a given value of $m_0, m_{\frac{1}{2}}$[3]. It is found that a solution may be obtained for h consistent with the top quark mass and $m_0, m_{\frac{1}{2}}$ consistent with the values needed to achieve gauge unification (the latter is sensitive to the point at which the new supersymmetric states switch on in the β-function). This very pleasing self-consistency check of the MSSM shows how the electroweak breaking scale is generated from the unification scale, realising the original aim in introducing supersymmetry. It also allows us to investigate the hierarchy problem for, from eq 32, one may see the Higgs mass squared, and hence the electroweak breaking scale, is generated largely by the squark mass and has its natural value proportional to this mass.

Dark matter abundance Finally there is also a prediction for the dark matter abundance made up by the LSP, the lightest supersymmetric particle which is stable in the MSSM. Since the spectrum of the supersymmetric state is determined by the soft supersymmetric breaking parameters and the radiative corrections, one can determine the character of the LSP and hence, knowing its interactions, determine its annihilation rate and thus its present day abundance.

The remarkable thing is that all four of these unification predictions work well for a common choice of the soft supersymmetry breaking parameters. Much work had been done on this topic over the years[42]. In Figure 11 I show the results of a combined fit to data for the "constrained" MSSM[42]. This is the MSSM in which the soft supersymmetry breaking parameters are constrained by a simultaneous fit to gauge coupling constant unification, to the value of Z mass coming from radiative electroweak breaking, to the Yukawa coupling constant unification, $m_b = m_{\tau,}$ to the dark matter abundance. In addition one requires consistency with phenomenological constraints, namely the SUSY spectrum should be consistent with the lower limits on SUSY states established experimentally and also with the branching ratio b→ $s\gamma$ (this turns out to be quite sensitive to radiative corrections involving supersymmetric states). Two regions of parameter space have been identifed where all of these constraints can be simultaneously satisfied. They correspond to low and high values of tan β. The allowed regions of the m_0, $\mathrm{m}_{1/2}$

[3] $m_{1/2}$ is the common gaugino mass at the unification scale.

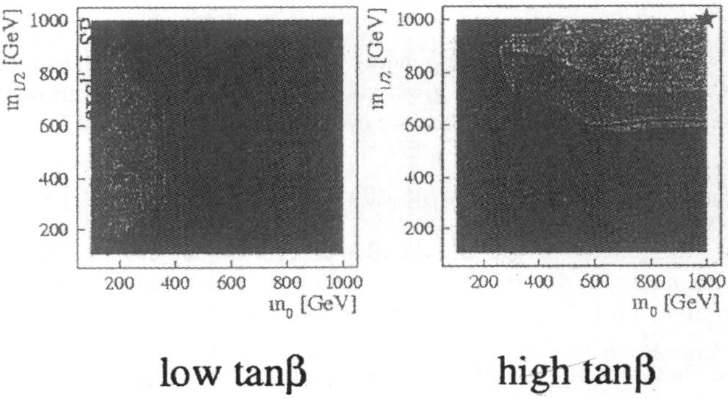

Figure 11: The allowed regions in the m_0 $m_{\frac{1}{2}}$ plane for the CMSSM fit discussed in the text. Also shown is the future LEP sensitivity.

plane are shown in Fig 11. In the low tan β solution the top Yukawa coupling is large, close to its infra-red fixed point value, M_t = 206 sinβ GeV. The Higgs mass is relatively light $M_H \lesssim 115$ GeV (cf the 132 GeV absolute upper bound in the MSSM). The LSP abundance is $\Omega h^2 = 0.42$, consistent with it being the dominant component of dark matter. with M_{LSP} = 80 GeV and composed mainly (99%) of bino. The large tan β solution extends the equality of the bottom and τ Yukawa couplings to equality with the top Yukawa coupling, giving $m_b = m_\tau = m_t$, as is suggested by many SO(10) grand unified models. Again the lightest Higgs state is relatively light, $M_H \approx 110 GeV$, and again the dark matter abundance is large. $\Omega h^2 = 0.19$, consistent with the LSP being the dominant dark matter component.

4.2 String unification

Attempts to go beyond the Standard Model usually focus on the unexplained multiplet patterns and the unification of gauge couplings that seem to require a stage of Grand Unification. It is quite remarkable that the observed quark and lepton states of the Standard Model transform as singlets or as the fundamental representation of the Standard Model gauge group. It is also quite remarkable that the quarks and leptons fill out complete representations of $SU(5)$ (and $SO(10)$) even though the low-energy gauge group is not $SU(5)$. Moreover the $SU(5)$ assignment of Standard Model states offers an immediate explanation for the left-handed nature of the weak interactions for both the quarks and leptons. However this symmetric view is somewhat spoilt by the sector responsible for spontaneous symmetry breaking. The Higgs doublet is the only representation of matter fields in the Standard Model spectrum that does not have its $SU(5)$

partners (additional $SU(2)$ singlet d quarks). In Grand Unified theories this omission must be explained and this gives rise to the doublet-triplet splitting problem, namely the need to give the additional d quarks a mass while leaving the Higgs doublets light. Solutions to this problem exist but typically require a large number of additional Higgs fields and a somewhat unconvincing and complicated set of interactions. In addition Grand Unified theories (GUTs) must be protected against the radiative coupling of the large Grand Unified scale to the low electroweak breaking scale - the hierarchy problem. Its solution requires a low energy stage of syupersymmetry. This leads to the MSSM, the supersymmetric extension of the Standard Model. Grand Unification of couplings plus the radiative corrections of the MSSM lead to the remarkably successful prediction for the ratio of gauge couplings at low energy provided the $U(1)$ normalization is chosen to be that given by $SU(5)$.

How is this unification picture changed in superstring theories? String theory is the only candidate we have for a unification of all the fundamental interactions *including* gravity. Of course at low energies one is looking at the four dimensional effective field theory which results from string theory defined at the Plack scale. Often this involves a stage of compactification of some of the space-time dimensions and in the low energy theory these act as internal degrees of freedom. The resulting 4D theory has several promising properties summarised in Table 8.

- The symmetries are specified by the underlying string. In the heterotic string case there is an $E_8 \otimes E_8$ gauge symmetry before compactification leading to an $E_8 \otimes E_6$ on compactification on a Calabi-Yau manifold. The resulting gauge group can easily accommodate the Standard Model but also can be larger. For example it can incorporate the Grand Unified groups SU(5), SO(10) or E_6. The only symmetries that are known to descend from the string are gauge symmetries, either continuous or discrete. Unfortunately there are many possible vacua in the compactified 4 dimensional theory and we do not know how to select between them.

- The multiplet structure is specified in a given string vacua. Usually this includes some number n_g of chiral families + some number n_V of vector like states which come in complex conjugate pairs. The latter are expected to acquire mass at a high scale through a stage of symmetry breaking below the compactification scale. Moreover, if the string theory is built from level-1 Kac Moody level theories the representation content of the theory is restricted offering an explanation to the question why only low lying representations of quark and leptons are observed.

- Of course the gauge group must be broken and, just as for Grand Unification, an explanation must be provided for the presence of light Higgs multiplets. However the string offers an elegant explanation for the doublet triplet splitting problem in the case there is *no* stage of Grand Unification below the Planck scale. In this case there is no symmetry reason demanding there should be partners to the Higgs thus finessing the doublet-triplet problem. However the string now has to explain why the quarks and leptons still come as complete $SU(5)$ representations; we shall discuss how this happens naturally in a class of string theories.

- String theory has only one parameter, the string scale which can be related to the Planck scale. All other parameters are determined in terms of the vacuum expectation values

GUTs	4D STRINGS
• GAUGE GROUP *e.g.* $SU(5) \supset SU(3) \otimes SU(2) \otimes U(1)$	 $G_{\text{“}Hidden\text{''}} \otimes G'_{\text{“}Visible\text{''}}$ G=E_8 or subgroup $G' = E_6 \; ... \; SU(3) \otimes SU(2) \otimes U(1)$ Unification of gauge couplings even without a GUT.
• MATTER $SU(5) : \overline{5} + 10 \to 1$ family $SO(10) : 16 = (1 + \overline{5} + 10)$	 Definite multiplet structure even without a GUT eg “level-1” : only non-exotic $SU(3) \otimes SU(2)$ reps allowed
• FAMILIES Larger gauge structure? ...not very convincing	 3 generation examples known
• MASS HIERARCHY ⇒ SUSY GUTS	 ⇒ (N=1) SUSY can persist from superstring (No need for coloured Higgs)
• LOW ENERGY STRUCTURE (N=1) SUSY B# ? L# ? ⇒ Discrete symmetry needed	 (N=1) SUSY B# ? L# ? ⇒ Discrete symmetry predicted
• PARAMETERS $g_i \to g_{GU}$ $m_b = m_\tau, ..$ Higgs sector? ...many undetermined parameters.	 In principle all predicted in terms of single parameter $M_P \equiv M_{Planck}$

Table 8: Comparison of GUT and 4D superstring structures.

(vevs) of moduli fields, Φ_M. For example the string coupling constant is determined by

$$g_i^{1/2}(M_{string}) = \frac{k_i}{g^2_{string}} \tag{33}$$

where k_i is a parameter associated with the particular string construction and

$$g^2_{string} = \frac{1}{< S >}. \tag{34}$$

Here S is a moduli field known as the dilaton. In the absence of supersymmetry breaking the moduli fields have no potential and any value for their vacuum expectation value gives a viable string theory. Once supersymmetry is broken, the moduli's vev will be fixed and the coupling will be determined as in eq 34. Similarly the Yukawa couplings will be determined as functions of additional moduli known as complex structure moduli. (Unfortunately we are still far from an understanding of the string vacuum structure and hence from a prediction of the string coupling.)

4.2.1 String unification of gravity

As mentioned above, the string makes an important additional prediction which goes beyond Grand Unification, namely it determines the unification scale in terms of the Planck scale. If this could be checked it would provide the first quantitative test of the unification of the strong, electromagnetic and weak forces with gravity which the superstring provides. We saw that preliminary indications are promising in that the scale of unification predicted by continuing the gauge couplings up in energy from the laboratory scale assuming the minimal supersymmetric Standard Model (MSSM) spectrum gives a very large scale of unification. From Figure 10 we see this is some $(1-3) \times 10^{16} GeV$, not so far from the Planck scale which sets the scale for the string prediction. In the case of the (weakly coupled) heterotic string the detailed prediction (see the next section) is [43]

$$M_{string} \approx g_{string} \times (5.2 \times 10^{17} GeV) \approx 3.6 \times 10^{17} GeV \tag{35}$$

which is only a factor of 20 above the "observed" gauge unification scale.

It is difficult to overemphasise the potential importance of this result. Our belief that there is a stage of Grand Unification of the strong, weak and electromagnetic interactions rests largely on the quantitative success of the unification of the associated couplings. The prediction of the unification scale would be the first indication of unification with gravity. While the prediction within a factor of 20 is encouraging, the residual discrepency leaves the question open. What could be the source of this discrepancy? This has been examined in some detail. There are various effects that can cause the prediction of the Grand Unified scale to vary [43]. One should remember that the Grand Unified scale is the argument of the log and thus, in order to get it correctly, one has to work to very high precision. Possible problems in determining this scale are :

- SUSY threshold effects. It seems that although these could make the factor of 20 difference this only can be achieved with a very peculiar supersymmetry spectrum at low energies in which the gluino is lighter than the Wino [44].

- A second possibility are threshold effects the Grand Unified scale [45]. One can easily arrange that small corrections add up to give a change in the prediction for the Grand Unified scale by a factor of 20. However, in my opinion this is rather unsatisfactory for the moment one adds extra states at high energy the predictive powers is reduced and the remarkable agreement of gauge unification looks less convincing.

- Another possibility that has been explored are string threshold effects which in a given string theory are calculable[46] and amount to including the effect of the heavy Kaluza Klein modes which are themselves split when the Grand Unified theory is broken. Although it is possible to construct theories in which they can also give you a change in the unification scale by a factor of 20, there is only a small class of theories in which this happens and they look rather unnatural.

- Perhaps the most plausible explanation follows because if one takes the factor of 20 discrepancy at face value the prediction indicates the origin of the problem. If unification does not occur until $3.6 \times 10^{17} GeV$ the value of the string coupling is so large as to mean the perturbative analysis breaks down. In this case one must redo the calculation determining the non-perturbative effects. Techniques to do this have been developed using the string dualities that relate strong to weak coupling string theories. In the case of the heterotic string, the strongly coupled version is known as M-theory. Given the importance of the result I will discuss in some detail how M-theory can resolve the discrepancy in the prediction of the string or Planck scale.

4.2.2 M-theory determination of the gauge unification scale

The prediction for the gauge unification scale in the weakly coupled heterotic string follows from the general form of the 4D Lagrangian

$$\begin{aligned} L_{eff} &= -\int d^{10}x\sqrt{g}\alpha_{10}^{-1}(\frac{4}{\alpha'^4}R + \frac{1}{\alpha'^3}TrF^2 + ...) \\ &= -\int d^4xV\sqrt{g}\alpha_{10}^{-1}(\frac{4}{\alpha'^4}R + \frac{1}{\alpha'^3}TrF^2 + ...) \end{aligned} \tag{36}$$

In this we may see that Newton's constant, G_N, and the value of running gauge couplings at the unification scale, α_{GUT}, are given in terms of the 10D string coupling α_{10}, the string tension $\alpha' \sim \frac{1}{M_{string}}$ and the volume of the 6D compactified space V by

$$G_N = \frac{\alpha_{10}\alpha'^4}{64\pi V}, \quad \alpha_{GUT} = \frac{\alpha_{10}\alpha'^3}{16\pi V} \tag{37}$$

For the case that α_{10} is small the volume V is approximately M_{string}^{-6} and one obtains eq(35) eliminating V between the two equations. However there is another way of exhibiting the implication of eq(37). If one uses the "measured" value for $V = O((1-3).10^{16}GeV^{-6})$ one may obtain the value of α_{10} instead. This gives an enormous value, quite inconsistent with the assumption of weak coupling that went into the derivation of eq(37). Thus the failure of the prediction of eq(35) is not surprising - it was the wrong calculation. Instead one should go to the strongly coupled heterotic string case.

It has long been known that there are five distinct classes of string theory, a fact that has caused concern because one might hope that the "Theory of Everything" is unique. However developments in string theory have shown that there is a rich class of duality symmetries relating various string theories[47, 48] so now the expectation is that all string theories are related to each other and to an underlying theory known as "M-theory". This theory lives in eleven dimensions and the various string theories may be viewed as different limits of this underlying theory. In addition it is now realised that eleven dimensional supergravity also belongs to this web of inter-related theories. A feature of the dualities relating different theories is that they often relate one theory in the weak coupling (perturbative) limit to another theory in the strong (nonperturbative) limit or one theory in the small compactification radius limit to another in the large compactification limit. By making use of these dualities one may gain insight into non-perturbative physics by performing calculations in the dual theory in which the perturbation series makes sense.

In the case of the strongly coupled heterotic string Horava and Witten showed that the theory is equivalent to M-theory. Although it is not possible to construct M-theory explicitly it is possible to write down the long-distance effective field theory coming from M-theory. It is just 11 dimensional supergravity compactified on a six-dimensional Calabi-Yau space times a line interval, $CY_3 \otimes S_1/Z_2$. The effective Lagrangian in this case is given by [49]

$$L_{eff} = -\frac{1}{2\pi\kappa_{11}^2}\int d^{11}x\sqrt{g}R - \frac{1}{8\pi(4\pi\kappa_{11}^2)^{2/3}}\int d^{10}x\sqrt{g}TrF^2 + ...) \tag{38}$$

giving

$$G_N = \frac{\kappa_{11}^2}{16\pi^2 V R_{11}}, \quad \alpha_{GUT} = \frac{(4\pi\kappa_{11}^2)^{2/3}}{2V} \tag{39}$$

The reason this form changes is that the gravitational fields now propagate in the bulk of the 11th dimension while the gauge fields live on the end points of the line integral (one E_8 on each fixed point) and propagate only in 10D. One may see that this changes the relation between Newton's constant and α_{GUT} through the appearance of the radius of the 11th dimension, R_{11}. Given the appearance of a new parameter ~~it is~~ clear one may adjust it to eliminate the discrepancy between scales found in the weakly coupled case. This gives [50]

$$\pi R_{11} = (5.10^{15} GeV)^{-1} \tag{40}$$

While it is encouraging that the discrepancy between scales has been eliminated, apparently the predictive power has been lost. However Witten [50] has shown that R_{11} cannot be arbitrarily large because the value of the gauge couplings now depend on it in a calculable way :

$$\frac{1}{g_{6,8}^2} = S \pm \alpha R_{11} \tag{41}$$

where $g_{6,8}$ refers to the E_6 and E_8 couplings of the two gauge groups at either end of the 11D line interval. Here α is a constant which for standard embedding is of O(1). Its precise value is model dependent. Now we see that R_{11} cannot be arbitrarily large because for $R_{11} = S/\alpha$ a coupling becomes infinite (at the same time the associated volume of the Calabi Yau space vanishes). Taking $\alpha = 1$, this implies a maximum value for R_{11} given by

$$\pi R_{11} \leq (5.10^{15} GeV)^{-1} \tag{42}$$

It is remarkable that the bound corresponds closely to the desired value given in eq(40) , suggesting that R_{11} might be driven to the extremum value (R_{11} is determined by a moduli whose vacuum expectation value must be determined by supersymmetry breaking effects.)

4.3 Strings and large extra dimensions - D-Branes

The original formulation of the heterotic superstring was in 10 space-time dimensions. Indeed most superstring theories have an underlying interpretation involving more than 4 space-time dimensions. It was only realised quite recently that strings necessarily contain D-branes[51, 52], structures which exist in (D+1) space-time dimensions less than or equal to the number of dimensions of the full string theory. Open string states may be confined to a lower dimensional space because their end points are constrained to lie on the D-branes through Dirichelet boundary conditions (hence the name D-brane).

In the heterotic string that we have discussed up until now, the Standard Model states, like the states related to gravity, are described by closed string states. However in a different class of string theories non-gravitational states may be described by open string states. If it is the case that Standard Model are open string states they may live in a lower dimension than the closed string states which, having no end points, are not confined to the D-brane. The graviton is necessarily a closed string state and so string theories naturally lead to the possibility that gravity may live in a larger number of dimensions than the states of the Standard Model. Of course having more space-time dimensions than four is in conflict with observation unless there is some form of compactification which leads at large distance scales to a world that is effectively four dimensional.

String theories have also raised further possibilities for the phenomenology of new large dimensions because, although open string states are confined to $D-$branes, the space dimension of ths $D-$brane may be more than 3. Thus even the Standard Model states may live in more than the 3 space dimensions of our low-energy world. Here I will briefly discss the phenomenological implications of such theories.

5 Phenomenology of large extra dimensions

In Section 3.3.1 we discussed how the tower of Kaluza Klein states provides the signal for large new dimensions. The phenomenological implications differ for the factorisable and non-factorisable cases. In the factorisable case there are very light Kaluza Klein excitations with mass gap given by eq(13). These have various experimental signatures:

5.1 Cavendish experiments

The Kaluza Klein states modify the Newtonian potential

$$V_{gravity}(r) = -\frac{G_N m_1 m_2}{r}(1 + 2\delta e^{-\frac{r}{R}} + ...)$$

The non-observation of departures from Newtonian gravity currently places the bounds $R < 0.8mm$ at 95% confidence level[53].

Collider		R/M(n = 2)	R/M(n = 4)	R/M(n = 6)
Present	SN1987A	$3.10^{-5}/50000$	$10^{-9}/1000$	$6.10^{-11}/100$
	LEP 2	$4.8.10^{-2}/1200$	$1.9.10^{-9}/730$	$6.8.10^{-12}/530$
	Tevatron	$5.5.10^{-2}/1140$	$1.4.10^{-9}/860$	$4.1.10^{-12}/780$
Future	LC	$1.2.10^{-3}/7700$	$1.2.10^{-10}/4500$	$6.5.10^{-13}/3100$
	LHC	$4.5.10^{-4}/12500$	$5.6.10^{-11}/7500$	$2.7.10^{-13}/6000$

Table 9: Current and future sensitivities to large extra dimensions from missing-energy experiments. All values for colliders are expressed as 95% of extra dimensions R (in cm) and the effective Planck scale M (in GeV). For the analysis of SN1987A, we give probable-confidence limits.

5.2 Astrophysical bounds

Cooling due to the emission of the Kaluza Klein (KK) tower of states proves to be quite a significant bound. The rate of the graviton tower production of an object at temperature T is proportional to $(TR)^\delta/M^2_{Planck,4} = T^\delta/M^2_{Planck,4+\delta}$. Here the first factor $(TR)^\delta$ simply counts the number of KK modes excited at a temperature T. Because this is a large number the net effect is only suppressed by the $(4+\delta)$ Planck scale i.e. gravitational effects become strong at this scale which may be as low as $1TeV$! Clearly the hotter the object the larger the effect. A supernova has a temperature of $O(30MeV)$ and one may use the observed phenomena of the supernova to limit the possible new effects due to the KK tower. This gives a limit of $O(1TeV)$ for the case $\delta > 2$ and $O(50TeV)$ for $\delta = 2$[54]. Similar results have been obtained by studying the effects on the diffuse $\gamma-$ray backgrond[55].

5.3 Laboratory probes

If the graviton KK modes are produced in the laboratory they will not be seen because of their gravitational strength coupling. However this does not mean their effect is unobservable. For example they may significantly change the cross section for $e^+e^- \longrightarrow \gamma+$ missing and $pp \longrightarrow jet+$ missing. The process $e^+e^- \longrightarrow \gamma + g^n$, summed over the graviton Kaluza Klein modes g^n which are energetically accessible, has a cross section given by $(ER)^\delta/M^2_{Planck,4} = E^\delta/M^{2+\delta}_{Planck,4+\delta}$. Once again the first factor counts the number of KK modes which can be excited. The main background to this process $e^+e^- \longrightarrow \gamma + \nu\overline{\nu}$. Provided one can use polarised electrons one can suppress the dominant t-channel W exchange part of the background. Further the cross section for tensor production has a characteristic angular dependence which will also help in establishing a signal One may estimate the reach of various colliders in probing new large dimensions[26, 27]. A summary is given in Table 9.

As discussed in Section 3.4.3 it is also possible to detect indirect effects of the graviton KK tower via modifications of Standard Model processes, for example via graviton KK exchange in the processes e^+e^-, $q\overline{q} \longrightarrow f\overline{f}$. Once again after summing over the KK one finds it is the $(4-\delta)$ Planck mass that determines the strength of the effect. However in this case the cross section is proportional to $M^{-4}_{Planck,4}$ with no dependence on δ whereas the graviton KK production cross section was proportional to $(E/M_{Planck,4+\delta})^\delta$. As a result they have different sensitivities to different numbers of new large dimensions so the processes give largely complementary bounds[56].

Limits on Mass Scale M_D for n Extra Dimensions

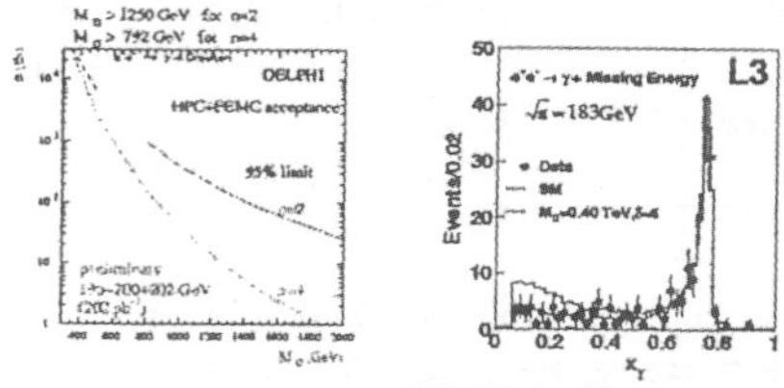

Efficiencies for signal from $e^+e^- \to \nu\bar{\nu}\gamma$ reweighted to expected signal differential cross-section

$\sqrt{s}$ (GeV)	$\int L$ (pb^{-1})	n = 2	3	4	5	6	7
189-202	410	1.10	0.86	0.70	0.60	0.52	
196-202	202	1.25		0.79			
189	176	1.02	0.81	0.67	0.58	0.51	0.45
189	177.3	1.09	0.86	0.71	0.61	0.53	0.47

M_D^{95} (TeV)

Figure 12:

Summary of Limits from Indirect Searches

	Λ_T^+/Λ_T^-	M_s (λ=+1/−1)		
Channel	ALEPH	OPAL	DELPHI	[illegible]
$\mu^+\mu^-$	680/630	600/630 *	725/592	690/560
$\tau^+\tau^-$	590/570	630/500	645/557	540/580
$q\bar{q}$	610/660	---	---	490/490
e^+e^-	1030/800	---	---	994/911
Fermions combined	1040/820	680/610	755/598	1000/840
$\gamma\gamma$	920/910	641/629 *	713/691	790/800
WW	---	---	---	790/680
ZZ	---	---	---	770/760
Bosons combined	890/770	641/629	713/691	890/820
Bosons + fermions	1120/840	---	---	1070/870

$\sqrt{s} \leq 189$ GeV published

$\sqrt{s} \leq 189$ GeV preliminary

$\sqrt{s} \leq 202$ GeV preliminary

$$\frac{|\lambda|}{M_S^4} = \frac{\pi}{2}\frac{1}{\Lambda_T^4}$$

Best limits from Bhabha scattering in which interference terms contributes significantly

* (OPAL limits from $\gamma\gamma$ are on $\left(\frac{2}{\pi}\right)^{1/4} M_s$)

Figure 13:

The results of LEP experimental searches for direct and indirect evidence of large new dimensions are summarised[57] in Figures 12 and 13.

5.4 Phenomenology of new large dimensions - non-factorisable metric

The case of non-factorisable metric can lead to quite a different phenomenology. As discussed in Section 3.3.1, due to the warp factor the size of the extra dimension can be quite small while still generating the hierarchy. As a result the mass of the KK modes need not be as low as for the factorisable case, eq(11). It is found in this case that the KK modes are roughly equally spaced with mass of order ϖk. Since $k \lesssim M_{Planck,4}$ the lightest mode will not be far below $1TeV$ in stark contrast to the factorisable case. Furthermore the coupling of this mode, rather than being proportional to $M_{Planck,4}^{-1}$ is in fact enhanced by the inverse warp factor giving a coupling $O(TeV^{-1})$. As a result the phenomenology is quite different. In e^+e^- collisions the new KK state will show up as a new resonant state with the final state angular distribution characteristic of tensor exchange giving a very clear signal. The phenomenology of more general configurations have been studied [59, 60] where it is argued that the phenomenology may be much richer, a combination of the factorisable metric phenomenology discussed with very light gravitational strength coupled KK states and the more massive, strongly coupled KK modes just discussed.

5.4.1 Small neutrino masses - large new dimensions

We have seen how Grand Unification provides an elegant explanation for the lightness of neutrinos. In the case of large new dimensions, however, the solution to the hierarchy problem is supposed to be due to the absence of any fundamental mass scale much larger than the electroweak breaking scale[7]! As a result it cannot use the see-saw mechanism, eq(30) involving the very large Grand Unified scale. Somewhat surprisingly, these theories provide an equally elegant reason for the lightness of neutrinos.

Factorisable Case. In these theories the smallness of Newton's coupling is due to flux spreading of the graviton in the additional dimension. If the right-handed neutrino is also able to propagates in the new large dimension its flux too will spread in the extra dimensions and this will lead to a suppression of the Yukawa coupling of the neutrino to the Higgs. As a result the neutrino mass will be naturally small, having a similar see-saw form to that in eq(30) suppressed by an inverse power of the four dimensional Planck mass[58]. To make this more explicit consider a five dimensional (5D) space with coordinates (x_i, y). The interaction of the right-handed bulk neutrino, $\nu_R(x.y)$, with the Standard Model lepton doublet and Higgs states, $l(x)$, $h(x)$, confined to a 4-D brane is given by

$$S = \lambda \int d^4x l(x) h(x) \nu_R(x, y = 0)$$

If the fifth dimension is compact (here, compactified on a circle) we may expand the neutrino wave function in terms of its Kaluza Klein tower of states

$$\nu_R(x, y) = \frac{1}{\sqrt{2\pi R M_*}} \sum_{n=-\infty}^{\infty} \nu_R^n(x) e^{iny/R}$$

where the normalisation factor has been chosen so that the states, $\nu_R^n(x)$, have canonical normalisation for their kinetic terms - this factor accounts for the flux spreading into the fifth dimension. Here R is the radius of compactification and M_* is the fundamental mass scale. The neutrino mass is given by

$$\begin{aligned} m_\nu &= \frac{\lambda < h >}{\sqrt{2\pi R M_*}} \\ &= \frac{\lambda < h > M_*}{M_{Planck,4}} \\ &\approx 10^{-4}\lambda eV.(\frac{M_*}{1TeV}) \end{aligned}$$

where we have used the fact that the 5D mass scale is related to the 4D Planck scale by $2\pi R M_* = M_{Planck,4}/M_*$. Thus we see the effect of flux spreading is to suppress the neutrino coupling in the same way as the gravitational coupling is suppressed since both propagate in the extra dimension(s). As a result one naturally obtains a see-saw mechanism suppressed by the large Planck mass even though there are no fundamental mass scales of this magnitude in the theory. For M_* of order $1TeV$, a scale that avoids the hierarchy problem, one readily obtains neutrino masses of the correct order.

Non-factorisable case

In this case the graviton wave function (the zero mode of the Kaluza Klein tower) is peaked on the Planck brane and masses on the Standard Model brane are suppressed by the warp factor corresponding to the fall-off of the graviton propagator, $M_i \sim M_{Planck}e^{-kr}$. If the right handed neutrino, ν_R, propagates in the bulk one finds it, too, has its lightest peaked on the Planck brane. As a result its Yukawa coupling to the lepton doublet and Higgs doublet fields living on the Standard Model brane is suppressed by the factor $e^{-m/k}$ where m is the ν_R mass in the bulk and k is the 5D curvature. One may try to use these factors to obtain a pattern of mass hierarchies [61] but one may wonder why the hierarchies needed are quite small when the natural expectation from an exponential factor of this type is a very large hierarchy.

Neutrino probe of the Kaluza-Klein tower Neutrinos can provide a very sensitive probe of the Kaluza-Klein tower. If solar neutrino oscillation is due to oscillation from the electron neutrino into a tower of KK states, if these are closely spaced, it is possible that more than one state to be produced. This gives rise to observable phenomena. In Figure 14 I show the recoil electron spectra compared to the SuperK result that would result from various choices of spacing between the KK levels. The best agreement is for a compactified radius of $0.1mm$ with three sterile neutrinos being significantly excited via the MSW mechanism as the neutrino escapes the solar core! Recent SNO results, however, disfavour a sterile neutrino interpretation unless the ^8B spectrum is significantly different from current estimates[63].

Propagation of Standard Model states in new dimensions Up to now we have assumed that Standard Model states do not propagate in extra dimensions. However in Section 4.3 we saw that in string theories it is possible for open string states to live in $D-$branes when space-time dimension is greater than 4. The first signal of such a new dimension states will be the lightest Kaluza Klein excitation of a Standard Model states. Since these have gauge interactions

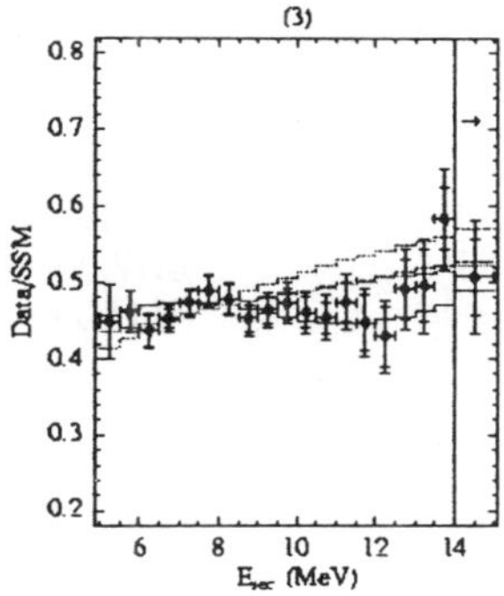

Figure 14: Solar neutrino recoil electron spectrum for various compactification radii : $(1/R) = 2 \cdot 10^{-2} eV$ (dotted line), $(1/R) = 5 \cdot 10^{-3} eV$ (darker solid line), $(1/R) = 2 \cdot 10^{-3} eV$(dashed line) and $(1/R) = 0.5 \cdot 10^{-3} eV$ (lighter solid line).

their production cross section will be large. In the case that they can be singly produced present experiment puts bounds on such states of order $1TeV$ and so the new dimensions must be $\leqslant O(1TeV^{-1})$. It may happen that, to a good approximation, the momentum in the new dimension is conserved. In this case the Kaluza Klein number n of eq.(12) will also be conserved to a good approximation and the KK states will only be pair produced. As a result the bound on the mass of such states is smaller, of order $0.5TeV$.

5.5 Gauge Unification as a test of large new dimensions

In Section 4.1.1 we discussed the remarkable accuracy of the prediction for gauge coupling unification that follows in the MSSM with a high unification scale. Can theories without a high unification scale achieve gauge unification? In [64] it was suggested that gauge unification could unify at a low scale because gauge couplings run like a power rather than a log if the number of space-time dimensions is greater than 4. Thus, if the beta functions above the compactification scale remain in the same as in the MSSM, the prediction of gauge unification will be as in the MSSM but the unification scale will be lower, perhaps as low as $1TeV$. While this is an appealing idea, it proves impossible to realise it in a way that preserves the MSSM *prediction*. The problem is that the MSSM beta function comes from a combination of the gauge sector and the Higgs/matter sector. If the SM fields propagate in extra dimensions, they will have Kaluza Klein excitations which, being massive, fill out $N = 2$ supermultiplets. A $N = 2$ vector supermultiplet combines a $N = 1$ chiral supermultiplet with a $N = 1$ vector supermultiplet, both in the same adjoint representation of the gauge group. These contribute to the one-loop beta function with opposite signs. On the other hand a $N = 2$ hypermultiplet

$$|V_{u_i d_j, CKM}| \approx \begin{bmatrix} 0.973-0.975 & 0.217-0.222 & 0.0023-0.004 \\ 0.21-0.24 & 0.9-1.2 & 0.038 \pm 0.041 \\ 0.006-0.01 & 0.026-0.04 & 0.84-1.14 \end{bmatrix}$$

$M_d(M_Z) = (1.8-5.3)MeV$	$M_u(M_Z) = (0.9-2.9)MeV$
$M_s(M_Z) = (35-100)MeV$	$M_c(M_Z) = (0.53-0.68)GeV$
$M_b(M_Z) = (2.8-3.0)GeV$	$M_t(M_Z) = 168-180GeV$
$Q = \frac{m_s/m_d}{\sqrt{1-(m_u/m_d)^2}}$	22.7 ± 0.08
m_u/m_d	0.553 ± 0.043
m_c/m_s	8.23 ± 1.5

Table 10: **Experimental data on quark masses and mixing angles.**

contains two $N=1$ chiral multiplets, both in the same representaion of the gauge group. Thus the $N=2$ hypermultiplet contributes twice that of a single $N=1$ multiplet. Because of the different structure of the vector and hyper multiplets their total contribution to the one-loop beta functions of the various gauge group factors is *not* proportional to the $N=1$ contribution [32]. Thus, in order to regain the MSSM prediction, additional states must be added to the theory. In this case the precision prediction of the MSSM must be viewed as a complete accident as the running above the string scale comes from a spectrum completely unrelated to that of the MSSM, something I find hard to accept given the remarkable precision of the prediction.

Of course our analysis does not preclude the existence of large new dimensions not associated with the propagation of Standard Model states. A particular example is the strongly coupled heterotic string in which gravity but not the Standard Model states propagate in the eleventh dimension. For the case the eleventh dimension is three orders of magnitude larger than the string length, corresponding to a compactification radius of $O(10^{-14}fm)$, the gauge unification scale may be reduced to that found in the MSSM. However, even in this case, the size of the extra dimensions is still severely constrained by the need to have a high unification scale to be $\geq 10^{-14}fm$, the minimum occurring for just one additional dimension as in the strongly coupled heterotic case.

6 Quark masses and mixing angles

As we have seen, unified theories offer answers to all the questions raised by the structure of the Standard Model, Section 2.1. We have also discussed how such theories give precise predictions for the gauge coupling of the Standard Model too. What about the remaining parameters, in particular those responsible for quark and lepton masses and mixing angles?

In Table 10 I present the data for the quark masses and mixing angles as currently known [33]. Note that what we know are the eigenvalues of the mass matrix and the CKM matrix which is a combination of the rotations matrices needed to diagonalize the left-handed components of the up and down quarks.

$$\mathrm{U}_{CKM} = V_L^{u\dagger} V_L^d \tag{43}$$

However, the full mass matrix requires a knowledge of both the left and right handed rotation matrices

$$M^{u,d} = V_R^{u,d^{T^\dagger}} M^{u,d}_{Diag} V_L^{u,d} \tag{44}$$

Thus measurement of the CKM matrix and the diagonal mass entries is insufficient to construct the full mass matrix. This is an embarrasment when trying to develop a theory of masses because masses in a field theory arise through the matrix of couplings of quarks to the Higgs boson. Upon spontaneous breakdown this leads to a mass matrix as the fundamental starting point. What then do we know about this mass matrix?

6.1 Perturbative analysis of mass matrix

Bjorken [65, 66, 67, 68] has introduced a systematic way to determine the mass matrix from the observed masses and mixing angles for the case of small off diagonal matrix elements. Although not inevitable, this is the preferred case for it leads to small mixing angles as are observed without requiring a cancellation between up and down rotations. In this case it is easy to show that the CKM matrix elements determine those elements of the mass matrix above the diagonal. Those below the diagonal are only weakly constrained. To see this we write the mass matrix in the form

$$M = M_D + \Delta M + m \tag{45}$$

in which M_D is a diagonal matrix with the mass eigenvalues along the diagonal, m refers to the small off diagonal elements and ΔM refers to the correction to the diagonal elements on going to the non-diagonal form. The mass matrices M and M_D are related by

$$V_L^+ M M^+ V_L = M_D^2 \qquad (V = V^{u,d} \quad M = M^{u,d}) \tag{46}$$

Substituting the form of eq(45) and premultiplying by V_L gives

$$(M_D + \Delta M + m)(M + \Delta M + m^+)V_L = V_L M_D^2 \tag{47}$$

Since the rotation matrices are ordered in powers of the small parameters m_{ij} by

$$V_{ij} = O(m) \quad V_{ii} = 1 + O(m^2) \tag{48}$$

we may solve eq(47) perturbatively to obtain two equations in leading non-vanishing order

$$2M_i\,\Delta M_i + \sum_j |m_{ij}|^2 + \sum_j (m_{ij}M_j + M_i m^x_{ji})V_{ji} = 0 \tag{49}$$

and

$$M_i^2 V_{ij} + \sum_k (m_{ik}M_k + M_i m^*_{ki})V_{kj} + \sum_k m_{ik}m^\tau_{jk} = V_{ij}M_j^2 \tag{50}$$

These may be solved for the off diagonal elements of the CKM matrix giving

$$\begin{aligned} V_{ij} &= -\frac{(m_{ij}M_j + m^*_{ji}M_i)}{(M_i^2 - M_j^2)} + \frac{(m_{ik}M_k + m^*_{ki}M_i)(m_{kj}M_j + m^*_{jk}M_k)}{(M_i^2 - M_j^2)(M_k^2 - M_j^2)} \\ &\quad -\frac{m_{ik}m^*_{jk}}{(M_i^2 - M_j^2)} + O(m^3) \end{aligned} \tag{51}$$

¿From eq (51) we may determine the contributions to the CKM coming from diagonalisation of the up and the down quarks. Combining these we have

$$
\begin{aligned}
V_{us}^{CKM} &= \frac{m_{12}^{d}}{M_s} - \frac{m_{12}^{u}}{M_c} + \frac{m_{23}^{d*} m_{13}^{u}}{M_b M_t} + \frac{m_{21}^{d*}}{M_s}\frac{M_d}{M_s} + \ldots \\
V_{cd}^{CKM} &= -\frac{m_{12}^{d*}}{M_s} + \frac{m_{12}^{u*}}{M_c} + \frac{m_{13}^{d*} m_{23}^{u}}{M_b M_t} + \ldots.
\end{aligned}
\tag{52}
$$

This form immediately leads to a constraint between the matrix elements. To leading order we have

$$
|V_{us}| \backsimeq |V_{cd}| \tag{53}
$$

This is in excellent agreement with the measurements $|V_{us}| = 0.217 - 0.222$, $|V_{cu}| = 0.21 - 0.24$ and supports the initial assumption of small off diagonal mass matrix elements. Note that to leading order the magnitude of the matrix elements is determined only by the mass matrix entries above the diagonal.

The same procedure applied to V_{cb}, V_{ts} gives

$$
\begin{aligned}
V_{cb}^{CKM} &= \frac{m_{23}^{d}}{M_b} - \frac{m_{23}^{u}}{M_t} + \ldots \\
V_{ts}^{CKM} &= -\frac{m_{23}^{d*}}{M_b} + \frac{m_{23}^{u*}}{M_t} + \ldots
\end{aligned}
\tag{54}
$$

so

$$
|V_{cb}| \backsimeq |V_{ts}| \tag{55}
$$

again in good agreement with experiment $|V_{cb}| = 0.038 - 0.041$, $|V_{ts}| \backsimeq 0.026 - 0.04$.

Finally one finds

$$
\begin{aligned}
V_{ub} &= \frac{m_{13}^{d}}{M_b} - \frac{m_{13}^{u}}{M_t} + \frac{m_{12}^{u}}{M_c}\left(\frac{m_{23}^{u}}{M_t} - \frac{m_{23}^{d}}{M_b}\right) + \ldots \\
V_{td} &= -\frac{m_{13}^{d*}}{M_b} + \frac{m_{13}^{u*}}{M_t} - \frac{m_{12}^{d*}}{M_s}\left(\frac{m_{23}^{u*}}{M_T} - \frac{m_{23}^{d*}}{M_b}\right) + \ldots
\end{aligned}
\tag{56}
$$

Again if the leading term dominates we have approximate equality of V_{13} and V_{31} in magnitude, but this is in disagreement with the observation (*cf* Table 10). If, on the other hand, we assume that the leading terms are zero *ie* $m_{13}^{d} = m_{13}^{u} = 0$, so called "texture zeros" [69], then the next term in eq(56) dominates and leads to

$$
\begin{aligned}
|V_{ub}| &= \sqrt{\frac{M_u}{M_c}}\,|V_{cb}| \\
|V_{td}| &= \sqrt{\frac{M_d}{M_s}}|V_{ts}|
\end{aligned}
\tag{57}
$$

Substituting the measured values given in Table 10 one finds these relations are in reasonably good agreement with experiment. Since presenting these lectures, new data has indicated [70] that there are corrections to eq(57) which require non-zero m^d_{13}, although smaller than the other mass matrix elements. As we shall discuss, family symmetries capable of explaining the approximate texture zeros also indicate that the texture zeros are not exactly zero and this new information will be valuable in testing these schemes. I shall return to a discussion of this point later.

6.2 Texture Zeros

We have seen that the assumption of the vanishing (in leading order) of two elements of the mass matrix ("texture zeros") leads immediately to a prediction for CKM matrix elements in excellent agreement with experiment. Are there any other possible texture zeros? Suppose the diagonal element $(M + \Delta M)_{11} = 0$ and further assume that the $(1,2)$, $(2,1)$ elements are symmetric. The sub-matrix for either the up or the down quarks has the form

$$\begin{pmatrix} 0 & m_{12} \\ m_{12} & m_{22} \end{pmatrix}$$

This has now only two parameters (the phase can be absorbed in a redefinition of the fermion fields). Hence there is a relation between its two mass eigenvalues and the rotation diagonalising it. One may readily see that this implies [55, 71, 72, 68]

$$\begin{aligned} |V_{us}| &= |\sqrt{\frac{M_d}{M_s}} - \sqrt{\frac{M_u}{M_c}}\, e^{i\sigma}| \qquad (58) \\ c.f.(0.218 - 0.224) &= |(0.16 - 0.33) - (0.047 - 0.07)e^{i\sigma}| \end{aligned}$$

Here σ is a phase that enters when combining the up and the down rotations which can be shown to be the Jarlskog CP violating phase that enters into CP violating processes[73, 74, 70]. The experimental comparison is also shown and one may see this prediction is in good agreement with experiment (due to the smallness of the up quark contribution the σ dependence is quite small).

Thus we see that experiment is in good agreement with the hypothesis of four texture zeros (counting pairs of off-diagonal zeros as one zero since we have assumed a symmetric form). We shall return to a more systematic discussion of texture zeros shortly. Note that as was the case for V_{12}, V_{21} the other CKM matrix elements principally depend on the elements of m *above* the diagonal so we are largely ignorant of the elements below the diagonal. In fact the only hint for the magnitude of these off-diagonal lower elements comes from the success of eq(58) which applies only if $m_{21} = m_{12}$. In what follows we will often follow this hint and assume that the full mass matrices are symmetric.

To summarise, we have found the data consistent with four texture zeros plus the assumption that the off-diagonal matrix elements are small. This simple structure determines 5 of the 6 CKM matrix elements (*ie* three of the four independent elements of a unitary matrix)

The mass matrix has the form

$$M^{d(u)} = \begin{pmatrix} 0 & m_{12} & 0 \\ m_{12} & m_{22} & m_{23} \\ 0 & m_{33} & m_{33} \end{pmatrix} \qquad (59)$$

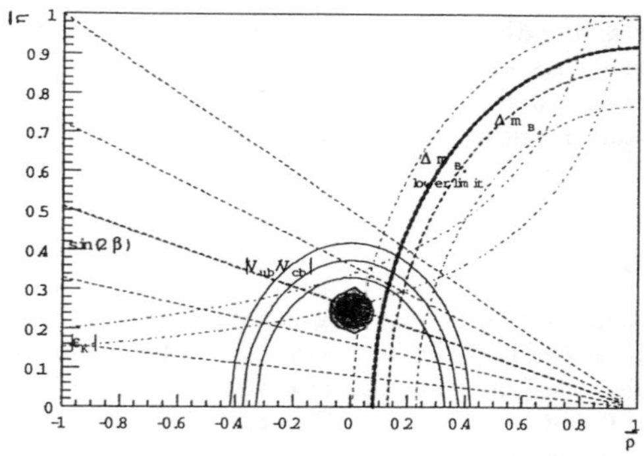

Figure 15: The fit for the relations $|V_{td}|/|V_{ts}| = \sqrt{m_d/m_s}$ and $|V_{ub}|/V_{cb}| = \sqrt{m_u/m_c}$. The experimental constraints + SM interpretation indicate that these relations are disfavoured.

The one remaining matrix element left to be determined is V_{cb} which is related to the mass matrix elements as in eq(51)

$$V_{cb} = \frac{m_{23}^d}{M_b} - \frac{m_{23}^u}{M_t} \tag{60}$$

We will return to a discussion of how such a structure might emerge from a field theory description of the fundamental interactions in a later section. First however we turn to a more systematic discussion of the possible texture zeros in the mass matrix.

6.2.1 Tests of the texture zero structure

Barbieri et al (BHR) [71] emphasised that these relations lead to a very tight determination of the CKM unitarity triangle. In terms of the re-scaled Wolfenstein parameters, $\bar{\rho} = c\rho$, $\bar{\eta} = c\eta$, $c = (1 - \lambda^2/2)$, the ratios $|V_{ub}|/|V_{cb}|$ and $|V_{td}|/|V_{ts}|$ are

$$\frac{|V_{ub}|}{|V_{cb}|} = \frac{\lambda}{c}\sqrt{\bar{\rho}^2 + \bar{\eta}^2} \qquad \frac{|V_{td}|}{|V_{ts}|} = \frac{\lambda}{c}\sqrt{(1-\bar{\rho})^2 + \bar{\eta}^2} \tag{61}$$

In the context of the Standard Model, the measurable quantities which give information on $\bar{\rho}$ and $\bar{\eta}$ are

(a) the ratio $|V_{ub}|/|V_{cb}|$ obtained from semi-leptonic decays of B mesons,

(b) Δm_{B_d} and Δm_{B_s} which are the mass differences in the $B_d^0 - \bar{B}^0{}_d$, $B_s^0 - \bar{B}^0{}_s$ systems,

(c) $|\epsilon_K|$ the parameter related to CP violation in the K, $\bar{K}$ system, and

(d) $\sin 2\beta$ (β is one of the angles in the unitarity triangle) obtained from CP asymmetries in various B decays.

In Figure 15 I show the results of a recent fit to all the data including the recent information coming from the $b-$factories BarBar and Belle. In the constraints on $\bar{\rho}$ and $\bar{\eta}$ are shown corresponding to the various measurements together with the prediction following from eq.(57). We see that, largely due to the improvement in the measurement of Δm_{B_s}, the new precision data is now inconsistent with an absolute zero in the $(1,3)$ position.

6.3 The quark mass matrices

Given that the exact texture zero prediction does not fit, what is the form of the quark mass matrices? A fit to the data allowing for a non-zero entry in the $(1,3)$ position gives [70]

$$\frac{M^u}{m_t} = \begin{pmatrix} 0 & b'\epsilon^3 & c'\epsilon^3 \\ ? & \epsilon^2 & a'\epsilon^2 \\ ? & ? & 1 \end{pmatrix} \tag{62}$$

and

$$\frac{M^d}{m_b} = \begin{pmatrix} 0 & b\bar{\epsilon}^3 & c\bar{\epsilon}^3 \\ ? & \bar{\epsilon}^2 & a\bar{\epsilon}^2 \\ ? & ? & 1 \end{pmatrix} \tag{63}$$

where the parameters below the diagonal are only weakly constrained because, as discussed above, measurement of the quark masses and the CKM matrix does not provide enough information to determine the full quark mass matrices. The parameters of the up quark mass matrix are given by $\epsilon = 0.05$, $b' \simeq 1$ while a' and c' are very weakly constrained. The parameters of the down quark mass matrix are much better determined with

$$\begin{aligned} \bar{\epsilon} &= 0.15 \pm 0.01 \quad b = 1.5 \pm 0.1 \quad a = 1.31 \pm 0.14 \\ |c| &= 0.4 \pm 0.02 \quad \psi = 24^0 \pm 3^0 \quad \text{or} & (64) \\ |c| &= 1.27 \pm 0.05 \quad \psi = -58^0 \pm 5^0 & (65) \end{aligned}$$

where $c = |c|\, e^{i\psi}$. The fact that the $(2,2)$ and $(2,3)$ matrix elements are very similar in magnitude is required by the smallness of V_{cb}. The recent data now also requires that the $(1,2)$ and $(1,3)$ are also quite similar [72, 71][4]. Note however that with the smaller solution for c the second texture zero of eq.(57) is still a good approximation, reminding us that texture zeros really only require texture small but not zero.

7 Family symmetry

Of course the underlying motivation for studying the detailed structure of the quark and lepton mass matrices is that they may lead to an insight about the structure beyond the Standard Model. In my opinion the hierarchical structure for the fermion mass matrices strongly suggests it originates from a spontaneously broken family symmetry. In this approach, when the family symmetry is exact, only the third generation will be massive corresponding to only the $(3,3)$ entry of the mass matrix being non-zero. When the symmetry is spontaneously broken, the zero elements are filled in at a level determined by the symmetry. Suppose a field θ which transforms non-trivially under the family symmetry acquires a vacuum expectation value, thus

[4] If one allows for an asymmetric form of the quark mass matrices with large entries below the diagonal it is still possible to have a $(1,3)$ zero[70]. Here we concentrate on symmetric structures which naturally accommodate the successful relation of eq.(58)

spontaneously breaking the family symmetry. The zero elements in the mass matrix will now become non-zero at some order in $<\theta>$. If only the 2-3 and 3-2 elements are allowed by the symmetry at order θ/M, where M is a mass scale to be determined, then a second fermion mass will be generated at $O((\theta/M)^2)$. In this way one may build up an hierarchy of masses.

$$\mathcal{M} \sim \begin{pmatrix} 0 & 0 & 0 \\ 0 & 0 & 0 \\ 0 & 0 & 1 \end{pmatrix} \rightarrow \begin{pmatrix} 0 & 0 & 0 \\ 0 & 0 & <\theta>/M \\ 0 & <\theta>/M & 1 \end{pmatrix} \tag{66}$$

7.1 Symmetry breaking in the mass matrix

An important question is how do these elements at $O(\theta/M)$ arise? A wide variety of models have been constructed in which the mechanism for communicating the breaking to the mass matrix, and generating the hierarchy takes on different forms:

7.1.1 Froggatt Nielsen mixing

A widely studied approach communicates symmetry breaking via an extension of the "see-saw" mechanism mixing light to heavy states - in this context it is known as the Froggatt Nielsen mechanism [75]. To illustrate the mechanism, suppose there is a vector-like pair of quark states X and $\overline{X}$ with mass M and carrying the same Standard Model quantum numbers as the c_R quark, but transforming differently under the family symmetry, so that the Yukawa coupling $h\overline{c_L}XH$ is allowed. Here H is the Standard Model Higgs responsible for giving quarks a mass. When H acquires a vacuum expectation value (vev), there will be mixing between $\overline{c_L}$ and $\overline{X}$. If in addition there is a gauge singlet field θ transforming non-trivially under the family symmetry so that the coupling $h'\overline{X}c_R\theta$ is allowed, then the mixing with heavy states will generate the mass matrix.

$$\begin{pmatrix} \overline{c_L} & \overline{X} \end{pmatrix} \begin{pmatrix} 0 & h<H> \\ h'<\theta> & M \end{pmatrix} \begin{pmatrix} c_R \\ X \end{pmatrix}$$

Diagonalising this gives a see-saw mass formula

$$m_c \simeq \frac{hh'<H><\theta>}{M} \tag{67}$$

This mass arises through mixing of the light with heavy quarks. Here these heavy quarks play the role of the messenger sector communicating the family symmetry breaking to the light quarks

A similar mechanism can generate the mass through mixing of the light Higgs with heavy messenger Higgs states. Suppose H_X, $\overline{H}_X$ are Higgs doublets with mass M. If H_X has family quantum numbers allowing the coupling $H\overline{H}_X\theta$, there will be mixing between H and H_X. If the family symmetry also allows the coupling $\overline{c_L}c_RH_X$, the light-heavy Higgs mixing induces a mass for the charm quark of the form given in eq.(67).

7.2 Identification of the Family symmetry

The nature of the spontaneously broken family symmetry is hard to identify because the available data on quark masses and mixings is insufficient uniquely to pin it down. The kinetic terms

	Q_i	u_i^c	d_i^c	L_i	e_i^c	ν_i^c	H_2	H_1
$U(1)_{FD}$	α_i	α_i	α_i	a_i	a_i	a_i	$-2\alpha_1$	$-2\alpha_1$

Table 11: $U(1)_{FD}$ symmetries.

and gauge interactions of the Standard Model has a very large family symmetry group, namely $U(3)^5$, where the $U(3)$ factors act on the left- and right- handed multiplets of quarks and leptons respectively. The group is extended to $U(3)^6$ if three right-handed neutrinos are added. Any family group should be contained in $U(3)^6$ but this leaves very many possibilities. In what follows I shall illustatrate the possibilities by discussing two characteristic possibilities. In the first I consider the simplest case of an Abelian family symmetry. Such symmetries abound in compactified string theories so are quite natural extensions of the Standard Model. The second is a non-Abelian family symmetry, a subgroup of the $U(3)^6$. In both cases I shall insist that there be texture zeros giving not just qualitative (order of magnitude) predictions but also *quantitative* predictions. These models require a symmetric mass matrix structure. As a result I will not discuss many, many models that have been developed which adequately describe the data. For a more general discussion and extensive references, see [68, 69] I will also consider how a family symmetry may be combined with a Grand Unified symmetry to relate quark and lepton masses. The discussion below assumes the effective theory is supersymmetric as only in this case are the radiative corrections associated with the Grand Unified symmetry consistent. However it is not difficult to develop (non-Grand-Unified) non-supersymmetric examples.

7.3 Abelian case

It turns out to be remarkably easy[76] to construct a model leading to the mass matrices in eqs. (62,63) through the introduction of an Abelian gauge symmetry, $U(1)$ (such additional symmetries abound in string theories). The most general charge assignment of the Standard Model states is given in Table 11. This follows since the need to preserve $SU(2)_L$ invariance requires (left-handed) up and down quarks (leptons) to have the same charge. This together with the requirement of symmetric matrices then requires that all quarks (leptons) of the same i-th generation transform with the same charge $\alpha_i(a_i)$. If the light Higgs, H_2, H_1, responsible for the up and down quark masses respectively have $U(1)$ charge so that only the (3,3) renormalisable Yukawa coupling to H_2, H_1 is allowed, only the (3,3) element of the associated mass matrix will be non-zero as desired. The remaining entries are generated when the $U(1)$ symmetry is broken. A particularly interesting example may be constructed in a supersymmetric extension of the Standard Model [76]. We assume this breaking is spontaneous via Standard Model singlet fields, θ, $\bar{\theta}$, with $U(1)_{FD}$ charge -1, +1 respectively, which acquire vacuum expectation values (vevs), $<\theta>$, $<\bar{\theta}>$, along a "D-flat" direction. After this breaking all entries in the mass matrix become non-zero. For example, the (3,2) entry in the up quark mass matrix appears at $O(\epsilon^{|\alpha_2-\alpha_1|})$ because U(1) charge conservation allows only a coupling $c^c t H_2(\theta/M_2)^{\alpha_2-\alpha_1}$, $\alpha_2>\alpha_1$ or $c^c t H_2(\bar{\theta}/M_2)^{\alpha_1-\alpha_2}$, $\alpha_1>\alpha_2$ and we have defined $\epsilon=(<\theta>/M_2)$ where M_2 is the unification mass scale which governs the higher dimension operators. As discussed in reference [76] one may expect a different scale, M_1 for the down quark mass matrices (it corresponds to mixing in the

H_2, H_1 sector with M_2, M_1 the masses of heavy H_2, H_1 fields). Thus we arrive at mass matrices of the form

$$\frac{M_u}{m_t} \approx \begin{pmatrix} h_{11}\rho_{11}\epsilon_a^{|2+6a|} & h_{12}\rho_{12}\epsilon_b^{|3a|} & h_{13}\rho_{13}\epsilon_a^{|1+3a|} \\ h_{21}\rho_{21}\epsilon_b^{|3a|} & h_{22}\rho_{22}\epsilon^2 & h_{23}\rho_{23}\epsilon^1 \\ h_{31}\rho_{31}\epsilon_a^{|1+3a|} & h_{32}\rho_{32}\epsilon^1 & h_{33} \end{pmatrix} \tag{68}$$

$$\frac{M_d}{m_b} \approx \begin{pmatrix} k_{11}\sigma_{11}\bar{\epsilon}_a^{|2+6a|} & k_{12}\sigma_{12}\bar{\epsilon}_b^{|3a|} & k_{13}\sigma_{13}\bar{\epsilon}_a^{|1+3a|} \\ k_{21}\sigma_{21}\bar{\epsilon}_b^{|3a|} & k_{22}\sigma_{22}\bar{\epsilon}^2 & k_{23}\sigma_{23}\bar{\epsilon}^1 \\ k_{31}\sigma_{31}\bar{\epsilon}_a^{|1+3a|} & k_{32}\sigma_{32}\bar{\epsilon}^1 & k_{33} \end{pmatrix} \tag{69}$$

where $\bar{\epsilon} = (\frac{<\theta>}{M_1})^{|\alpha_2-\alpha_1|}$, $\epsilon = (\frac{<\theta>}{M_2})^{|\alpha_2-\alpha_1|}$, and $a = (2\alpha_1 - \alpha_2 - \alpha_3)/3(\alpha_2 - \alpha_1)$. For $-3a > 1$ $\epsilon_a = \epsilon_b = \epsilon$ and $\bar{\epsilon}_a = \bar{\epsilon}_b = \bar{\epsilon}$. In this case it is easy to check that there are *no* texture zeros because all matrix elements contribute at leading order to the masses and mixing angles. For $1 > -3a > 0$, ϵ_a, $\bar{\epsilon}_a$ change and are given by $\bar{\epsilon_a} = (\frac{<\bar{\theta}>}{M_1})^{|\alpha_2-\alpha_1|}$, $\epsilon_a = (\frac{<\bar{\theta}>}{M_2})^{|\alpha_2-\alpha_1|}$. In this case texture zeros in the (1,1) and (1,3) positions *automatically* appear for small $< \bar{\theta} >$. However the (1,2) matrix element is too large (cf Table 11). For $a > 0$ however $\bar{\epsilon}_{a,b} = (\frac{<\bar{\theta}>}{M_1})^{|\alpha_2-\alpha_1|}$, $\epsilon_{a,b} = (\frac{<\bar{\theta}>}{M_2})^{|\alpha_2-\alpha_1|}$, the texture zeros in the (1,1) and (1,3) positions persist, and the (1,2) matrix element can be of the correct magnitude.

Note that the family symmetry does not make the small elements exactly zero so it predicts only approximate texture zeros. Indeed, fixing the parameter $a = 1$ to obtain the measured magnitude of the (1,2) matrix element one finds that the (1,1) elements occurs at $O(\varepsilon^8)$ and $O(\bar{\epsilon}^8)$ for the up and down mass matrices respectively. This is so small that eq.(55) is valid to a high degree of accuracy if $\rho_{12} = \rho_{21}$. On the other hand the (1,3) matrix element is predicted to occur at $O(\epsilon^4)$, $O(\bar{\epsilon}^4)$ for the up and down matrices respectively and, as discussed above, a term of this order (with a coefficient 1/2) is sufficient to correct the prediction for $\left|\frac{V_{ub}}{V_{cb}}\right|$ following from the assumption of an exact texture zero, c.f.eqs.(62) and (63).

7.4 Non-Abelian family symmetry

Although the Abelian symmetry is able to describe many of the features of the quark mass matrix it does fail in some respects. The best fit prefers the (1,3) element to occur at $O(\epsilon^3)$, $O(\bar{\epsilon}^3)$, i.e. close to the (1,2) matrix elements. Moreover the measured value of V_{cb} requires the (2,2) and (2,3) matrix elements of the down quark mass matrices should be of the same magnitude, of $O(\bar{\epsilon}^2)$. This is in contradiction to the predictions of the Abelian family symmetry, unless one appeals to the unknown coefficients of $O(1)$. The most plausible way to get such relations for matrix elements involving different family members is to invoke a non-Abelian family symmetry [67].

The candidate family symmetry group, namely those symmetries which commute with the gauge interactions of the Standard Model, is quite large. For the quarks it is $U(3)^3$, corresponding to a separate $U(3)$ factor for the left-handed doublets and each of the right-handed $SU(2)$ singlet fields respectively. If we restrict the choice to one with the same transformation for the left-handed and the charge conjugate right-handed quarks, this symmetry is reduced to a diagonal $U(3)$ or one of its subgroups. It is straightforward to see how this symmetry can lead to the desired relation in the quark mass matrix. To be definite, let us consider the case of $SU(3)$. The family symmetry assignments for the left-handed quarks and leptons are:

$$\psi_i \in (Q_i, L_i) \sim \mathbf{3}, \quad \psi_i^c \in (U_i^c, D_i^c, E_i^c, N_i^c) \sim \mathbf{3} \tag{70}$$

where $i = 1,2,3$ are $SU(3)$ labels. To build a viable model we need spontaneous breaking

$$SU(3) \longrightarrow SU(2) \longrightarrow \text{Nothing} \tag{71}$$

Note that whenever we write $SU(3)$ it will always refer to the new family symmetry and $SU(2)$ will always refer to its subgroup. These should not be confused with the SM gauge group factors with which the family symmetry commutes.

The Higgs fields responsible for the above symmetry breaking are the $SU(3)$ antitriplets (but SM singlets) ϕ_3^i, ϕ_{23}^i, which develop vacuum expectation values (VEVs)

$$<\phi_3>= \begin{pmatrix} 0 \\ 0 \\ a_3 \end{pmatrix}, \quad <\phi_{23}>= \begin{pmatrix} 0 \\ b \\ b \end{pmatrix}. \tag{72}$$

Note that $<\phi_3>$ can always be rotated into the third position using $SU(3)$. The alignment of the $<\phi_{23}>$ VEVs is non-trivial but can be fixed by the scalar potential constrained by the $SU(3)$ family symmetry[78].

The leading contributions to the Yukawa matrices arise from operators which are quadratic in ϕ_3^i and ϕ_{23}^i

$$\left(\frac{1}{M^2}\psi_i\phi_{23}^i\psi_j^c\phi_{23}^j + \frac{1}{M^2}\psi_i\phi_3^i\psi_j^c\phi_3^j\right) H_\alpha \tag{73}$$

where H_α are the two Higgs doublets of the MSSM, which are $SU(3)$ singlets, but carry the usual electroweak quantum numbers. If $\psi^c = U^c, N^c$ then $\alpha = 2$, while if $\psi^c = D^c, E^c$ then $\alpha = 1, 2.I$ This gives the mass matrix

$$\frac{M}{m_3} = \begin{pmatrix} 0 & 0 & 0 \\ 0 & \epsilon^2 & \epsilon^2 \\ 0 & \epsilon^2 & 1+\epsilon^2 \end{pmatrix} \tag{74}$$

where we have taken $a_3 \simeq M >> b$ and $\epsilon = b/M$. As in the Abelian case the expansion parameter is expected to differ between the up and the down sectors because of the different messenger masses. Note that the equality of the $(2,2)$ and $(2,3)$ matrix elements is ensured by the family symmetry.

The remaining matrix elements are filled in by higher order operators. In order to generate Yukawa entries in the first row and column, we need to introduce the $SU(3)$ triplet (SM singlet) Higgs $\overline{\phi}_{3,i}$, $\overline{\phi}_{23,i}$, VEVs which, in fact, are necessary in a supersmmetric theory to cancel the large vacuum energy coming from D-terms :

$$<\overline{\phi}_3>= \begin{pmatrix} 0 \\ 0 \\ a_3 \end{pmatrix}, \quad <\overline{\phi}_{23}>= \begin{pmatrix} 0 \\ b \\ -b \end{pmatrix}. \tag{75}$$

The operators responsible for the first row and column of the Yukawa matrices are then

$$\frac{1}{M^5}(\epsilon^{ijk}\psi_i\overline{\phi_{23}}_j\psi_k^c)H_\alpha(\phi_3^l\overline{\phi}_{23,l})^2 \rightarrow \frac{\epsilon^2}{M}(\epsilon^{ijk}\psi_i\overline{\phi}_{23,j}\psi_k^c)H_\alpha \tag{76}$$

$$\frac{1}{M^{16}}(\epsilon^{ijk}\psi_i\overline{\phi_3}_j\overline{\phi_{23}}_k)(\epsilon^{lmn}\psi_l^c\overline{\phi_3}_m\overline{\phi_{23}}_n)H_\alpha(\phi_3^p\overline{\phi}_{23,p})^6 \rightarrow \frac{\epsilon^6}{M^4}(\epsilon^{ijk}\psi_i\overline{\phi}_{3,j}\overline{\phi}_{23,k})(\epsilon^{lmn}\psi_l^c\overline{\phi}_{3,m}\overline{\phi}_{23,n})H_\alpha \tag{77}$$

The reason these operators only appear at a high order is due to an additional Abelian symmetry which restricts the allowed terms[78].Together with the operators of eq.(73) these lead to the mass matrix

$$\frac{M}{m_3} = \begin{pmatrix} O(\epsilon^8) & \lambda\epsilon^3 & \lambda\epsilon^3 \\ -\lambda\epsilon^3 & \epsilon^2 & \epsilon^2 \\ -\lambda\epsilon^3 & \epsilon^2 & 1 \end{pmatrix} \tag{78}$$

This is of the form of the "experimental" mass matrices of eqs.(62) and (63), with small corrections in higher order in the ϵ expansion allowing a perfect fit to the coefficients. It demonstrates how a non-Abelian symmetry can order the hierarchy of quark masses and mixings and, in addition, explain the closeness of some of the matrix elements. Of course the next question to ask is does the same symmetry give a good description of lepton masses and in particular explain the large neutrino mixing angles?

8 Charged lepton masses.

The main issue we address here is whether the structure of the quark masses can be related to that of the leptons. We will discuss the implications that follow if we adopt the most symmetric possibility in which the family symmetry properties of the charged leptons is the same as that of the down quarks. This is required if there is an underlying GUT such as $SU(5)$ or $SO(10)$.

We first consider the charged leptons. Given the same family properties the form of the charged lepton mass matrix will be the same as that of the down quark matrices in eq(63) up to coefficients of $O(1)$. If there is an underlying GUT the coefficients too may be related. As mentioned in Section 4.1.4 the relation $m_b = m_\tau$ at the unification scale is in good agreement with the measured masses. From eq.(24) we see that such an equality applies in $SU(5)$ if the Higgs responsible for the third generation masses transforms as a $\overline{5}$ of $SU(5)$. In $SO(10)$ equality applies if the Higgs belongs to a 10 or 120 representation. In the former case one also expects equality between the top quark and the third generation Dirac neutrino mass, something we explore below.

What about the two lighter generations? In this case we must address the question whether the expansion parameters are related. From a phenomenological point of view, note that the relation $Det[M^d] = Det[M^\tau]$ at the unification scale is also in good agreement with the experimental measurements and this requires that the expansion parameters be the same. If there is an underlying lepton-quark GUT symmetry the expansion parameters should be the same because the down quarks and leptons are both $T_{R,3} = -1/2$ states and both acquire their mass from the Higgs doublet H_d in a supersymmetric theory. Equality of the expansion parameter is also required if there is an underlying GUT such as $SU(5)$ or $SO(10)$ because the down quarks and leptons fall in the same representation. The success of the relation $Det[M^d] = Det[M^\tau]$ is also consistent with equality of the coefficients determining the $(1,2)$ matrix elements, something that follows if the Higgs responsible for this element transforms as a $\overline{5}$ of $SU(5)$. However it is not possible to have identical forms for the charged lepton and down quark mass matrices for the relations $m_s = m_\mu$ and $m_d = m_e$ are is in gross disagreement with experiment after taking account of the radiative correction on going from high to low scale, giving approximately a factor of 3 increase in the quark masses. As pointed out by Georgi and Jarlskog[79] this discrepancy is readily explained if there is an underlying GUT through the appearance of Clebsch Gordon factors in the matrix element coefficients. In particular from eq.(25) we see that if the Higgs responsible for the $(2,2)$ matrix element should belong to a $\overline{45}$ of $SU(5)$ the lepton coupling is a factor -3 times the down quark coupling. In this case the relations for the light generations are modified to give $m_s = m_\mu/3$ and $m_d = 3m_e$. Including the radiative corrections needed to determine the masses at laboratory scales, these relations are in excellent agreement with the measured masses. If there is also an underlying non-Abelian family symmetry relating the matrix elements as in eq(7.4), the $(2,2)$ and $(2,3)$ matrix elements are related and so it is necessary that the Higgs field responsible for the $(2,3)$ matrix element be also in the $\overline{45}$ of $SU(5)$. As a result the form of the lepton mass matrix is given by

$$M^l \simeq \begin{pmatrix} \bar{\epsilon}^8 & \bar{\epsilon}^3 & \bar{\epsilon}^3 \\ -\bar{\epsilon}^3 & -3\bar{\epsilon}^2 & -3\bar{\epsilon}^2 \\ -\bar{\epsilon}^3 & -3\bar{\epsilon}^2 & 1 \end{pmatrix} \tag{79}$$

Note that in order to realise this scheme there must be a very particular origin to the effective Lagrangian given in eq(73). The effective operators may be generated through quark or Higgs mixing with states carrying different family symmetry quantum numbers. In order to generate the factor 3 it is necessary that the operators responsible for these terms be dominately given Higgs mixing of the $\overline{5}$ Higgs responsible for the $(3,3)$ element with the $\overline{45}$ generating the factor–3 enhanced terms.

The mass matrix of eq(79) gives excellent relation between the down quarks and charged lepton masses of all three generations. Due to the approximate texture zero in the $(1,1)$ position it also implies a contribution to the $(1,2)$ mixing angle given by

$$V^l_{\nu_e\mu} \simeq \sqrt{\frac{m_e}{m_\mu}} \tag{80}$$

As we shall discuss this is significant in solar neutrino oscillation if the SMA solution applies. On the other hand, due to the smallness of the $(1,3)$ matrix element, the contribution to the $(2,3)$ matrix element is small

$$V^d_{\nu_\mu\tau} \simeq \frac{m_\mu}{m_\tau} \tag{81}$$

This is the analogue to the relation in the down quark sector $V_{cb} \simeq m_s/m_b$, which is in good agreement with the small value observed. However it is too small to explain the large mixing angle observed in atmospheric neutrino oscillation. This discrepancy is at the heart of the apparent conflict between quark and lepton masses and we turn now to the discussion how tis conflict may be resolved.

9 Neutrino masses

9.1 Dirac mass

Following the discussion in Section 7 we immediately see that the family symmetry properties of the left- and right- handed neutrinos must be the same as those of the charged leptons. As a result the form of the neutrino Dirac mass matrix coupling the doublet neutrinos to the singlet (right-handed) neutrinos should be the same as eq(79) although the expansion parameter may be different. If there is an underlying lepton-quark ($SO(10)$) GUT symmetry the expansion parameters should be the same because the up quarks and neutrinos are both $T_{R,3} = +1/2$ states and both acquire their mass from the Higgs doublet H_u in a supersymmetric theory. Thus we take the Dirac neutrino mass matrix of the form

$$\frac{M_D}{m_3} = \begin{pmatrix} O(\epsilon^8) & \epsilon^3 & \epsilon^3 \\ -\epsilon^3 & -3\epsilon^2 & -3\epsilon^2 \\ -\epsilon^3 & -3\epsilon^2 & 1 \end{pmatrix} \tag{82}$$

9.2 Majorana mass

Of course it is necessary to determine the Majorana mass matrix before one can determine the effective neutrino mass matrix via the see-saw formula of eq(29). Although the family symmetry properties of the right-handed neutrinos are related to those of the charged leptons it is not possible to use this information unambiguously to determine the structure of the Majorana mass matrix. The reason is twofold. Firstly the Majorana masses are only generated via new $\Delta L = 2$ lepton number violating Higgs vevs and it is also necessary to specify the representation content of these fields. Secondly the Majorana mass matrix involves the coupling of identical fermions and so antisymmetric terms allowed in the Dirac mass matrix will not arise in the Majorana matrix. Given these uncertainties we will adopt the pragmatic approach and consider whether it is possible to explain the large mixing in the neutrino sector with a "reasonable" form for the Majorana matrix. By "reasonable", we mean that large mixing should not be due to a detailed correlation between the Dirac and Majorana mass matrix elements because this would require a correlation between the $\Delta L = 2$ and $\Delta L = 0$ symmetry breaking sectors for which there is no obvious symmetry. In practice we will assume an hierarchical structure for the Majorana mass matrix. Such an hierarchical structure is to be expected because the underlying family symmetry is the same as applies to the Dirac matrix which leads to a structure ordered by the expansion parameter $\varepsilon_M = \theta/M'$. If a single symmetry breaking field, θ, dominates there will no possibility of degenaracy between matrix elements. Here we consider the case in which $\varepsilon_M << \epsilon$, when the mixing in the neutrino sector is dominated by the mixing coming from the Dirac mass. In this case the matrix can be diagonalised by small rotations of $O(\varepsilon_M)$ giving

$$M = \begin{pmatrix} m_1 & & \\ & m_2 & \\ & & m_3 \end{pmatrix} \tag{83}$$

The important feature of this form is that the basis of mass eigenstates is very close to that used in writing the Dirac matrix. This follows from the fact that the (broken) family symmetry properties of the Matrix elements right-handed neutrinos are the same as those of the Dirac neutrino case and the mass eigenstates will thus be the family symmetry eigenstates up to corrections of $O(\varepsilon_M)$.

9.3 The light neutrino mass matrix

We are now able to determine the masses and mixing angles of the light neutrinos using eqs(82,83) in eq(29). The effective Lagrangian associated with the see-saw mass is of the form

$$\mathcal{L} = \frac{\epsilon^6}{m_1}(\nu_\mu+\nu_\tau+O(\epsilon)(\nu_\mu-\nu_\tau))^2+\frac{\epsilon^4}{m_2}(3\nu_\mu+3\nu_\tau-\epsilon\nu_e+O(\epsilon)(\nu_\mu-\nu_\tau))^2+\frac{1}{m_3}(\nu_\tau-3\epsilon^2\nu_\nu+\epsilon^3\nu_e)^2 \tag{84}$$

It is now straightforward to determine the viable forms of the mass matrix.

9.3.1 Maximal mixing

Consider first the heaviest neutrino which should have near maximal mixing to explain atmospheric neutrino oscillation. One sees from eq(84) that, surprisingly, this is easy to achieve. In particular, if $\frac{\epsilon^6}{m_1} > \frac{\epsilon^4}{m_2}, \frac{1}{m_3}$ the heaviest eigenstate will be $\nu_1 \propto \nu_\mu+\nu_\tau$.. Thus ν_1 is very close to a maximal mixed state of muon and tau neutrinos. It is also straightforward to compute the next lightest neutrino state. If $\frac{1}{m_3} > \frac{\epsilon^6}{m_2}$ the third generation Majorana exchange dominates and the eigenstate ν_2 is proportional to $\nu_\tau - \nu_\mu + O(\epsilon)\nu_e$. Thus the mixing angle $V_{\nu_e\mu}$ relevant to solar neutrino oscillation will be dominated by the contribution from the charged lepton sector given in eq(80). This is of the correct magnitude to fit the SMA solution.

9.3.2 Bi-maximal mixing

The question naturally arises whether the near bi-maximal mixing case with a large mixing angle in the solar oscillations can also be consistent with this extended symmetry. In the case $\frac{1}{m_3} < \frac{\epsilon^6}{m_2}$ the second generation Majorana exchange dominates the determination of the next lightest neutrino state. In leading order this term just gives a contribution to the heaviest eigenstate mass. Including the subleading terms one finds the next lightest state $\nu_2 \propto (x(\nu_\mu - \nu_\tau) + y\nu_e)$ where x and y are of $O(1)$. Thus even (near) bi-maximal mixing is consistent with the form of eq.(82) of the same form as the quark mass matrices! As we have seen the reason for this is the see-saw mechanism which can relate small quark mixing to large neutrino mixing. Thus, somewhat surprisingly, the measured fermion masses and mixings are consistent with an underlying (non-Abelian) family symmetry plus a GUT symmetry unifying quarks and leptons.

10 Summary and Conclusions

In these lectures I have tried to review the most promising attempts to go Beyond the Standard Model and answer the questions posed in Section 2.1. The overall situation is summarised in Table 12. We saw that the first three questions were answered if the Standard Model is an effective field theory following from some scale of new physics far above the electroweak breaking scale. However to maintain this interpretation it is necessary that the Standard Model be supplemented by some new physics to avoid the hierarchy problem and to keep light the

	Technicolour	**Large extra dimensions**	**SUSY**
M_W/M	?	✓	✓
Precision tests : decoupling	×	✓?	✓
$(g-2)_\mu$	×?	×?	✓
Light Higgs	×	×(✓$_{SUSY}$)	✓
$SU(2) \times U(1) \to U(1)_{EM}$	✓	?	✓
Direct searches	×?	×?	×?
Gauge Unification	?	×	✓
Standard Model parameters	?	?	✓

Table 12: Score sheet for models solving the hierarchy problem

Higgs boson needed for electrowek symmetry breaking. The candidate explanations for this are compositeness, supersymmetry and large new dimensions.

The remaining questions of Section 2.1 seem to require a further stage of unification. The gauge group and quark and lepton multiplet structure are suggestive of a GUT either below or above the string scale. The Higgs structure does not fit so obviously into this pattern but can be accommodated in a conventional GUT and appears quite natural in a compactified string theory. Gauge coupling unification works amazingly well for high scale unification but not for the case there are large new dimensions. In string theories the unification scale is related to the string or Planck scale offering a simple explanation for high scale unification.

The Standard Model parameters determining electroweak breaking fit very nicely in a SUSY unified scheme with radiative breaking following from an underlying soft mass unification. The remaining parameters determining the quark, charged lepton and neutrino masses fit well with an underlying GUT combined with a spontaneously broken (non-Abelai) family symmetry. Due to the see-saw mechanism the large neutrino mixing angle(s) are perfectly consistent with small quark mixing angles. Of course there are many alternative possibilities but this simple example illustrates that there is a good prospect for an underlying unified theory to be the origin of all the parameters responsible for the masses and mixing angles of the fermions.

While all this evidence is circumstantial it does make a very persuasive case favouring the SUSY solution to the hierarchy problem with a unification scale close to the usual Planck scale. If these ideas are correct much more evidence will be available at LHC energies when the SUSY threshold is passed.

References

[1] LEP EWWG, http://lepewwg.web.cern.ch/LEPEWWG/plots/summer2000/

[2] See, for example, S. Weinberg,"Conceptual foundations of the unified theory of weak and electromagnetic interactions", Les Prix Nobel 1979. See, for example, S. Weinberg, "Conceptual foundations of the unified theory of weak and electromagnetic interactions", Les Prix Nobel 1979.

[3] For reviews see:
J. Wess and J. Bagger, Supersymmetry and Supergravity (Princeton University Press,

1993);
P. Fayet, Unification of the Fundamental Particle Interactions, eds. S. Ferrara, J. Ellis and P. Van Nieuwenhuizen (Plenum Press, New York, 1980), p. 587;
G.G. Ross, Grand Unified Theories, Benjamin/Cummings (1984);
H.-P. Nilles, Phys. Rep. 110 (1984) 1 ;
H.E. Haber and G.L. Kane, *Phys. Rep.* **117**, 75 (1985);
A.B. Lahanas and D.V. Nanopoulos, *Phys. Rep.* **145**, 1 (1987).
S. Martin, "A Supersymmetry Primer", hep-ph/9709356
D.I. Kazakov, "Beyond the Standard Model, : In search of Supersymmetry", Lectures given at European School of High-Energy Physics, Caramulo, Portugal, 20 Aug - 2 Sep 2000, [hep-ph/0012288]

[4] J. Ellis, K. Enqvist, D.V. Nanopoulos and F. Zwirner, *Mod. Phys. Lett.* **A1**, 57 (1987); R. Barbieri and G.F. Giudice, *Nucl. Phys.* **B306**, 63 (1988); S. Dimopoulos and G.F. Giudice, *Phys. Lett.* **B357**, 573 (1995).

[5] G. Anderson and D.J. Castano, *Phys. Lett.* **B347**, 300 (1995) and *Phys. Rev.* **D52**, 1693 (1995).

[6] L. Giusti, A. Romanino and A. Strumia, 1998, Natural ranges of supersymmetric signals, *Nucl. Phys.* **B550**, 3 [hep-ph/9811386].

[7] N. Arkani-Hamed, S. Dimopoulos and G. Dvali, *Phys. Lett.* **B429**, 263 (1998); *Phys. Rev.* **D59**, 086004 (1999); I. Antoniadis, N. Arkani-Hamed, S. Dimopoulos and G. Dvali, *Phys. Lett.* **B426**, 257 (1998).

[8] L. Randall and R. Sundrum, *Phys. Rev. Lett.* **83**, 3370 (1999); hep-th/9906064.

[9] M.E. Peskin and T. Takeuchi, *Phys. Rev. Lett.* **65**, 964 (1990), *Phys. Rev.* **D46**, 381 (1992).

[10] G. Altarelli and R. Barbieri, *Phys. Lett.* **B253**, 161 (1991); G. Altarelli, R. Barbieri, and S. Jadach, *Nucl. Phys.* **B369**, 3 (1992).

[11] M. L. Swartz, hep-ex/9912026, to appear in the Proceedings of the XIX International Symposium on Lepton and Photon Interactions at High Energies.

[12] E. Eichten and K. Lane, *Phys. Lett.* **B90**, 125 (1980). E. Farhi and L. Susskind, *Phys. Rev.* **D20**, 3404 (1979).

[13] C.T. Hill, *Phys. Lett.* **B266**, 419 (1991).

[14] B.A. Dobrescu and C.T. Hill, *Phys. Rev. Lett.* **81**, 2634 (1998) [hep-ph/9712319]; B. A. Dobrescu, hep-ph/9903407.

[15] R.S. Chivukula, B.A. Dobrescu, H. Georgi and C. T. Hill, *Phys. Rev.* **D59**, 075003 (1999) [hep-ph/9809470].

[16] M.K. Gaillard and B.W. Lee, *Phys. Rev.* **D10**, (1974) 897; M.K. Gaillard, B.W. Lee and R. Schrock, *Phys. Rev.* **D13**, 2674 (1976).

[17] For a recent review see R. Sekhar Chivukula, "Lectures on Technicolour and Compositeness", lectures presented at TASI 2000, Boulder (2000) hep-ph/0011264.

[18] A. Masiero, L. Silvestrini, Two Lectures on FCNC and CP Violation in Supersymmetry, Lectures given by A. Masiero at the International School on Subnuclear Physics, 35th Course: "Highlights: 50 Years Later", Erice, Italy, 26 August-4 September 1997 and at the International School of Physics Enrico Fermi, Course CXXVVII: "Heavy flavour physics: a probe of Nature's grand design", Varenna, Italy, 8-18 July 1997

[19] J.L. Hewett, *Phys. Rev. Lett.* **82**, 4765 (1999), hep-ph/9811356. T. Han, J.D. Lykken and R. Zhang, *Phys. Rev.* **D59**, 105006 (1999), hep-ph/9811350.

[20] F. Gabbiani and A. Masiero, *Nucl. Phys.* **B322**, 235 (1989); J. Hagelin, S. Kelley and T. Tanaka, *Nucl. Phys.* **B415**, 293 (1994); *Mod. Phys. Lett.* **A8**, 2737 (1993); D. Choudhury, F. Eberlein, A. Konig, J. Louis and S. Pokorski, *Phys. Lett.* **B342**, 80 (1995); F. Gabbiani, E. Gabrielli, A. Masierio and L. Schwestrini, Rome preprint ROM2F/96/21; hep-ph/9604387.

[21] J.A. Casas, J.R. Espinosa, M. Quiros, *Phys. Lett.* **B342**, 171 (1995); *Nucl. Phys.* **B436**, 3-29 (1995). ERRATUM-ibid.B439:466-468,1995;

G. Altarelli, G. Isidori, *Phys. Lett.* **B337**, 141-144, (1994). (2987309)

[22] A.H. Chamseddine, R. Arnowitt and P. Nath, *Phys. Rev. Lett.* **49**, 970 (1982), *Nucl. Phys.* **B227**, 121 (1983). R. Barbieri, S. Ferrara and C.A. Savoy, *Phys. Lett.* **B119**, 343 (1982). L. Hall, J. Lykken and S. Weinberg, *Phys. Rev.* **D27**, 2359 (1983).

[23] M. Dine, A.E. Nelson and Y. Shirman, it Phys. Rev. **D51**, 1362 (1995), hep-ph/9408384; M. Dine, A.E. Nelson, Y. Nir and Y. Shirman, *Phys. Rev.* **D53**, 2658 (1996), hep-ph/9507378.

[24] G.F. Giudice, M.A. Luty, H. Murayama and R. Rattazzi, *JHEP* **9812**, 027 (1998), hep-ph/9810442. 9810442;

L. Randall and R. Sundrum, breaking," *Nucl. Phys.* **B557**, 79 (1999), hep-th/9810155.

[25] M.E. Peskin, theoretical summary lecture for EPS HEP99, hep-ph/0002041

[26] G.F. Giudice, R. Rattazzi and J.D. Wells, *Nucl. Phys.* **B544**, 3 (1999) [hep-ph/9811291].

[27] E.A. Mirabelli, M. Perelstein and M.E. Peskin, *Phys. Rev. Lett.* **82**, 2236 (1999) [hep-ph/9811337].

[28] H. Georgi and S.L. Glashow, *Phys. Rev. Lett.* **32**, 438 (1974): *Nucl. Phys.* **B193**, 150 (1981).

[29] H. Georgi, H.R. Quinn, S. Weinberg, *Phys. Rev. Lett.* **33**, 451 (1974).

[30] J. Ellis, S. Kelley, D.V. Nanopoulos, *Phys. Lett.* **B260**, 131 (1991);
U. Amaldi, W. deBoer, H. Furstenau, *Phys. Lett.* **B260**, 447 (1991);
P. Langacker, M. Luo, Univ. of Pennsylvania preprint, UPR-0466T (1991).

[31] L.E. Ibanez, G.G. Ross, *Phys. Lett.* **B105**, 439 (1982);
M. Einhorn, D.R.T. Jones, *Nucl. Phys.* **B196**, 475 (1982).
S. Dimopoulos, S. Raby, F. Wilczek, *Phys. Rev.* **D24**, 1681 (1981).

[32] D. Ghilencea, G.G. Ross, hep-ph/0102306.

[33] C. Caso *et al.*, physics," *Eur. Phys. J.* **C3**, 1 (1998).
H. Leutwyler, *Phys. Lett.* **B378**, 313 (1996) [hep-ph/9602366]. S. Narison, in *Nucl. Phys. Proc. Suppl.* **86**, 242 (2000) [hep-ph/9911454]. A. Ali and D. London, hep-ph/0002167.

[34] M. Gell-Mann, P. Ramond and R. Slansky, proceedings of the Supergravity Stony Brook Workshop, New York, 1979, ed. by P. Van Nieuwenhuizen and D. Freedman (North-Holland, Amsterdam).

[35] For a review and the original references, see H. Murayama, proceedings of the 28th International conference on High Energy physics, Warsaw (1996) World Scientific edited by Z. Ajduk and A.K. Wroblewski.

[36] N. Sakai and T. Yanagida, *Nucl. Phys.* **B197**, 533 (1982).

[37] V. Lucas and S. Raby, Ohio State Preprint OHSTPY-HEP-T-95-028.

[38] J. Ellis, J. Lopez and D.V. Nanopoulos, CERN Preprint CERN-TH/95/260 (1995) (hep-ph/9519246).

[39] L.E. Ibanez and G.G. Ross, *Nucl. Phys.* **B368**, 3 (1992).

[40] A. Buras, J. Ellis, M.K. Gaillard and D.V. Nanopoulos, *Nucl. Phys.***B135**, 66 (1978);
D.V. Nanopoulos and D.A. Ross, *Nucl. Phys.* **B157**, 273 (1970);
H. Arason, D.J. Castano, B.E. Keszthelyi, S. Mikaelian, E.J. Piard, P. Ramond and B.D. Wright, *Phys. Rev. Lett.* **67**, 2933 (1991);
L. Ibanez and C. Lopez, *Nucl. Phys.* **B233**, 511 (1984);
S. Dimopoulos, L. Hall and S. Raby, *Phys. Rev. Lett.* **68**, 1984 (1992); *Phys. Rev.* **D45**, 4192 (1992).

[41] L.E. Ibáñez and C. López, *Phys. Lett.***B126**, 54 (1983); *Nucl. Phys.* **B233**, 511 (1984).

[42] For a recent review and extensive references, see D.I. Kazakov, "Beyond the Standard Model, : In search of Supersymmetry", Lectures given at European School of High-Energy Physics, Caramulo, Portugal, 20 Aug - 2 Sep 2000, [hep-ph/0012288].

[43] For an extensive review and further references see K.R. Dienes, *Phys. Rept.* **287**, 447 (1997).

[44] L. Roszkowski and M. Shifman, *Phys. Rev.* **D53**, 404 (1996).

[45] S. Weinberg, *Phys. Lett.* **B91**, 51 (1980); L.J. Hall, *Nucl. Phys.* **B178**, 75 (1981); C.H. Llewellyn-Smith, G.G. Ross, J. Wheater, *Nucl. Phys.* **B177**, 263 (1981); R. Barberi and L. Hall, *Phys. Rev. Lett.* **68**, 752 (1992).

[46] V. Kaplunovsky, *Nucl. Phys.* **B307**, 145 (1998); 382 (1992) E436; L.J. Dixon, V.S. Kaplunovsky and J. Louis, *Nucl. Phys.* **B329**, 27 (1990); 355 (1991) 649; J.P. Derendinger, S. Ferrara, C. Kounnas and F. Zwirner, *Nucl. Phys.* **B372**, 145 (1992); *Phys. Lett.* **B271**, 307 (1991); G. Lopez Cardoso and B.A. Ovrut, *Nucl. Phys.* **B369**, 351 (1992); I. Antoniadis, K.S. Narain and T.R. Taylor, *Phys. Lett.* **B267**, 37 (1991); I. Antoniadis, E. Gava and K.S. Narain, *Phys. Lett.* **B283**, 209 (1992); 393 (1992) 93; I. Antoniadis, E. Gava, K.S. Narain and T.R. Taylor, preprint NUB-3057 (1992); L.E. Ibanez, D. Lust and G.G. Ross, *Phys. Lett.* **B272**, 251 (1991); L. Ibanez and D. Lust, *Nucl. Phys.* **B382**, 305 (1992); I. Antoniadis, J. Ellis, R. Lacaze and D.V. Nanopoulos, *Phys. Lett.* **B268**, 188 (1991); G. Cardoso, D. Lust and T. Mohaupt, HUB-IEP-95/50, hep-th/9412209.

[47] C.M. Hull and P.K. Townsend, *Nucl. Phys.* **B438**, 109 (1995) [hep-th/9410167].

[48] E. Witten, *Nucl. Phys.* **B443**, 85 (1995) [hep-th/9503124].

[49] P. Horava, E. Witten, *Nucl. Phys.* **B475**, 94 (1995); P. Horava, *Phys. Rev.* **D54**, 7561 (1996).

[50] E. Witten, *Nucl. Phys.* **B471**, 135 (1996).

[51] J. Polchinski, *Phys. Rev. Lett.* **75**, 4724 (1995) [hep-th/9510017].

[52] J. Polchinski, hep-th/9611050.

[53] V.P. Mitrofanov and O.I. Ponomareva, *Sov. Phys. JETP 67*, 1963 (1988).

[54] Schuyler Cullen, Maxim Perelstein, *Phys. Rev. Lett.* **83**, 268-271 (1999).

[55] Lawrence J. Hall, David Smith,"Cosmological Constraints on Theories with Large Extra Dimensions", hep-ph/990426.

[56] J.L. Hewett, *Phys. Rev. Lett.* **82**, 4765 (1999).

[57] P. Krieger, SUSY2K, June 2000, CERN [http://wwwth.cern.ch/susy2k/susy2kfinalprog.html].

[58] K.R. Dienes, E. Dudas and T. Gherghetta, *Nucl. Phys.* **B557**, 25 (1999) [hep-ph/9811428]; N. Arkani-Hamed, S. Dimopoulos, G. Dvali and J. March-Russell, hep-ph/9811448.

[59] J. Lykken, L. Randall, hep-th/9908076.

[60] I.I. Kogan, S. Mouslopoulos, A. Papazoglou, G.G. Ross and J. Santiago, hep-ph/9912552.

[61] Y. Grossman and M. Neubert, *Phys. Lett.* **B474**, 361 (2000) [hep-ph/9912408].

[62] A. Lukas, P. Ramond, A. Romanino and G.G. Ross, *Phys. Lett.* **B495**, 136-146 (2000) [hep-ph/0008049]; *JHEP* **0104**, 010 (2001) [hep-ph/0011295]

[63] Q. Ahmad *et al*, nucl-ex/0106015.

[64] K. Dienes, E. Dudas, T. Gherghetta, *Nuclear Physics* **B537**, 47 (1999).

[65] J.D. Bjorken, lectures presented at Oxford 1996, available on Oxford WWW home page.

[66] L.J. Hall and A. Rasin, *Phys. Lett.* **B315**, 164 (1993).

[67] G.G. Ross, lectures presented at TASI 2000, Boulder (2000).

[68] H. Fritzsch, *Nucl. Phys.* **B155**, 189 (1979).

[69] F. Wilczek and A. Zee, *Phys. Lett.* **B70**, 418 (1977). H. Fritzsch, *Phys. Lett.* **B70**, 436 (1977). For a recent review on quark masses and extensive references on textures see: H. Fritzsch and Z. Xing, *Prog. Part. Nucl.*

[70] R.G. Roberts, A. Romanino, G.G. Ross and L. Velasco-Sevilla, hep-ph/0104088.

[71] R. Barbieri, L.J. Hall and A. Romanino, *Nucl. Phys.* **B551**, 93 (1999).

[72] P. Ramond, R.G. Roberts and G.G. Ross, *Nucl. Phys.* **B406**, 19 (1993).

[73] L.J. Hall, hep-ph/9303217, LBL33677, talk given at 16th Texas symposium on relativistic astrophysics and 3rd symposium on particles, strings and cosmology, Berkeley 1991.

[74] H. Fritzsch and Z. Zing, *Phys. Lett.* **B413**, 396 (1997).

[75] C.D. Froggatt and H.B. Nielsen, *Nucl. Phys.* **B147**, 277 (1979).

[76] L. Ibanez and G.G. Ross, *Phys. Lett.* **B332**, 100 (1994) [hep-ph/9403338].

[77] P. Binetruy and P. Ramond, *Phys. Lett.* **B350**, 49 (1995);
V. Jain and R. Schrock, *Phys. Lett.* **B352**, (1995);
E. Dudas, S. Pokorski and C.A. Savoy, *Phys. Lett.* **B356**, 45 (1995);
Y. Nir, *Phys. Lett.* **B345**, 107 (1995).

[78] S. King and G.G. Ross, preprint SHEP 01-21; OUTP 0146P (2001) [hep-ph/0108112].

[79] H. Georgi and C. Jarlskog, *Phys. Lett.* **B86**, 297 (1979); H. Georgi and D.V. Nanopoulos, *Nucl. Phys.* **B159**, 16 (1979); J. Harvey, P. Ramond and D.B. Reiss, *Phys. Lett.* **B92**, 309 (1980); *Nucl. Phys.* **B199**, 223 (1982).

RUNII PERSPECTIVES AT CDF: TOP, HIGGS AND SEARCHES

PATRIZIA AZZI

INFN Padova, Via Marzolo 8, 35100 Padova, Italy
E-mail: azzi@pd.infn.it

The CDFII experiment is ready for the new data taking period at the Tevatron scheduled to start in March 2001. The main features of the upgraded detector, in particular the new tracking system are reviewed. The impact of the detector and accelerator performances on top physics measurements is described. The sensitivity to a SM Higgs boson as a function of integrated luminosity has been evaluated and a new approach for the search of more exotic phenomena is presented.

1 The Tevatron and CDFII upgrade

Until the start up of LHC, the Fermilab Tevatron Collider remains the high energy frontier of particle physics. The luminosity enhancement provided by the Main Injector dramatically increases the discovery reach and, in conjunction with the upgrade of the collider detectors, moves the experimental program into a regime of *precision* hadron collider physics. Precision capability at the energy frontier will allow simultaneous attack on the open questions of high energy physics, including: characterization of the properties of the top quark, and the direct search for the Higgs boson and new phenomena, that are the subject of this paper. The high luminosity Tevatron of RunII requires extensive changes to the experimental apparatus. Ten years of experience with CDF and Tevatron physics have been applied in the design of an upgraded CDF detector with powerful new features[1].

1.1 *The Tevatron upgrade*

For RunII The Tevatron accelerator will increase its center-of-mass collision energy from 1.8 to 2 TeV and achieve a design peak luminosity of $2\times10^{32}\,\mathrm{cm}^{-2}\mathrm{s}^{-1}$. The RunII is divided in two data taking periods: RunIIa (IIb) where $\mathcal{L}_{int} = 1 \approx 2\,\mathrm{fb}^{-1} (10 \approx 15\,\mathrm{fb}^{-1})$. Due to the higher instantaneous luminosity conditions the number of multiple interactions in a single beam crossing might become large, see Fig. 1.

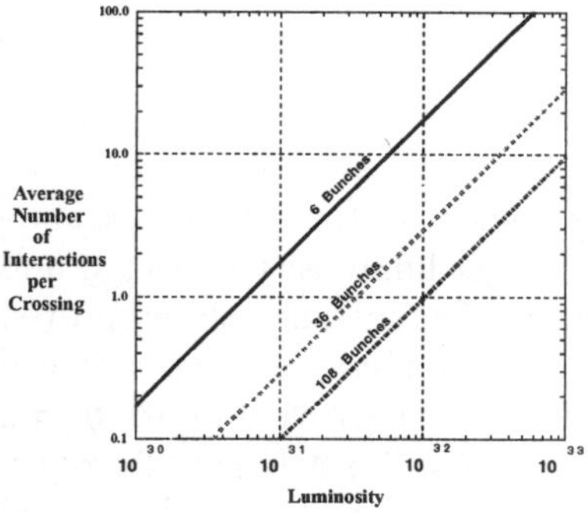

Figure 1: Average number of multiple interactions

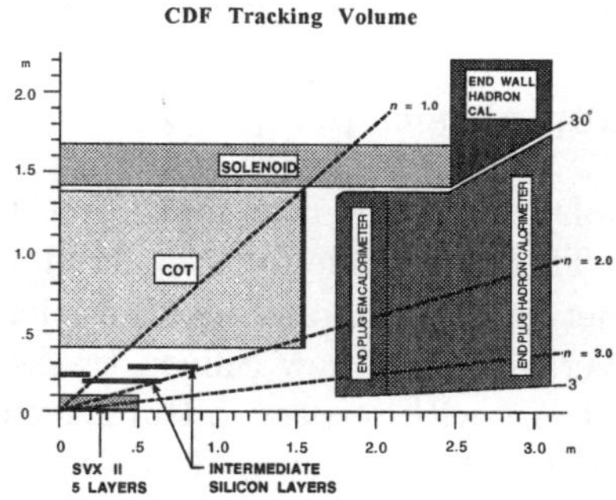

Figure 2: The CDFII detector upgrade

1.2 The CDFII Tracking System

In order to sustain the higher luminosity and beam crossing time both the central tracking chamber (CTC) and the silicon vertex detector (SVX) have been replaced, see Fig. 2. The new central tracking chamber (COT) has a maximum drift time of 100 ns obtained reducing the cell size and using a fast gas mixture. It is organized in eight superlayers, four axial and four stereo, that combined with the single hit resolution of 200 μm provide excellent pattern recognition capabilities.

The space inside the COT, has been equipped with a silicon tracker system composed by three different detectors. This new silicon system has an increased z (length $\simeq 1\,m$) and η coverage (up to $|\eta| < 2$). The presence of double sided layers allows 3D track reconstruction with an overall resolution on the impact parameter of $\sigma_\phi \leq 30\,\mu$m and $\sigma_z \leq 60\mu$m.

The innermost layer (L00), a single sided silicon layer made with rad-hard technology, is positioned on the beam pipe at a radius of 2.5 cm: it guarantees a good and uniform impact parameter resolution in the transverse plane. The region of $3 < r < 10$ cm is occupied by SVXII: five layers of double sided silicon detectors equipped with axial, small angle stereo (1.2^o) and 90° readout strips. Finally, the region $10 < r < 20$ cm is occupied by ISL: these two additional layer of double sided silicon with axial and small angle stereo (1.2^o) readout strengthen the linking of COT tracks with silicon information and allow for stand alone silicon tracking in the larger η region not covered by the COT.

Moreover, tracking capabilities at the trigger level have also been improved. CDFII is equipped with a fast track reconstruction processor at Level 1 (XFT) and displaced track triggering at Level 2 (SVT), obtained combining XFT tracks with SVXII $r - phi$ information.

2 Top physics perspectives

The plan for RunII is to make precision measurement of top physics parameters. This is possible by a combination of the larger luminosity expected and the better selection and identification efficiencies. For instance the improved z coverage of the new Silicon tracker implies a 50% increase in geometrical acceptance with respect to RunI. Extending track reconstruction up to $|\eta| = 2$ allows to recover about 30% of b's in $t\bar{t}$ events. Moreover, a stronger fake rejection is achievable exploiting 3D secondary vertex reconstruction. The overall event selection efficiency is also increased by the improved muon system coverage and the possibility of electron ID at large η. The net result is that we will be able to improve the purity of the already large $t\bar{t}$ samples, obtain sizable samples of single top events (likely less pure), and also collect large control samples that will allow a significant reduction of the systematic uncertainties.

2.1 Top production and decay

At the Tevatron, $t\bar{t}$ pairs are produced strongly through the process $q\bar{q}, gg \rightarrow t\bar{t}$ with $\sigma_{t\bar{t}} = 6.8\,\text{pb}$ for $m_t = 175\,\text{GeV/c}^2$ at $\sqrt{s} = 2.0\,\text{TeV}$. There is a 40% increase in the production cross section due to the increase in center-of-mass energy from 1.8 to 2 TeV accompanied by an increase of the contribution from the gg process that becomes 15% of the total.

However, top can also be singly produced through weak processes such as W-gluon fusion, $\sigma(Wg \rightarrow tb + X) = 1.44\,\text{pb}$, and virtual W exchange, $\sigma(W^* \rightarrow tb + X) = 0.74\,\text{pb}$. The W^* production process is particularly interesting for the measurement of V_{tb} since it has smaller uncertainties in the cross section. It might be possible to isolate this channel exploiting the fact that polarization information is preserved in the top decay products and can be used for event selection and reduction of combinatoric in mass reconstruction.

The dominant top decay mode is through the weak process $t \rightarrow Wb$ (vertex V_{tb} as in the CKM matrix) and the final states are usually named depending on the W('s) decay.

2.2 Measurements in the $t\bar{t}$ sample

The fundamental measurements that will be performed in $t\bar{t}$ events are the top quark mass and cross section, see Table 1. The mass is a fundamental parameter of the Standard Model and is needed to determine the ttH coupling: for $\delta m(W) = 20\,\text{MeV/c}^2$ and $\delta m(t) = 2\,\text{GeV/c}^2$, the Higgs mass is constrained

Table 1: Summary of expected precision of measurements from $t\bar{t}$ events in RunII

Run I value	RunIIa ($2\,\text{fb}^{-1}$)	RunIIb ($10\,\text{fb}^{-1}$)
$M(t) = 176 \pm 6.5$ GeV/c^2	$\delta M(t) = 3\,\text{GeV/c}^2$	$\delta M(t)\ 1\,\text{GeV/c}^2$
$\sigma(t\bar{t}) = 6.5^{1.7}_{1.4}\,\text{pb}$	$\delta\sigma(t\bar{t}) = 9\,\%$	$\delta\sigma(t\bar{t}) = 6\,\%$

to within 50% of its value. The experimental handles that will be used to improve the precision of the measurement are the requirement of double tags to reduce the overall background and the combinatorial, and the fact that data driven systematic scale with $1/\sqrt{N}$ (jet energy scale, gluon radiation).

The larger statistics and better selection efficiency will also allow to improve other measurement and perform new ones:

- The branching ratio $B(t \to (b)W)$ is obtained from the ratio of double tagged to single tagged events. The precision on the indirect V_{tb} measurement extracted from this BR will go down to 3% for $\mathcal{L}_{int} = 1\,\text{fb}^{-1}$.

- The branching ratio $B(t \to b(W))$ is obtained from the ratio of dilepton to single lepton events. This branching ratio can be used to set a limit on the charged Higgs mass through the process $t \to bH^+$: $10\,\text{fb}^{-1}$ of data will allow to exclude $m(H^+) < m(W)$ for any value of $tan\beta$.

- The measurement of the W helicity is possible since the top quark decays before hadronizing thus preserving the helicity information. The SM prediction is $B(t \to bW_0) = 69.2\%$ for $m(t) = 170\,\text{GeV/c}^2$. With $\mathcal{L}_{int} = 10\,\text{fb}^{-1}$ of RunIIb, $V - A$ and $V + A$ contribution can be probed at the % level.

- Rare decays: not only the FCNC decays limits will go down to $10^{-2} - 10^{-3}$ @95%C.L. but the CKM suppressed decays $t \to Ws, Wd$ might be visible.

2.3 Measurements in the single top sample

This is a whole new world: single top samples will be smaller and also less clean. The main analysis strategy is to rely on the identification of the high p_T isolated lepton ($\ell = e, \mu$) and the missing energy from the W decay accompanied by the b tagging information in order to reduce the background and solve ambiguities in the mass reconstruction. The main measurements will be the production cross section(s), the mass, and the direct V_{tb} coupling, see Table 2.

Table 2: Summary of expected precision of measurements from single top events in RunII

Value	RunIIa ($1\,\mathrm{fb}^{-1}$)	RunIIb ($10\,\mathrm{fb}^{-1}$)
$\delta\sigma(tbX)$	26%	10%
$\delta\Gamma(t \to Wb)$	28%	12%
$\delta\|V_{tb}\|$	14%	6%

Table 3: Higgs production cross section at the Tevatron

m(H)[GeV/c^2]	$\sigma_H[fb]$	$\sigma_W H[fb]$	$\sigma_W H[fb]$	$\sigma_H qq[fb]$
110	900	224	128	100
120	704	165	96.5	85.4
130	558	124	73.5	73.2

3 Standard Model Higgs search

The dominant Standard Model Higgs production processes at the Tevatron are: gluon fusion, Higgs*strahlung* and vector boson fusion. The cross sections at $\sqrt{s} = 2\,\mathrm{TeV}$ for the interesting Higgs mass range are reported in Table 3. Higgs decay modes are strong function of Higgs mass, the two dominant ones are $H \to b\bar{b}$ at low Higgs masses and $H \to WW$ for $m(H) > 140\,\mathrm{GeV/c^2}$.

The Run I searches looked for the process of associated production with a vector boson in the various channels, defined depending on the vector boson decay: lepton plus jets, missing energy plus jets and multijets. All the analyses relied heavily on b identification techniques in order to reduce the background from boson plus jets events. The overall combined limit was more than an order of magnitude bigger than the Standard Model prediction. The situation in RunII will still be difficult because the production cross section is limited and the event signature is less striking from $t\bar{t}$: the number of jets is smaller and the invariant mass is lower. However, new tools are being developed that might be crucial for Higgs search. First of all the improvement in b identification: larger overall geometrical acceptance, increased efficiency and better purity (due to 3D vertexing). Then, the improved overall lepton identification efficiency and geometric acceptance and the development of specific trigger paths for Higgs (displaced track trigger plus jets, or missing energy, multijet) will allow to optimize the collection of events for all the final states. Finally, RunI studies on $Z \to b\bar{b}$ events show that significant improvements in the $b\bar{b}$ invariant mass resolution are achievable if specific jet energy corrections are used. A summary of RunII sensitivity[2] as a function of Higgs mass, combining CDF and D0 data, is shown in Table 4 and Fig. 3.

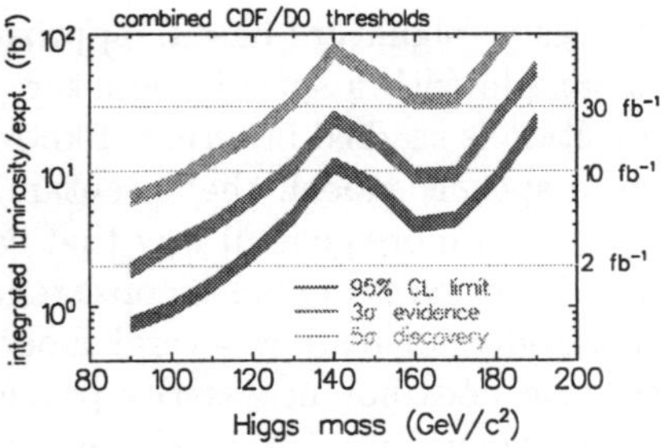

Figure 3: Combined CDF and D0 threshold for Standard Model Higgs search

Threshold	RunII (CDF+D0)	
	Mass [GeV/c^2]	$\mathcal{L}_{int}$ [fb^{-1}]
95% C.L. excl.	< 190	10
3σ evid.	< 180	20
5σ disc.	< 120	30

Table 4: Summary of Higgs sensitivities at Tevatron RunII

Studies have also shown that additional sensitivity to Standard Model Higgs can come from the $Ht\bar{t}$ production channel that while having a very small production cross section, also has an absolutely striking signature. The selection efficiency for these events is extremely high and the background is manageable. For instance, it has been shown[3] that a 2.8σ evidence is achievable for $m(H) = 120\,\text{GeV/c}^2$ and $\mathcal{L}_{int} = 15\,\text{fb}^{-1}$. 12 $t\bar{t}b\bar{b}$ and 2 $t\bar{t}c\bar{c}$ background events. A 2.8σ evidence is achievable in RunIIb[3].

4 New particle searches

The improvements described above to the accelerator and the detector make the RunII a very promising place for exciting new physics discoveries. This might also be a good time to re-think the approach to new particle searches. First of all the Standard Model has been extremely successful in the description of the high energy physics data. Secondly, a new detector will need time to be understood and validated with known Standard Model measurements before data can be used for new particle searches. Two co-existing approaches to searches need to be pursued both in a a balanced way: the traditional "model driven" analysis and a broader "signature based" approach. The "model driven" approach starts from the choice of a specific model and is based on the optimization of the analysis' selection using a Monte Carlo simulation of the desired signal. Background estimates are then made and limits are calculated. This procedure gives the best optimization for the analysis, however as soon as the model is outdated so is the analysis.

For exploring a wider range of possibilities a "signature based" approach is advisable: in its implementation a data sample with a specific signature is chosen (i.e. $\gamma\gamma$ + X) and the sample composition is studied in terms of known processes. Instead of publishing a limit on a specific model, the acceptances and cross section information are made public in a more general way that can be used by theorists or other experiments. If an excess or an inconsistency is found, this approach leaves open the possibility of testing several models quickly. Clearly this implies that the selection does not necessarily provide the best optimization. The idea is then to implement in a balanced way both approaches in order to optimize the time spent in analysis and the utilization of selected data samples and background estimate while guaranteeing timely but high quality results.

5 Conclusions

The top, Higgs and exotic physics program that awaits CDFII is extremely exciting. The amount of data that will be collected in RunII plus the improvements to the detector will allow not only more precise measurement of the top production cross section and mass, but also will give the opportunity to study single top production and possibly make the observation of the Higgs boson. Moreover, new tools and strategies are being considered in order to be ready for any hint of new physics that might be out there.

References

1. "The CDF II Detector Technical Design Report", FERMILAB-Pub-96/390-E .
2. M. Carena *et al.*, "Report of the Higgs Working Group of the Tevatron Run 2 SUSY/Higgs Workshop", hep-ph/0010338, Fermilab, Nov 19-21, 1998.
3. J. Goldstein *et al.*, "$p\bar{p} \rightarrow t\bar{t}H$: A Discovery mode for the Higgs boson at the Tevatron", *Phys. Rev. Lett.* **86**, 1694-1697 (2001).

PHYSICS at TESLA

G.A. BLAIR

Department of Physics, Royal Holloway, University of London, Egham, Surrey. TW20 0EX, UK

E-mail: g.blair@rhul.ac.uk

The physics at a 500–800 GeV electron positron linear collider, TESLA, is reviewed. The machine parameters that impact directly on the physics are discussed and a few key performance goals for a detector at TESLA are given. Emphasis is placed on precision measurements in the Higgs and top sectors and on extrapolation to high energy scales in the supersymmetric scenario.

1 Introduction

A high energy linear collider (LC) operating in the energy regime 0.5–$\sim$ 1 TeV will provide a programme of precision measurements and possibly new discoveries in an energy regime matching that of the LHC. Worldwide studies [1] have been exploring the physics potential of the LC, representing machine designs from Japan (JLC), America (NLC) and Europe (TESLA). A path to multi-TeV machines is also proposed by CERN (CLIC). The TESLA design is well advanced and a Technical Design Report (TDR) [2] was published in March 2001 that covers the machine [3], the detector and the physics.

2 Experimental Environment

The advantage of a linear collider (LC) over a hadronic one (LHC) centres on the facts that the centre of mass (cms) energy of the colliding particles is tunable (given by the energy of the colliding beams) and that the events themselves are much cleaner. The LC has the additional advantage that it can provide polarised beams, which can be used to explore the spin structure of fundamental couplings, to reduce systematic errors in precision measurements, to provide an effective luminosity gain in some channels such as the production of chiral supersymmetric particles (sparticles) and, importantly, to reduce dramatically the backgrounds to some processes from W-pair production.

Both the LHC and the LC will be needed to explore fully the TeV scale. Some states that are invisible at the LHC, such as an invisible Higgs, would be both discovered and have their properties measured at the LC. Other states that may be discovered at the LHC will, if their mass is not too high, be studied with high precision at the LC. The benefits apply in both directions;

for instance if supersymmetry (SUSY) is realised in nature then the LHC will almost certainly discover it and will tell the LC where to concentrate its threshold scans. However LHC analysis are limited by the overall scale, set by the lightest SUSY particle (LSP) mass. A precision measurement of the LSP mass at the LC would then feed back directly to the LHC analysis. This type of feedback would clearly be greatly beneficial to both programmes and is a strong argument for concurrent running of the LHC and the LC.

2.1 Machine Parameters

TESLA will provide tunable e^+e^- collisions at cms energies ranging from the Z^0 pole up to about 0.8 TeV. The machine offers very high luminosity of 3–5$\times 10^{34}$ cm^{-2}s^{-1}. This performance permits a general analysis to assume a total integrated luminosity of order 500 fb^{-1} and specific high-importance analysis (such as the Higgs self coupling, discussed below) to assume an integrated luminosity of order 1 ab^{-1}. Electron polarisation of $\sim$80% is planned and positron polarisation of up to $\sim$60% is also envisaged (there will be some trade-off between luminosity and positron polarisation). The polarisation can be measured to a precision of about 0.5%.

Although the nominal beam energies will be known to the 10^{-4} level, there are additional effects influencing the energy of individual events. One process, already familiar at LEP and SLC energies, is initial state radiation. An additional effect at the LC is beamstrahlung, where the high fields in the colliding bunches cause the emission of large numbers of photons, some of which will in turn produce e^+e^- pairs. Beamstrahlung results in a further widening of the effective electron/positron energy spectrum at the level of 3.3–4.7%.

The TESLA design proposes head-on collisions of electrons with positrons in one interaction region (IR). A second IR is also envisaged with a finite opening angle between the incoming beams. This second IR could be dedicated to $\gamma\gamma$ collisions and would need a specialised detector and final focus layout. In addition to the e^+e^- and $\gamma\gamma$ options, it would also be possible to run in e^-e^- or $e^-\gamma$ modes.

2.2 A Detector for TESLA

The detector at TESLA must be able to reconstruct high multiplicity multi-jet events as well as be able to measure precisely high-energy single particles. The detector must be hermetic and the TDR design allows energy measurements down to about 27 mrad with tagging of high energy electrons/positrons down to about 4 mrad. The low angle region is important for a range of physics processes; these include multi-jet events where at least one jet will be in the forward direction, SUSY processes - especially when the mass differences between the sparticles are small, and also for measuring (or vetoing) the large numbers of $\gamma\gamma$ events. Tracking at low angles is also important to be able to measure the centre of mass energy spectrum using acolinear Bhabha events. The detector design at low angles, and in the vertex detector region, is constrained by the presence of large numbers of e^+e^- pairs produced by the beamstrahlung photons mentioned in Sec. 2.1. A higher magnetic field is important to contain these particles at low-angle trajectories and a 4T detector field is presently being considered.

In a typical hadronic event, the individual particle energies are typically of order 1–2 GeV, so the quality of reconstruction hinges on the ability to measure very precisely the positions, directions and energies of lower-energy particles. The process whereby individual particles are reconstructed by combining information optimally from both tracking and calorimetry is called energy flow and the goal performance is set at $\delta E \sim 0.3\sqrt{E(\text{GeV})}$, a factor of two improvement on that achieved at LEP/SLC. This performance will require very high resolution and high granularity calorimetry, coupled with very high performance tracking.

In addition to measuring well the lower momentum tracks, it is also important to be able to make an excellent measurement of high momentum tracks, for instance to reconstruct the Higgs mass from recoil against the Z^0 boson (see Sec. 3.1). A global tracking performance of $\delta(1/p_T) \sim 5\times10^{-5}$ GeVc^{-1} is required. Excellent vertexing, for flavour identification, is essential to a wide range of physics processes, including measuring the branching ratios of the Higgs boson(s) and reconstructing the complicated cascade decay chains of SUSY. A high performance vertex detector is also essential to help with pattern recognition in high-density jets and to provide additional high-precision spatial measurements, which are essential to the global tracking performance.

3 Physics Processes

The physics programme at the LC is very rich; a comprehensive overview cannot be given here and only a few highlights from the Higgs and SUSY studies are presented in the following sections. A whole set of interesting measurements can also be made in the top sector, for instance the top quark mass can be measured to a precision where theoretical errors dominate (at $\delta m_t \sim 100$ MeV). Constraining the strong coupling α_s to the world average would give a top mass precision of order 40 MeV, an order of magnitude improvement on the LHC value [4]. High luminosity running at lower energies (the Z^0 pole and the W-pair threshold) can lead to typically an order of magnitude improvement

Table 1: Precisions on electroweak parameters from running TESLA at lower energies.

	LEP/SLC/Tevatron	TESLA
$\sin^2\theta^{\ell}_{\rm eff}$	0.23146 ± 0.00017	± 0.000013
lineshape observables:		
M_Z	91.1875 ± 0.0021 GeV	± 0.0021 GeV
$\alpha_s(M_{Z^2})$	0.1183 ± 0.0027	± 0.0009
$\Delta\rho_\ell$	$(0.55 \pm 0.10) \cdot 10^{-2}$	$\pm 0.05 \cdot 10^{-2}$
N_ν	2.984 ± 0.008	± 0.004
heavy flavours:		
$\mathcal{A}_b$	0.898 ± 0.015	± 0.001
R^0_b	0.21653 ± 0.00069	± 0.00014
M_W	80.436 ± 0.036 GeV	± 0.006 GeV

on the current measurements of electroweak parameters, as shown in Table 1 [2].

3.1 Higgs Physics

Understanding electroweak symmetry breaking lies at the heart of the LC physics programme. This means proving that the Higgs boson has spin-0 and has couplings to the standard model fermions that are proportional to the mass of the fermion, that the couplings to the gauge bosons are those expected from the standard model and that the Higgs potential has the right form, as given by the value of its self couplings.

The Higgs boson can be extensively studied at the LC. Its mass can be measured to high precision, for example a 120 GeV standard model Higgs mass can be measured to 40 MeV by combining all channels and to 70 MeV from using the Z^0-recoil alone, a method that will work even if the Higgs decays invisibly. In addition, the Higgs tri-linear self coupling can be measured to a statistical precision of about 22% if its mass is 120 GeV [6].

The Higgs spin, parity and charge conjugation quantum numbers can also be measured model-independently as can its branching ratios, for which the prospects are shown in Figure 1 [5].

Additional information about the Higgs can also be obtained from increased precision measurements at the LC of the electroweak parameters, such as those listed in Table 1. For example, from such precision measurements the Higgs mass can indirectly be constrained at the level of 5% [7]. Comparing this result to that obtained from direct measurements will provide a very powerful test of the Higgs sector.

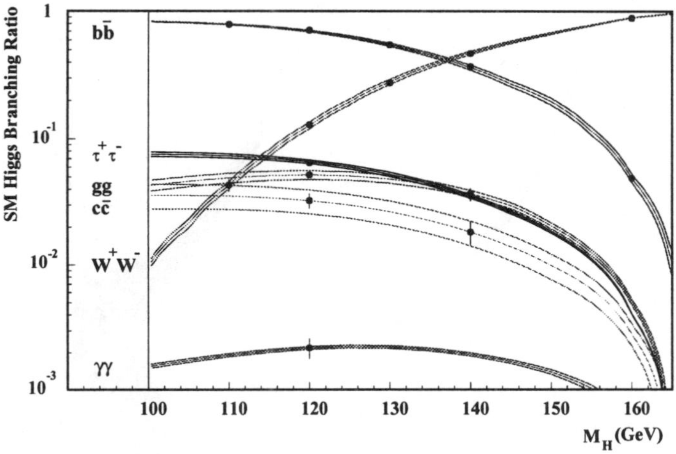

Figure 1: The precisions that can be obtained on the Higgs branching ratios for a range of Higgs masses. The points are the prospective experimental errors, the lines are the standard model uncertainties.

3.2 Supersymmetry

If SUSY is realised in nature, discovering it will require more than simply observing events with missing energy/momentum that are consistent with a given scenario. It will require measuring the couplings of the sparticles, testing that they are the same as those of the particles and also that the sparticles have the correct spin.

The current data favour a light Higgs boson and SUSY provides a mechanism that can accommodate such a light Higgs in a natural manner. The prospects for SUSY thus look good at present and, if SUSY is indeed realised in nature, the LC is ideal for making precision measurements of sparticle masses and cross sections. This is achieved by measuring the end-points of the energy spectra of the daughter particles and by performing threshold scans. Using such methods and allowing 100 fb^{-1} per threshold leads typically to per-mille precision on the sparticle masses [8].

After measuring the masses at the TeV scale, it is then possible to extrapolate the theory to higher energies by running the renormalisation group equations. In this way, the underlying structure of nature is reconstructed in a model-independent way [9]. Applying this procedure to a minimal supergravity model leads to Figure 2 where the very fine lines are obtained from the LC measurements of the colourless sparticles and the broader ones assume LHC

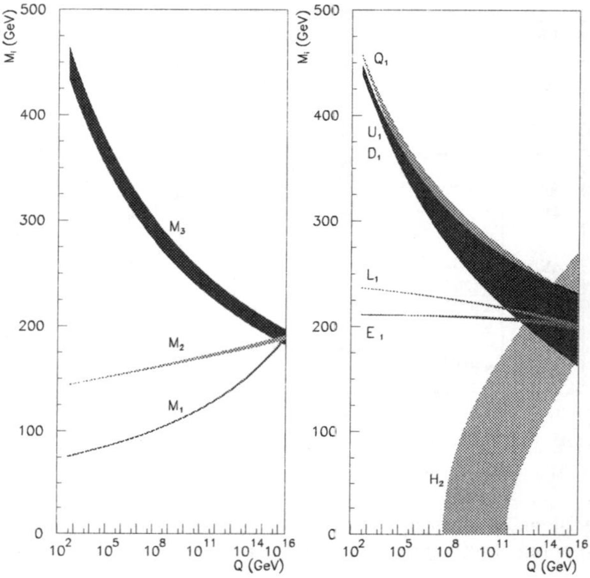

Figure 2: Model-independent extrapolations to high energies of SUSY soft breaking terms. The left hand plot shows the extrapolations of the gaugino mass parameters and the right hand plot the scalar mass parameters. The underlying model is minimal supergravity with $\tan\beta = 30$, M_0 =200 GeV,$M_{\frac{1}{2}}$ =190 GeV, A_0 =550 GeV, sign(μ)=-1.

precisions for the coloured states (which tend to be heavier). In this case, the observation that the curves meet at a common point would indicate that gravity plays a direct role in the breaking of SUSY.

4 Summary

The physics and detector studies reported here are only a small subset of a large amount of work by many people. More details can be found in the TDR [2] and in a complete set of LC-Notes [10], including work from all the international regions. It is clear that the LC presents a very exciting physics potential and much benefit to both the LHC and the LC programmes could be obtained from a significant time overlap in the running of these machines. The nature of electroweak symmetry breaking, the precision measurement of (s)particle properties and detailed extrapolations to high energy scales may yield model-independent answers to very fundamental questions.

Acknowledgments

Financial support from DESY and from the British Council ARC programme is gratefully acknowledged.

References

1. The central web page of the Worldwide Study of the Physics and Detectors for Future Linear e^+e^- Colliders: http://lcwws.physics.yale.edu/lc/
2. The TESLA Technical Design Report, March 2001. http://tesla.desy.de/new_pages/TDR_CD/start.html
3. W. Decking. "The TESLA Linear Collider Design", these proceedings.
4. D. Peralta, M. Martinez, R. Miquel. Proc. of LCWS99, Sitges (1999).
5. M. Battaglia. Proceedings of LCWS99 Sitges (1999).
6. A. Djouadi, W. Kilian, M. Mühlleitner, P.M. Zerwas *EPJC* **10**, 27 (1999); C. Castanier, P. Gay, P. Lutz, J. Orloff. LC Note LC-PHSM-2000-061.
7. S. Heinemeyer, Th. Mannel, G. Weiglein. Proc. of LCWS99, Sitges (1999). hep-ph/9909538. J. Erler *et al., Phys. Lett.* **B486**, 125 2000.
8. U. Martyn, G.A. Blair. Proc. of LCWS99, Sitges (1999). hep-ph/0009186.
9. G.A. Blair, W. Porod, P.M. Zerwas. *Phys. Rev.* **D63**, 17703 (2001).
10. The international LC notes are located at: http://www.desy.de/ lcnotes/LCnotes_welcome.html .

RADIATION CONCERNS IN HIGH-ENERGY PHYSICS AND THE SWITCHED CAPACITOR ARRAY CONTROLLER IN ATLAS

N.J. BUCHANAN[1] and D.M. GINGRICH[1,2]

[1] *Centre for Subatomic Research, University of Alberta, Edmonton, AB. T6G 2N5*
E-mail: buchanan@phys.ualberta.ca

[2] *TRIUMF Laboratory, 4004 Westbrook Mall, Vancouver, B.C. V6T 2A3*

The increased energy and luminosity of the Large Hadron Collider, which is now under construction, have placed new demands on detector designers. The radiation levels within the detectors have become sufficiently high that a large portion of the research and development resources must be used to ensure the detectors will function in these radiation environments. The switched capacitor array controller in the ATLAS detector will be used as an example of the challenges presented to experimental physicists who must design detectors which will operate in radiation fields. Some of the work involved in ensuring this device is radiation tolerant will be presented.

1 Introduction

Concerns relating to radiation levels have traditionally been restricted to those working in the nuclear energy, defense, radiology and oncology, or aerospace communities. The nuclear industry, energy and defense, have both had to contend with high total ionizing dose environments, while the nuclear energy industry has had the added problem of neutron radiation in their environments. The aerospace industry has found radiation concerns with both high total ionizing dose and radiation from charged particles, such as those found in the Van Allen belts.

When particle colliders reached the 10 MeV energy regime, induced radioactivity in surrounding materials occurs[1]. In these early accelerators, the injection and extraction points, along with beam stops, were most susceptible to activation and the primary concern was to ensure the safety of the people working near these regions.

Three developments led to the increased concern of the radiation environments at accelerators. The first development was the increase in both luminosity and centre-of-mass energy in modern colliders, which can be seen in Figure 1[2]. The second development which led to increased radiation concerns was the use of semiconductor detectors. These detectors provided advantages over traditional tracking detectors but are located near to the interaction re-

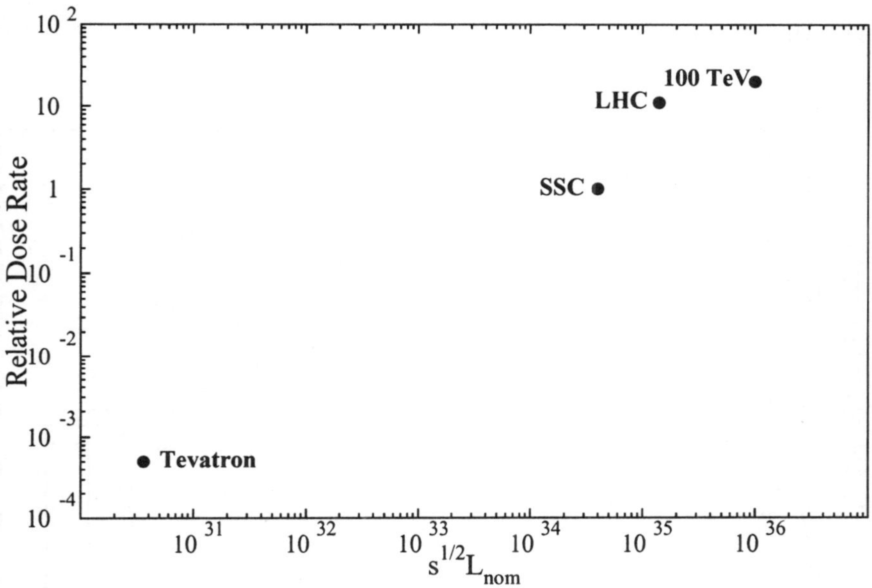

Figure 1: The relative dose rate as a function of the product of centre-of-mass energy and luminosity for the Tevatron (prior to upgrades), proposed Superconducting Super-Collider, Large Hadron Collider, and a 100 TeV machine running at LHC luminosity.

gion, which made their operation problematic[3,4]. Care had to be taken to ensure that signals could be differentiated from noise and that the detector would function over the projected lifetime of the detector. Lastly, the concern over noise due to the transmission of signals has motivated the placement of electronic components near, or even within, the detectors.

2 Types of Radiation Damage at Colliders

In high-energy accelerators there will be both hadronic and electromagnetic showers radiating out from the interaction region. The electromagnetic showers will contain photons, positrons, electrons, and muons. Hadronic showers will contain charged and neutral hadrons, and can initiate electromagnetic showers as well. Interactions of secondary particles with detector materials will clearly result in a mixed radiation field. Figure 2 shows the radiation energy spectra for various particles approximately 4 m from the interaction region inside the

ATLAS detector[a].

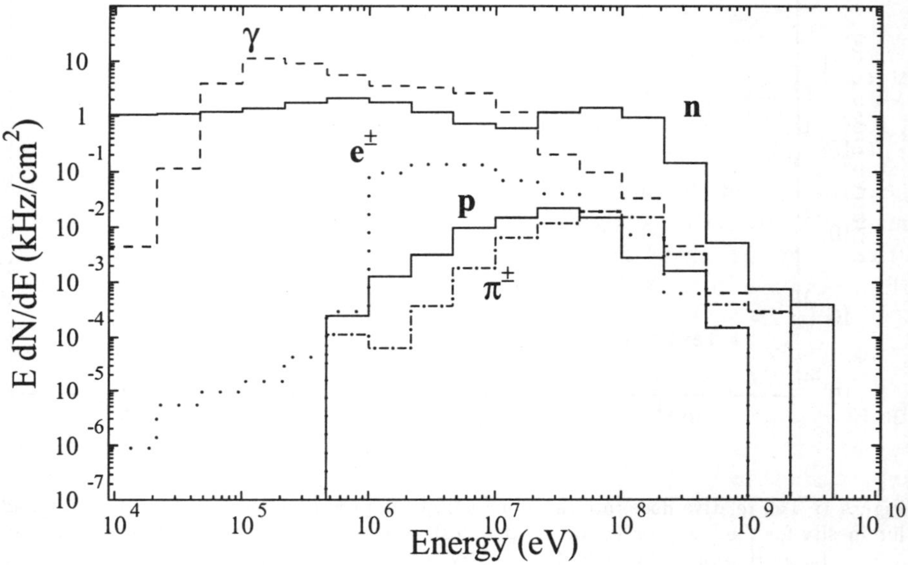

Figure 2: The particle energy spectra for the region of the ATLAS detector containing the readout electronics crates for the liquid argon calorimeter (approximately 4 m from the interaction region).

The primary effect of electrons and photons is the deposition of total ionizing dose within materials. Ionic displacements, where the lattice ions are moved from their original position, can also be caused by electrons and photons. Such displacements affect the materials to a much lesser degree. Above 10 MeV, photons can also initiate nuclear (photo-nuclear) reactions. Hadrons cause nuclear reactions, such as spallation, often resulting in the fragmentation of the target nucleus. Charged hadrons can contribute to total ionizing dose and neutral hadrons can interact with the nucleus at extremely low energies.

The type of damage caused by radiation varies with its type. Electromagnetic radiation is particularly damaging to electronics containing field effect transistors. The charge deposited can get trapped within the devices and change their characteristics[5,6]. Hadrons and fragments coming from nuclear interactions can cause temporary and permanent damage to electronics, as

[a] The particle energy spectra were calculated by Mike Shupe of the University of Arizona.

well as, damage to non-electronic materials used in the detector[7,8,9,10]. The radiation induced activation of nuclei within the detector materials can cause background events within the detector.

3 Radiation Concerns at ATLAS

The ATLAS detector will operate in the Large Hadron Collider (LHC), at CERN. The LHC is a proton-proton collider with a centre-of-mass energy of 14 TeV and a luminosity of $10^{34}\text{cm}^{-2}\text{s}^{-1}$. There will be a high radiation field throughout the detector and in the surrounding hall[11]. ATLAS will be made up of three sub-detector systems, an inner tracker, made up in part by semiconducting materials, a calorimeter system, and a muon detector. Of the three systems, the inner detector will be subject to the highest radiation field. The radiation field will be due to the high flux of charged and neutral hadrons from the interaction region and neutron backscatter from the calorimeter. The inner detector was designed to withstand the damage to semiconductor devices. In addition to damage, designers needed to address the issue of background in the detector, for example, a thermal neutron captured by a boron dopant in the silicon tracker could give a signal as much as 30 times greater than a minimum ionizing particle from a physics event[12].

The main concern for the calorimeters is the radiation effects on the readout electronics. The electronics must be located in the detector to reduce the coherent noise between readout channels. The balance between cost and reliable operation has driven intense research into radiation tolerant electronics for use in this detector system[13]. Other aspects of the calorimetry such as the radiation induced pollution of the liquid argon[b] have been explored[10].

The muon detectors are the outermost detector system located in ATLAS and the primary concern is that the radiation field from shower tails which penetrate the calorimeter will contribute to the background events.

4 The Switched Capacitor Array Controller in the ATLAS Liquid Argon Calorimeter

As an example of the procedure involved in determining the radiation tolerance of an electronic component for a high-energy physics experiment, the controller for the switched capacitor array will be examined. The controller is to be used in the liquid argon calorimeter in ATLAS. The controller will control data-flow in the front-end readout buffers for the liquid argon calorimeter prior to digitization. There are strict requirements on the speed and power consumption

[b]Some of the calorimeters in ATLAS will utilize liquid argon as an ionizing medium.

of the device, which limit the choice of candidate devices. The switched capacitor array controller will reside in the space between the tile calorimeters (approximately 4 m from the interaction region). Over the projected 10 years of operation of the experiment, the controller must survive 85 krad of total ionizing dose, 8×10^{12} neutrons/cm^2, and 3.8×10^{12} hadrons/cm^2.

To begin the device selection procedure we have tested a prototype controller implemented in a commercial Field Programmable Gate Array. We examined the device's performance under irradiation with photons and protons to simulate the required total ionizing dose and proton fluence at ATLAS respectively. The absorbed ionization causes an increase in power supply current until the device fails[14], while the protons cause temporary changes in state of transistors within the device leading to the corruption of stored data or circuit malfunction[7].

For the total ionizing dose test, a cobalt source was used to irradiate the device, which was operated in the same manner as it would be in ATLAS. Figure 3 shows that after about 20 krad of total ionizing dose was absorbed by the device the power supply current began to increase. After the current began to rise, reaching about 6 times its preirradiated value, errors occurred in the operation of the circuit.

Protons will cause parts of a microelectronic device to fail for a brief time. Unlike the total ionizing radiation, proton induced errors can be cleared and occur on a proton by proton basis and are thus called single event effects (for example, single event upsets). To determine the effect of protons on the prototype controller, it was placed in a proton beam of the Proton Irradiation Facility at TRIUMF in Vancouver. The cross-section for upsets (or probability of upset) as a function of proton energy was determined (Figure 4). The measured cross-section in Figure 4 was multiplied by the proton energy spectrum in Figure 2 and integrated over all energies. From this, the upset rate was predicted. The corresponding upset rate is given in Table 1.

In order to estimate the upset rates for neutrons and pions, the nuclear scattering cross-sections for protons, neutrons, and pions on silicon were used. As the single-event upset cross-section for a particle is proportional to the nuclear scattering cross-section for that particle in silicon, the SEU cross-sections for different particles should scale in the the same manner as their nuclear cross-sections. The estimated upset rates due to neutrons and pions are listed in Table 1.

The rates per device and for all 3072 devices to be used in ATLAS are included. The upset rates for this prototype device would be such that the time between errors would be on the order of minutes in ATLAS, making it inadequate. Work on developing a radiation tolerant device for the controller

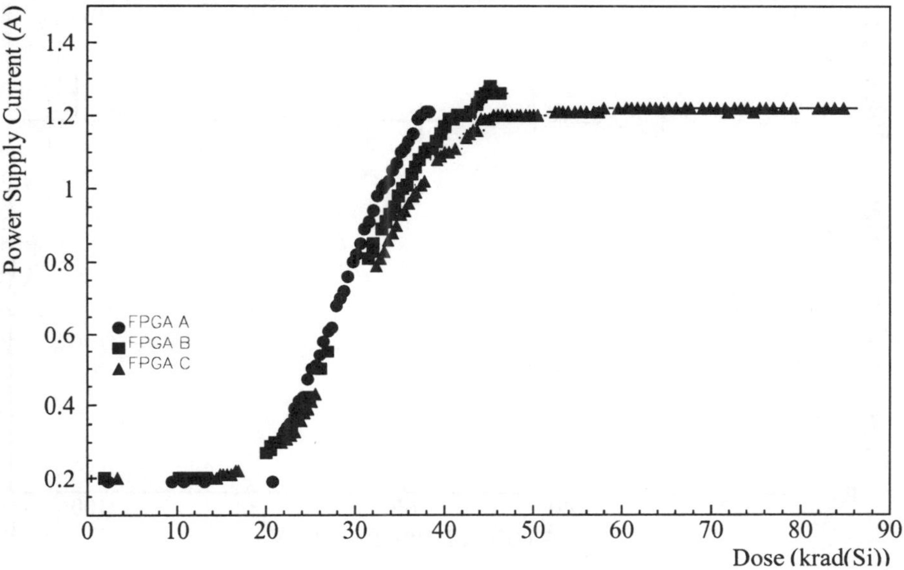

Figure 3: The power supply current of the switched capacitor array controller prototype (a Xilinx XC4036XLA FPGA) versus total dose, while under gamma irradiation.

Particle	Rate for device (upset/s)	Rate for ATLAS (upset/s)
Proton	3.3×10^{-7}	1.0×10^{-3}
Neutron	2.0×10^{-5}	6.1×10^{-2}
Pion	3.8×10^{-7}	1.2×10^{-3}

Table 1: Particle induced upset rates for the switched capacitor array controller prototype (a Xilinx XC4036XLA FPGA) if exposed to the predicted ATLAS radiation field.

is ongoing. We are developing and testing a radiation hardened application specific integrated circuit. A version using very fine microelectronic feature sizes is also being investigated.

5 Conclusion

With the increase in energy and luminosity of colliders has come the problem of high-radiation fields in and around them. Over the time period of a

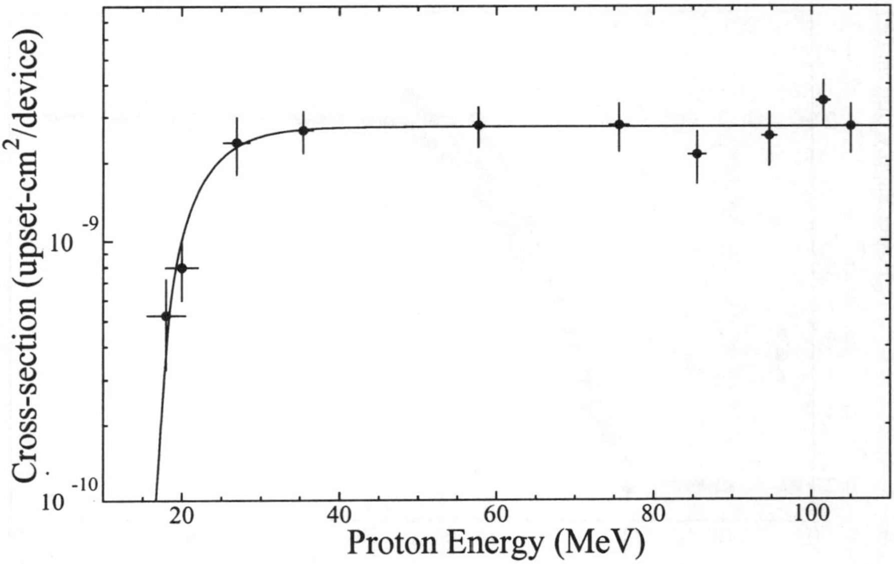

Figure 4: The proton induced single-event upset cross-section of the switched capacitor array controller prototype (a Xilinx XC4036XLA FPGA) versus proton energy under proton irradiation.

decade, the radiation dose rates have increased by several orders of magnitude. The ATLAS detector at the Large Hadron Collider demonstrates the level of constraint the radiation environment can impose on many aspects of detector design. While the switched capacitor array controller to be used in ATLAS provides a clear example of the problems and challenges associated with radiation tolerant detector components, it is only the beginning of the challenges facing future high-energy physics electronics design.

References

1. Gollon P.J., *IEEE Trans. Nuc. Sci.* **NS-23**, 1395 (1976).
2. Groom D., *Nucl Instr. Methods* **A279**, 1 (1989).
3. Khludkov S.S. *et al*, *J. Phys. D: Appl. Phys.* **29**, 1559 (1996).
4. Matthews J.A. *et al*, *Nucl Instr. Methods* **A381**, 338 (1996).
5. Ma T.P. and Dressendorfer P.V., Ionization Radiation Effects in MOS Devices and Circuits, John Wiley and Sons (1989).

6. Oldham T.R., Ionizing Radiation Effects in MOS Oxides, World Scientific (1999).
7. Buchanan N.J. and Gingrich D.M., *ATLAS Internal Note* **ATL-LARG-2001-009** (2001).
8. Andrieux M-L. *et al*, *Nucl Instr. Methods* **A456 (3)** 342 (2001).
9. Leroy C. *et al*, *ATLAS Internal Note* **ATL-LARG-1999-018** (1999).
10. Martin Ph. *et al*, *ATLAS Internal Note* **ATL-LARG-NO-052** (1996).
11. Battistoni G. *et al*, *ATLAS Internal Note* **ATL-GEN-NO-010** (1994).
12. Moszczynski A. *et al*, *ATLAS Internal Note* **ATL-INDET-99-009** (1996).
13. Dentan M., *ATLAS Internal Note* **ATC-TE-QA-0001** (2001).
14. MacQueen D.M. *et al*, *To be published in Nucl Instr. Methods* **A** (2001).

MEASUREMENT OF DIRECT CP VIOLATION BY NA48 EXPERIMENT AT CERN

G. COLLAZUOL
INFN e Università di Firenze
Largo E.Fermi 2, 50125 Firenze, Italy
E-mail: collazuol@fi.infn.it

on behalf of the NA48 Collaboration:
Cagliari, Cambridge, CERN, Dubna, Edinburgh, Ferrara, Firenze, Mainz, Orsay, Perugia, Pisa, Saclay, Siegen, Torino, Warszawa, Wien

The NA48 experiment at CERN has performed a measurement of the CP violating parameter $\mathrm{Re}(\varepsilon'/\varepsilon)$. In 1999 NA48 published [1] a result based on the data taken in 1997 $\mathrm{Re}(\varepsilon'/\varepsilon) = (18.5 \pm 7.3) \times 10^{-4}$. This paper describes the preliminary result from the analysis of data taken in 1998: $\mathrm{Re}(\varepsilon'/\varepsilon) = (14.0 \pm 4.3) \times 10^{-4}$.

1 Introduction

A long-standing question in High Energy Physics has been the origin of the phenomenon of CP violation. This was discovered [2] in the neutral kaon system in 1964, by the observation of K_L decays into two pions. The dominant effect is due to the asymmetric mixing of K^0 and $\bar{K^0}$ eigenstates. The deviation of the physical states K_S and K_L from the pure CP eigenstates, is parameterized by ε. CP violation can also occur directly in the two pion decay of the kaons, due to different amplitudes for K^0 and $\bar{K^0}$ to decay into final states with isospin 0 and 2. The difference is parameterized by ε'.

In the Standard Model, CP violation arises from the single phase occurring in the CKM quark mixing matrix. Direct CP violation is predicted in the range from 2 to 20×10^{-4}.

The NA48 experiment [5] aims at the determination of $\mathrm{Re}(\varepsilon'/\varepsilon)$ through the measurement of the double ratio R of the widths of neutral kaon decays into two pions:

$$R = \frac{\Gamma(K_L \to \pi^0\pi^0)/\Gamma(K_S \to \pi^0\pi^0)}{\Gamma(K_L \to \pi^+\pi^-)/\Gamma(K_S \to \pi^+\pi^-)} \approx 1 - 6 \cdot \mathrm{Re}(\varepsilon'/\varepsilon). \qquad (1)$$

The aimed accuracy on $\mathrm{Re}(\varepsilon'/\varepsilon)$ is $\approx 2 \cdot 10^{-4}$.

Previous measurements were reported by NA31 [3] giving a result greater than zero: $\mathrm{Re}(\varepsilon'/\varepsilon) = (23.0 \pm 6.5) \times 10^{-4}$, and E731 [4] giving $\mathrm{Re}(\varepsilon'/\varepsilon) = (7.4 \pm 5.9) \times 10^{-4}$.

2 The NA48 experiment

The NA48 experimental setup consists of:

i) Two beams, of K_L and K_S (see Figure 1), which are simultaneous and almost collinear. The beams are derived from 450 GeV/c protons with a flux of 1.5×10^{12} particles per pulse (ppp) accelerated by the CERN SPS. The K_L beam is produced by a target 126 m upstream from the decay region. A fraction of the non–interacting primary protons (3×10^7 ppp) is transported to the K_S production target. The K_S collimator selects secondary neutral particles at 4.2 mrad production angle, which is chosen in order to make the K_S and K_L decay spectra similar in the fiducial energy range $[70 \div 170]$ GeV. The kaon beams are vertically separated by ≈ 7 cm at the end K_S collimators and converge at the electromagnetic calorimeter with an angle 0.6 mrad.

ii) A detector, which measures the crossing time of the protons hitting the K_S target, allowing to recognize the K_S decay fragments.

iii) A magnetic spectrometer (resolution: $\frac{\sigma_p}{p} = 0.48\% \oplus 0.009\, p_{[\mathrm{GeV/c}]}\%$), which measures the kaon decays into charged pions. Two scintillator hodoscope planes measure the event time of the charged events, which is used by the tagging procedure.

iv) A quasi-homogeneous and fully active liquid Krypton ionization chamber, which measures the kaon decays into neutral pions. The calorimeter (resolution: $\frac{\sigma(E)}{E} = \frac{3.2\%}{\sqrt{E_{[\mathrm{GeV}]}}} \oplus \frac{10\%}{E_{[\mathrm{GeV}]}} \oplus 0.5\%$) is segmented in projective towers and provides excellent space uniformity. Short pulse shaping time provides high rate capability, and accurate timing information.

v) A hadron calorimeter followed by a muon filter.

The measurement proceeds by counting the number of events for each of the four decay modes, and carrying the double ratio. The simultaneous collection of the four decay modes in the same fiducial volume, and by the same detector, leads to the cancellation of acceptance and efficiency effects, as they are very similar for K_S and K_L decays. The systematic second order effects, residual from imperfect cancellation, are accurately measured, and used to correct the result.

The experiment is capable of standing very high rate (≈ 1 MHz) of instantaneous particle flux, in order to take data with an intense K_L beam. Very fast trigger electronics and read-out pipelined systems are devised to minimize

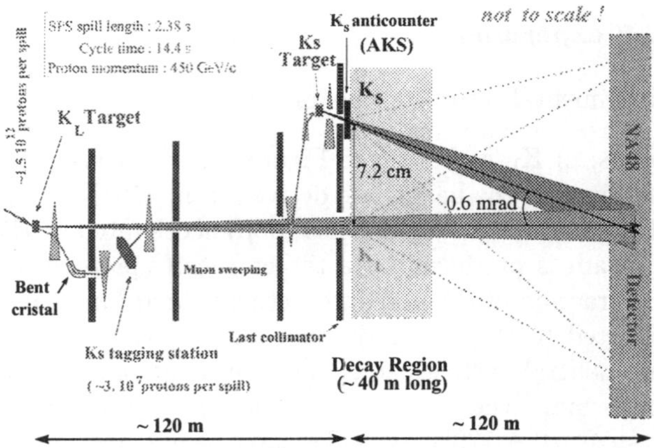

Figure 1: The NA48 experimental setup: the two simultaneous and almost collinear kaon beamlines converge in the same center at the central detector longitudinal position.

the systematic effects related to dead-time. The sensitivity to variations of the detection efficiency with time and rate are also minimized. The trigger for $\pi^0\pi^0$ decays is fully pipelined and based on the data from the calorimeter. It computes the number of clusters in horizontal and vertical projections, the energy, the center of energy and the proper lifetime of the kaon. The input rate is reduced from $\approx 1\,\text{MHz}$ to $\approx 2\,\text{kHz}$ by preserving $\sim$99.9% efficiency. The trigger for $\pi^+\pi^-$ decays is implemented in two levels. At the first level information from fast detectors (hodoscope, calorimeters) is used to decrease the rate from $\approx 1\,\text{MHz}$ to $\approx 100\,\text{kHz}$. At the second level the proper lifetime and invariant mass of the kaon are computed by using spectrometer information. The output rate is less than $2\,\text{kHz}$, with an inefficiency at the level of $\approx$2.5%. The dead time induced by the drift chamber read-out (about 25%), and by the charged trigger ($< 5\%$) is applied to neutral events to symmetrize any induced effect.

The experiment is designed to provide a background rejection factor of about 10^{-5}, which is achieved by very high resolutions in energy and position measured by the calorimeter and the magnetic spectrometer, and by the very efficient veto counters.

3 Data Analysis

The analysis closely follows the one performed on the 1997 data. More details can therefore be found in the Ref.[1].

3.1 *Event Selection and Background Subtraction*

To select $\pi^+\pi^-$ decays and reject the background from the abundant K_L semileptonic decays, events are rejected where a track is found to be consistent with being either an electron, or a muon. For good $\pi^+\pi^-$ events, the invariant mass $m_{\pi\pi}$ should be compatible with the kaon mass within $\pm 3\sigma$ (typical mass resolution: $2.5\,\mathrm{MeV/c^2}$). The two pion tracks must be coplanar with the kaon momentum: the kaon transverse momentum p_t is required to be $p_t^2 < 2 \times 10^{-4}\mathrm{GeV^2/c^2}$. The variable $A = |\frac{p_+ - p_-}{p_+ + p_-}|$ (p_+ and p_- are the pion track momenta) is used to reject Λ decays, and events with tracks near the beam hole. The residual semileptonic background in the K_L candidates is estimated, and the related correction to the double ratio R is $(19 \pm 3) \times 10^4$.

To select $\pi^0\pi^0$ decays, four photons are detected in the calorimeter. The longitudinal decay vertex position relative to the calorimeter is reconstructed from the measured energies and positions of the four photons, assuming that their invariant mass is the kaon mass. The invariant masses m_1 and m_2 of the photon pairs are computed and compared to the nominal π^0 mass (m_{π^0}). A χ^2 variable is used to find the best photon pairing, and to rejects the background from K_L into three pions, with undetected photons. The residual background is estimated. The related correction to R is $(6.6 \pm 2.0) \times 10^4$.

A small fraction of decays in the K_L beam comes from K_S decays originating from scattering on the collimators and regeneration. This kind of events are removed from the $\pi^+\pi^-$ sample by the coplanarity selection, but are kept in the neutral sample. A correction to R at the level of $(-9.9 \pm 3.0) \times 10^{-4}$ accounts for this.

3.2 K_S *and* K_L *Identification*

The event is identified as K_S if the tagger detects a proton within 2 ns off the event time, otherwise the event is identified as K_L .

Proton detection inefficiencies, or errors in measuring time, cause coincidences to be lost, and true K_S events to be classified as K_L . The probability for this to happen is measured at the level of $(1.95 \pm 0.05) \times 10^{-4}$, by using $\pi^+\pi^-$ events: they are identified by the vertical position of the vertex reconstructed by the spectrometer. The difference of tagging inefficiency in neutral and charged decays is measured at the level of 0.5×10^{-4}. The corresponding correction to the double ratio R is: $(0 \pm 3) \times 10^{-4}$.

Accidental coincidences with protons crossing the tagger may cause a K_L decay to be wrongly identified as K_S . This probability is measured for $\pi^+\pi^-$ events to be $(11.05 \pm 0.01)\%$. The difference of accidental tagging probability in charged and neutral events is at the level of $(0.5 \pm 4) \times 10^{-4}$. The corresponding correction to R is: $(1 \pm 8) \times 10^{-4}$.

3.3 Fiducial Volume and Energy Scale

The decay region is defined as a function of the kaon proper time: it covers the interval from 0 to 3.5 in K_S lifetime units. For K_S events the beginning of the decay region is accurately defined by a counter (AKS) which is used to veto decays occurring upstream. In the neutral mode, the reconstruction of the decay vertex position relies on the measurement of photon energies and positions by the calorimeter. To set the overall energy scale, decays of K_S into $\pi^0\pi^0$ are used. The energy scale is adjusted so that the sharp rise in the reconstructed decay vertex position created by the use of the AKS veto is at the known geometrical position of the AKS.

The accuracy of the energy scale setting is estimated at the level of 3×10^{-4}. The calorimeter non linearity is mainly studied using K_{e3}decays. Non linearities are limited at the level of 0.2%. The total systematic uncertainty on R due to the reconstruction of kaon energy and vertex position in neutral events is found to be 10×10^{-4}.

The systematic uncertainty on the longitudinal vertex position in the charged $\pi^+\pi^-$ decay mode results in an error on R at the level of 2×10^{-4}.

3.4 Acceptance Correction

The longitudinal distributions of the decay vertices are very different for K_S and K_L decays. In the calculation of the double ratio, K_L events are weighted with a function of the kaon proper time, to make similar the K_L and the K_S vertex distributions. Thus the K_S and K_L acceptances are made very similar in both charged and neutral modes, and there is no need for a very accurate simulation to estimate the acceptance correction. The Monte Carlo sample is five times larger than the data sample. The acceptance correction is evaluated and applied in each energy bin. The overall correction to the double ratio is: $(31 \pm 6(\text{stat}) \pm 6(\text{syst})) \times 10^{-4}$.

3.5 Accidental Activity

The effects of accidental activity in the detector are minimized by simultaneously collecting data in the four decay channels, and by applying off-line, to both $\pi^+\pi^-$ and $\pi^0\pi^0$ events, the dead-time conditions which are recorded event by event.

The residual effects are estimated by various methods[a], including overlaying K_S and K_L events with randomly triggered events. The noise induced by the drift chambers read-out doubles in the overlay procedure and has to be taken into account. The K_S /K_L ratio is constant with a R.M.S. of $\approx$ 10%,

[a] A detailed study of the effects of accidental activity can be found in Ref.[6]

Table 1: Corrections and systematic uncertainties to the double ratio R in the preliminary 1998 result. All corrections to R are almost smaller than 10^{-3}, in agreement with the goal of the experiment. In order to obtain the effect on $\mathrm{Re}(\varepsilon'/\varepsilon)$ the numbers must be multiplied by a factor of 1/6.

Source	1998(prel.) ΔR (Units = 10^{-4})	
	corr.	uncert.
Charged trigger	-1	± 11
Accidental tagging	+1	± 8
Tagging efficiency	–	± 3
Neutral rec. syst.	–	± 10
Charged vertex	+2	± 2
Acceptance	+31	± 9
Neutral BKG	-7	± 2
Charged BKG	+19	± 3
Beam scattering	-10	± 3
Accid. activity	+2	± 12
Total	+37	± 24

and the difference in accidental activity seen by $\mathrm{K_S}$ and $\mathrm{K_L}$ is about 1%: this allows to set an upper limit on the accidental effects on R at the level of 10×10^{-4}.

3.6 Result

The number of selected events in each channel used in the double ratio measurement from 1998 data are: 7.46×10^6 of $K_S \rightarrow \pi^+\pi^-$, 4.87×10^6 of $K_L \rightarrow \pi^+\pi^-$, 1.80×10^6 of $K_S \rightarrow \pi^0\pi^0$, 1.14×10^6 of $K_L \rightarrow \pi^0\pi^0$. The effects of the corrections applied to the double ratio are listed in Table 1.

The preliminary result (see Figure 2) for the direct CP violating parameter based on 1998 data is $\mathrm{Re}(\varepsilon'/\varepsilon) = (12.2 \pm 2.9 \pm 4.0) \times 10^{-4}$, where the first error is statistical and the second is systematic. Part of the systematic error is actually limited by the statistics of the used control samples.

4 Conclusions

Combining this preliminary measurement with the one published by NA48 from the 1997 data, the result is found: $\mathrm{Re}(\varepsilon'/\varepsilon) = (14.0 \pm 4.3) \times 10^{-4}$. The

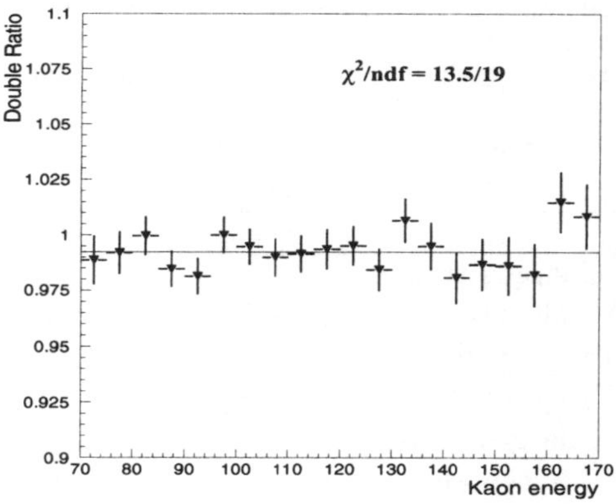

Figure 2: The double ratio R measured in the energy bins, after all corrections are applied.

resulting CP violating asymmetry is:

$$\frac{\Gamma(K_0 \to \pi^+\pi^-) - \Gamma(\bar{K}_0 \to \pi^+\pi^-)}{\Gamma(K_0 \to \pi^+\pi^-) + \Gamma(\bar{K}_0 \to \pi^+\pi^-)} = (4.5 \pm 1.4) \cdot 10^{-6}. \tag{2}$$

The analysis of the 1998 data is continuing and systematics are being reduced. The analysis of the 1999 sample, whose statistic is twice larger, is also in progress. The 2000 run, performed without the spectrometer, is being used for systematic checks related to the neutral mode. During summer 2001 there will be the last NA48 data taking for Re(ε'/ε) physics.

References

1. V. Fanti et al. A new measurement of direct CP violation in two pion decays of the neutral kaon. *Phys. Lett.* **B465**, 335 (1999). hep-ex/9909022.
2. J.H. Christenson, J.W. Cronin, V.L. Fitch, and R. Turlay. Evidence for the 2 π Decay of the K(2)0 Meson. *Phys. Rev. Lett.* **13**, 138 (1964).
3. G.D. Barr et al. A new measurement of direct CP violation in the neutral kaon system. *Phys. Lett.* **B317**, 233–242 (1993).
4. L.K. Gibbons et al. Measurement of the CP violation parameter *Re*(ε'/ε). *Phys. Rev. Lett.* **70**, 1203–1206 (1993).
5. G.D. Barr et al. Proposal for a precision measurement of ε'/ε in CP violating $K^0 \to 2\pi$ decays. CERN-SPSC/90-22, July 1990.
6. G. Collazuol. *Rate effects in the measurement of the direct CP violation with the NA48 experiment at cern.* PhD thesis, Università di Firenze, 2000. http://cern.ch/gcollazu/Phd-thesis.ps.gz.

SEARCHES FOR MSSM NEUTRAL HIGGS BOSONS AT LEP

S. CUCCIARELLI
University of Perugia, Via A. Pascoli 06123 Perugia, Italy
E-mail: susanna.cucciarelli@cern.ch

This article describes the analysis methods and the results of the searches for the MSSM neutral Higgs bosons at LEP. The analysis strategies used on the L3 experiment are described. The analyses are carried out on 216.6 pb^{-1} data collected by the L3 detector at centre-of-mass energies from 200 GeV to 209 GeV. No signal is observed. Combining the analyses from the LEP experiments, preliminary lower mass limits are given as a function of the tanβ parameter.

1 The Neutral MSSM Higgs Bosons

The Minimal Supersymmetric Standard Model (MSSM) [1, 2, 3] requires two Higgs doublets. This gives rise to five Higgs bosons: a charged scalar pair, two neutral CP-even, the lightest of which is called h, and a neutral CP-odd, A. The two most important production mechanisms of the neutral Higgs bosons in e^+e^- collisions are:

$$e^+e^- \to Z^* \to hZ \tag{1}$$

$$e^+e^- \to Z^* \to hA. \tag{2}$$

The cross section of the process (1) is smaller than that for the similar production of the Higgs boson in the Standard Model. This process is dominant at low values of tanβ, where tanβ is the ratio of the two Higgs vacuum expectation values. The pair-production of Higgs bosons (2) takes over at high values of tanβ.

The sensitivity to the production of MSSM neutral Higgs bosons is improved by combining all the data from the four LEP experiments and with the results of the searches at lower centre-of-mass energies.

2 L3 Analysis Strategies

For the hA production, the following decay modes are considered: hA $\to$ $b\bar{b}b\bar{b}$, hA $\to$ $b\bar{b}\tau^+\tau^-$ and hA $\to$ $\tau^+\tau^-b\bar{b}$. In the case of hZ, four event topologies covering approximately 98% of possible final states, are considered: $q\bar{q}q\bar{q}$, $q\bar{q}\nu\bar{\nu}$, $q\bar{q}l^+l^-(l = e, \mu, \tau)$ and $\tau^+\tau^-q\bar{q}$. The searches in channels with hadronic decays of the h boson are optimised for the dominant h $\to$ $b\bar{b}$ decay channel. Most of the analysis for hZ are the same as devised for the Standard Model

Higgs search[4]. Nevertheless dedicated analyses are performed to search for hA→ $b\bar{b}b\bar{b}$ and hA → $b\bar{b}\tau^+\tau^-$ channels.

2.1 Hadronic Channel

The signature of both the hA → $b\bar{b}b\bar{b}$ and hZ → $b\bar{b}q\bar{q}$ final states is four hadronic jets and the presence of b-hadrons. The dominant backgrounds come from $q\bar{q}(\gamma)$ production and hadronic decays of W and Z pairs. A common selection is applied to both the hA and hZ searches. It mainly reduces the two-photon interaction background while keeping the signal efficiency high ($\sim 93\%$), then a neural network is used to build a discriminating variable for hA and hZ searches separately.

High multiplicity events are selected and their visible energy, E_{vis}, is required to be greater than $0.6\sqrt{s}$ and less that $1.4\sqrt{s}$. Events with perpendicular imbalance greater than $0.35E_{vis}$ or with a lepton whose energy exceeds 65 GeV are rejected to suppress semileptonic W pair decays. Initial state radiation events are further suppressed by requiring $P^{L}_{mis}/(m_{vis} - m_Z) < 0.4$, where P^{L}_{mis} is the longitudinal component of the missing momentum, m_{vis} is the visible mass and m_Z the Z boson mass. The remaining events are then forced into four jets using the DURHAM algorithm[5] and a kinematic fit requiring energy and momentum conservation (4C) is performed. The number of expected and observed events in the data after this selection are listed in Table 1. At this level of the analysis, discriminating variables are combined

$\mathcal{L}_{int}$ (pb^{-1})	$\sqrt{s}$(GeV)	DATA	BKG	$q\bar{q}$	W^+W^-	ZZ
10.2	200.0-205.5	162	133.1	54.2	73.1	5.7
83.0	205.5-206.5	1130	1073.0	433.9	591.8	47.1
123.6	206.6-209.0	1679	1597.0	642.7	883.5	71.0

Table 1: Number of events observed and expected after the selection on the hadronic channels.

into a neural network. The input variables include the probability that the jets of the event contain b-quarks[7] (see Figure 1), called B_{Tag}, event shape variables and mass information. The mass information is summarised in a χ^2 variable defined as:

$$\chi^2(m_X, m_h) = (\Sigma_{ij}^{min} - (m_X + m_{\rm h}))^2 w_\Sigma + (\Delta_{ij}^{min} - |m_X - m_{\rm h}|)^2 w_\Delta \ , \quad (3)$$

where m_X is either m_Z or m_A. The pairing used gives the minimum value for the difference squared $(\Delta_{ij} - |m_X - m_{\rm h}|)^2$ between the measured and expected

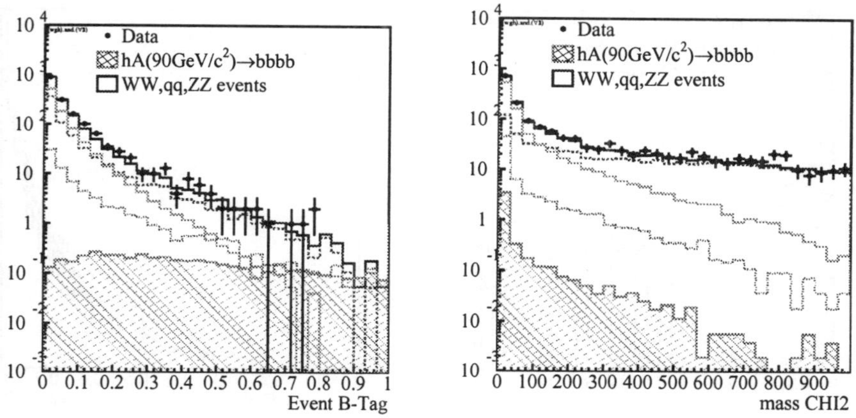

Figure 1: Examples of B_{Tag} (on the left) and $\chi^2(m_X, m_h)$ (on the right) distributions. The points represent the data, the dark line is the total backgrounds, the gray line the WW events, the dashed line the q$\bar{q}$ events and the dotted line the ZZ events. The hatched histogram is the hA $\to b\bar{b}b\bar{b}$ expected signal for $m_h = m_A = 90$ GeV.

dijet mass differences; Σ_{ij}^{min} is the corresponding sum of the dijet masses. The weights, w_Σ and w_Δ, are derived from the mass resolutions. In this way, the shape of the neural network output is made almost independent of the mass hypothesis. The polar angle of the Higgs boson, Θ, is also used as input for the hA analysis. It gives additional separation between the hA signal and W^+W^- background, due to the different spins of the W and the Higgs bosons.

The Neural Network has three output variables which correspond to the signal, O_{Higgs}, the q$\bar{q}(\gamma)$ final state, O_{qq}, and the W^+W^- final state hypotheses, O_{WW}. The discriminating variable is then obtained from the combination $O_{Higgs} \times (1 - O_{qq}) \times (1 - O_{WW})$. Two different neural networks are used for the hA and hZ analyses, and the events are classified as hA or hZ according to the largest value of the discriminating variable. An example of the discriminating variable distribution of hA analysis is shown in Figure 2. The highest signal sensitivity region corresponds to large values of the discriminating variable.

2.2 Semi-Leptonic Channel

The signatures of hA$\to b\bar{b}\tau^+\tau^-$ [a], hZ $\to b\bar{b}\tau^+\tau^-$ or hZ $\to \tau^+\tau^- q\bar{q}$ events are a pair of taus accompanied by two hadronic jets. For each of the channels hA and hZ an analysis is optimised based either on the tau identification or on the

[a] Both the decay modes (h$\to$ b$\bar{\text{b}}$, A$\to \tau^+\tau^-$) and (h$\to \tau^+\tau^-$, A$\to$ b$\bar{\text{b}}$) are considered.

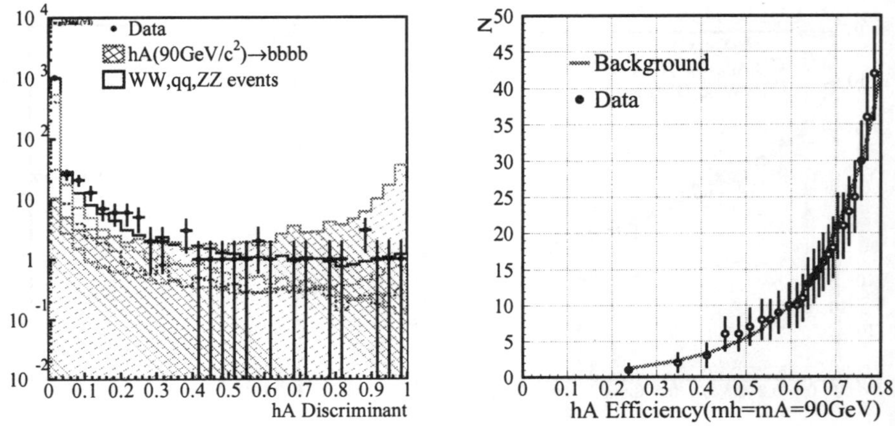

Figure 2: On the left, an example of the discriminating variable distribution in the hA $\rightarrow b\bar{b}b\bar{b}$ analysis and on the right, the number of the data (dots) and the expected backgrounds (line) versus the signal efficiencies.

event topology by requiring four jets with two of them being narrow and of low multiplicity. The main background results from W-pair decays containing taus. The invariant mass of the hadronic jets, $m_{\rm qq}$, must be between 25 GeV and 125 GeV. The ratio of the sum of the energies of the tau decay products over the sum of the energies of the jets is required to be less than one and the value of the missing momentum vector in the rest frame of the Higgs must be less than 40 GeV. Finally, the cosine of the polar angle of the Higgs boson, $|\cos\Theta|$, has to be less than 0.8. After the selection, 28 events are observed and 25.7 are expected from the total Standard Model background. The signal efficiency for $m_h = m_A = 90$ GeV is 34.7%.

For each event class j (ZZ, $\rm W^+W^-$, $\rm q\bar{q}$, $\rm Ze^+e^-$, hA, hZ), a probability function, f^i_j, is constructed, where i denotes the variables considered. These are the $\rm B_{Tag}$ for each hadronic jet, $m_{\rm qq}$ and $m_{\tau\tau}$. The probability, p^i_j, of an event to belong to class j, based on the value of the variable i, is then defined as

$$p^i_j = \frac{f^i_j}{\sum_k f^i_k} \,. \tag{4}$$

Finally, the probabilities for the individual variables are combined by calculat-

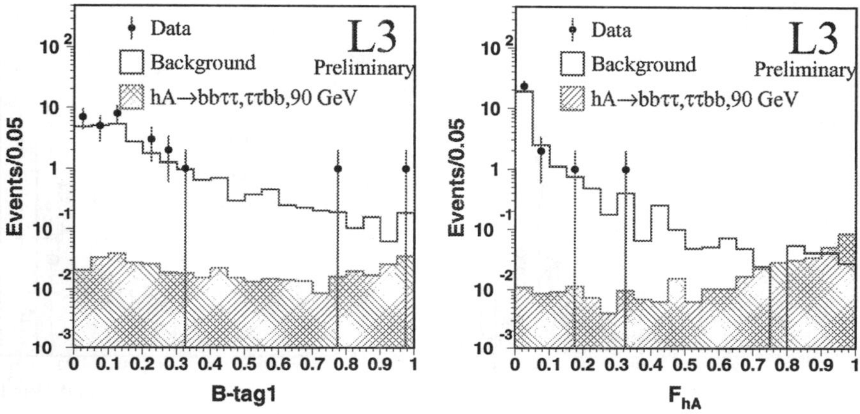

Figure 3: The B_{Tag} of the most energetic jet (on the left) and the discriminating variable, F_{hA}, (on the right) distributions into the hA→ $b\bar{b}\tau^+\tau^-$ analysis. The points represent the data, the open histogram is the total expected background and the hatched histogram is the expected signal for $m_h = m_A = 904$ GeV.

ing the likelihood that the event belongs to either signal class:

$$F_{\rm hA} = \frac{\prod_i p^i_{\rm hA}}{\sum_k \prod_i p^i_k} \qquad \text{and} \qquad F_{\rm hZ} = \frac{\prod_i p^i_{\rm hZ}}{\sum_k \prod_i p^i_k} \ . \tag{5}$$

Events retained by both the hA and the hZ selections, are classified according to highest value of the likelihood, as hA or hZ candidates. An example of the distribution of the discriminating variable for the hA search is shown in Figure 3.

3 Preliminary LEP Combined Results

A good agreement is observed between the data and expected background in all the analyses performed by the LEP experiments. Therefore no evidence of the production of the h and A bosons is found. Mass spectra of the most significant candidates in the hA $\rightarrow$ $b\bar{b}b\bar{b}$ and hA $\rightarrow$ $b\bar{b}\tau^+\tau^-$ search channels are shown in Figure 4.

The results of the search for the hA and the hZ production are interpreted in the framework of the constrained MSSM (CMSSM) assuming unification of the scalar fermion masses, unification of the gaugino masses and unification of the trilinear Higgs-fermion couplings at the GUT scale. This choice has little impact on the phenomenology of the Higgs bosons but reduces significantly

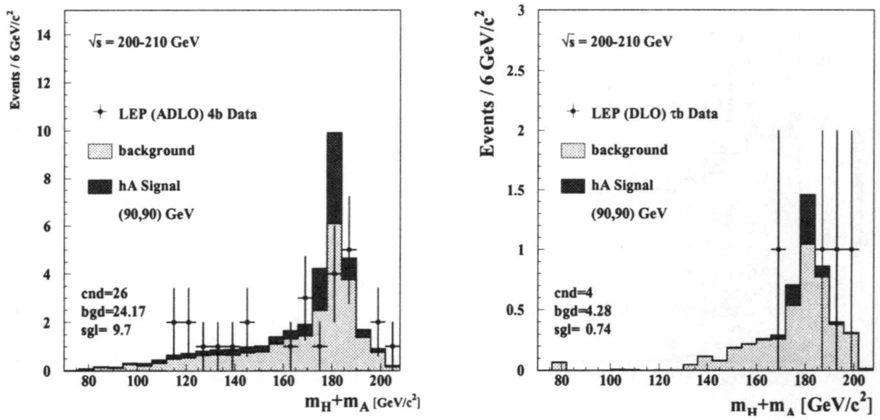

Figure 4: Distributions of the reconstructed mass $m_h + m_A$ on the hA $\rightarrow b\bar{b}b\bar{b}$ (on the left) and hA$\rightarrow b\bar{b}\tau^+\tau^-$ (on the right) channels for the most significant candidates. The points are the data from the four LEP experiments, the light grey histogram is the total background and the dark grey histogram is the signal+background distribution.

the number of free parameters. The remaining free parameters are tanβ, m_A, the gaugino mass parameter, M_2, the universal scalar fermion mass, m_0, the common scalar quark trilinear coupling, A, and the Higgs mixing parameter, μ.

Two benchmark scenarios [6] are considered. In the first one, termed "maximal mixing", the CMSSM parameters are chosen such that m_h acquires its maximal value for any given value of m_A and tanβ ; it is the scenario which gives the most conservative mass limits. The second scenario corresponds to vanishing mixing in the scalar top sector and is referred to as "minimal mixing".

The CMSSM parameters are chosen as follows: $m_0 = 1$ TeV, $\mu = -200$ GeV, $M_2 = 200$ GeV. The mass of the top quark is fixed to the measured value. The maximal mixing scenario is realised at $X_t = A - \mu \cot\beta = \sqrt{6}$ TeV, where X_t is the parameter which controls the mixing in the scalar top sector. The minimal mixing corresponds to $X_t = 0$. Keeping these values fixed, a scan over the two remaining independent parameters, tanβ and m_A, is performed in each mixing scheme in the ranges: $0.5 \leq \tan\beta \leq 30$ and 10 GeV $\leq m_A \leq 1$ TeV.

To set exclusion limits on the CMSSM parameters, the confidence level, CL, that the expected signal is absent in the data, is calculated [8] for each

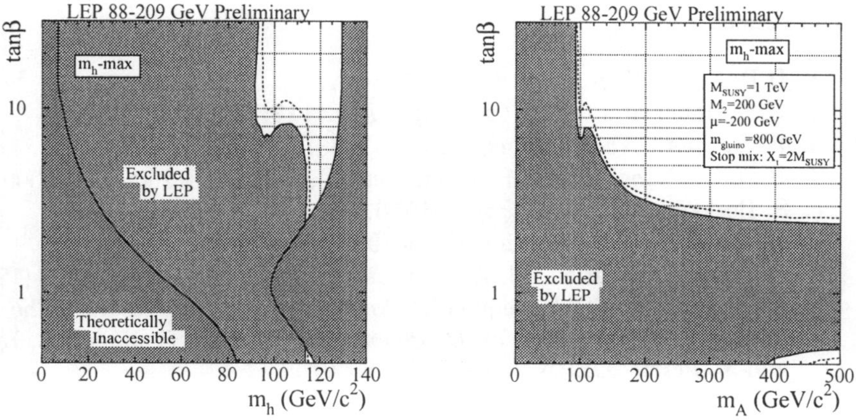

Figure 5: Exclusion plots in the (tanβ, m_h) and (tanβ, m_A) planes at the 95% CL for the maximal mixing scenarios. This is the scenario which gives the most conservative mass limits.

point tanβ, m_A of the scan. Figure 5 shows the region of the (tanβ, m_h) plane and (tanβ, m_A) plane excluded by LEP experiments for the maximal mixing scenario.

For the CMSSM parameters considered and assuming tanβ greater than 0.5, this results in preliminary lower mass limits at the 95% CL of:

$$m_h > 89.9 \text{ GeV}, \quad m_A > 90.5 \text{ GeV},$$

which compare to the median expected limits in the absence of a signal of $m_h > 93.8$ GeV and $m_A > 90.5$ GeV.

For $0.52 < \tan\beta < 2.25$ values of m_A up to 1 TeV are ruled out for any mixing scenario allowing to exclude this tanβ region in the CMSSM, for the measured value of the top mass.

Acknowledgements

We acknowledge the efforts of the engineers and technicians who have participated in the construction and maintenance of L3 and express our gratitude to the CERN accelerator divisions for the superb performance of LEP.

References

1. H.P. Nilles, *Phys. Rep.* **110**, 1 (1984).
2. H.E. Haber and G.L. Kane, *Phys. Rep.* **117**, 75 (1985).
3. R. Barbieri, *Nuovo Cimento* **11**, 1 (1988).
4. By L3 Collaboration (M. Acciarri *et al.*),CERN-EP/2000-146, 2000.
5. S. Bethke *et al.*, *Nucl. Phys.* **B370**, 310 (1992).
6. M. Carena *et al.*, Preprint hep-ph/9912223 (1999).
7. By L3 Collaboration (M. Acciarri *et al.*), *Phys. Lett.* **B411**, 373 (1997).
8. A. Read, "Modified Frequentist Analysis of Search Results (The CLs Method)" in *Workshop on Confidence Limits*, eds. F. James, L. Lyons and Y. Perrin, CERN 2000-05, 81 (2000).

THE TESLA LINEAR COLLIDER DESIGN

W. DECKING (for the TESLA Collaboration)
DESY, Notkestraße 85, 22603 Hamburg, Germany
E-mail: winfried.decking@desy.de

The international TESLA collaboration proposes a 500-800 GeV center-of-mass linear collider based on superconducting 1.3 GHz cavities. This technology promises a luminosity of $3.4 \times 10^{34}\,\mathrm{cm}^{-2}\mathrm{s}^{-1}$ at 500 GeV. Results from the cavity R&D program at the TESLA test facility and details of the collider design will be presented.

1 Introduction

A lepton collider with an initial center-of-mass energy of 500 GeV will be the perfect tool for precision measurements of the Higgs system and SUSY particle properties. [1] It is in many ways complementary to the LHC (presently under construction at CERN) and there is a broad consensus in the particle physics community that a linear collider should be the next accelerator facility to be built.

In the TESLA collaboration, centered at DESY, about 40 institutes from Armenia, China, Finland, France, Germany, Italy, Poland, Russia, Switzerland and the USA participate in the design and technical R&D for a linear collider based on superconducting niobium accelerating structures operated at 2K with a resonance frequency of 1.3 GHz. The combination of high power transfer efficiency with small emittance dilution in the low-frequency linac makes this choice of technology ideally suited for an optimum performance in terms of the achievable luminosity and operation stability.

The general scaling law for linear collider luminosity (assuming flat beams) is:

$$L \propto \frac{\eta P_{AC}}{E_{cm}} \sqrt{\frac{\delta_E}{\varepsilon_y}} \tag{1}$$

with P_{AC} the site electrical power, η the P_{AC} to beam-power transfer efficiency, E_{cm} the center-of-mass energy, δ_E the beamstrahlung energy loss and ε_y the normalized vertical emittance. With E_{cm} given, δ_E limited to a few percent due to background and beam transport considerations and a reasonable site-power of around 100 MW it becomes obvious that the only way to achieve high luminosity is to increase the power transfer efficiency and reduce the vertical beam emittance.

Superconducting cavity technology is the ideal choice to realize both goals:

- The low power dissipation in the cavity walls allows for a very high ($\approx 20\%$) power transfer efficiency.

- A major source of beam emittance deterioration are the beam induced electromagnetic fields (wakefields). The strength of these wakefields scales with the third power of the cavity resonance frequency. The relatively small frequency of the TESLA cavities ensures small wakefields.

The accelerating gradient of superconducting cavities is limited by the critical microwave magnetic fields on the cavity surface. For the TESLA cavity geometry this theoretical limit is above 50 MV/m. In practice superconductivity breaks down much earlier due to material imperfections and uncleanliness. The challenge has been to produce reliable and achievable cavities with a gradient as close as possible to the theoretical limit. The TESLA collaboration has thus started a R&D program in 1992 with the goal of increasing the average accelerating gradient by a factor of 4 to 5 while simultaneously reducing the cost per MV/m by the same factor. The success of this program will be presented in section 2.

The Technical Design Report [2] which includes the scientific case, technical feasibility and cost estimates for the proposed linear collider has been published in spring 2001. The baseline design and further options as well as the expected performance are summarized in section 3.

2 Superconducting Cavity Development

A typical 9-cell cavity is shown in Figure 1. These cavities are produced by industry and undergo further treatment and tests at the TESLA test facility (TTF), built at DESY. This facility includes clean rooms for preparation and assembly, chemical treatment, ultra-pure high-pressure water rinsing and heat treatment of niobium cavities. Single cavities undergo tests with continuous wave excitation in a vertical cryostat and in pulsed excitation on a horizontal test stand. A summary of cavity performance on the vertical test stand is given in Figure 2(a). Little difference has been observed between cavities tested both on the vertical and horizontal test stand. The improvement both in average gradient and quality is obvious. The TESLA 500 GeV goal of 23.4 MV/m is reached by the cavities of the third production series on average.

The ongoing R&D program focuses on simplifications in the cavity production and treatment and on pushing towards higher gradients. Recent developments in electro-polishing, done in collaboration between CERN, KEK, Saclay and DESY, have yielded very high accelerating gradients of 35...43 MV/m in

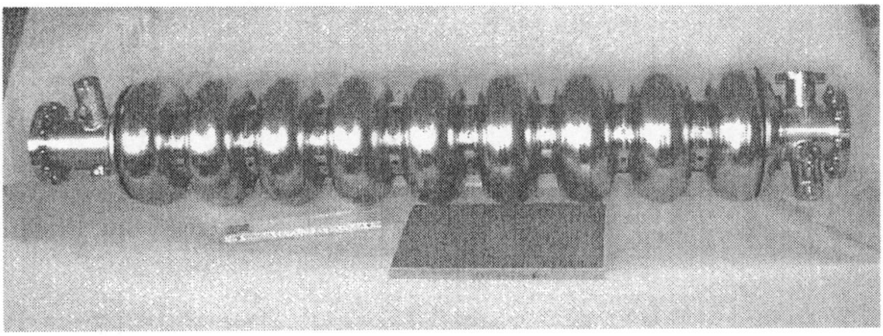

Figure 1: Photo of a 9-cell niobium cavity. The total length of the device is 1.25 m.

several single cell resonators and 33 MV/m in a standard TESLA 9-cell resonator. A hydroformed single cell cavity - this technique avoids welding - has shown a gradient of 42 MV/m.

After treatment and test the cavities are installed in an approximately 12 m long cryomodule which houses eight 9-cell cavities. The average gradient of the five cryomodules produced so far is shown in Figure 2(b).

A full integrated system test with beam is done at the TTF linac. The linac went into operation in May 1997 and up to date the first three accelerator modules have been tested with beam with a measured average gradient of 16MV/m, 20.5MV/m and 22MV/m.

3 Linear Collider Layout

The baseline design of the TESLA linear collider is optimized for e^+e^- collisions with $E_{cm} = 500\,\text{GeV}$ and a luminosity of $3.4 \times 10^{34}\,\text{cm}^{-2}\text{s}^{-1}$. A parameter list is given in Table 1. Notable is the long bunch train with a large bunch spacing, being only possible through the use of superconducting RF technology.

A second interaction region can be used for $\gamma\gamma$ or e^-e^- collisions with parameters as given in Table 1. For $\gamma\gamma$ collisions the horizontal beam size can be decreased because of the absence of beamstrahlung. Only the effective luminosity for collisions with $E_{cm} > 400\,\text{GeV}$ is given. The luminosity for the e^-e^- option is smaller than in the e^-e^+ case because the beams are defocused by the space charge fields of the accommodating equally charged bunch.

The progress on s.c. cavity R&D indicates that the accelerating gradient can be pushed beyond the required value of 23.4 MV/m. The sub-systems of the facility have been laid out to accommodate an energy upgrade to $E_{cm} =$

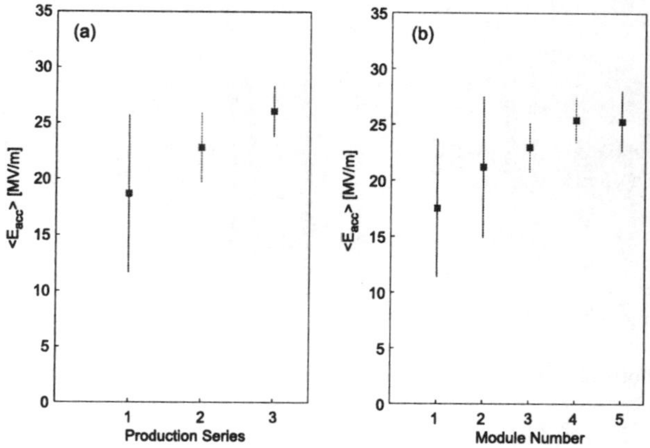

Figure 2: Average accelerating gradients measured in the vertical test cryostat of the cavities in the three production series since 1995 (left); and of the cavities installed in the first five cryogenic modules for the TTF linac (right).

800 GeV. Recent design improvements could lead to an increase of the filling factor (the ratio of active accelerating structure length to total machine length) by 6 %, yielding a necessary accelerating gradient of 35 MV/m at 800 GeV. The 800 GeV machine parameters, assuming an upgrade of the cryogenic plants and of the RF-system are shown in Table 1.

Additional to the collider options, the TESLA accelerator can be used for fixed target experiments: Low-intensity polarized electron pulses can be accelerated in parallel to the collider beam and used for polarized target experiments. [3] If built next to the DESY site the first part of the linac can be used as an injector for the HERA electron ring - operated as a pulse stretcher -, which could deliver continuous beam at 15-25 GeV for nuclear physics experiments. The linac could also be connected with the HERA proton ring and would allow for e^-p collisions with three to five times larger center-off-mass energy than at present HERA, but reduced luminosity.

Figure 3 shows a schematic overview of the TESLA linear collider. A brief description of the major subsystems will be given in the following.

The electron beam is generated in a polarized laser-driven gun and then pre-accelerated to 5 GeV before injection in the damping ring.

The positron source has to provide 5.6×10^{13} positrons per pulse. It is based on the concept of high-energy photon conversion into e^+e^- pairs in a thin target. The photons are generated by the high-energy electron beam which is sent through a planar undulator *before* the interaction point. This layout avoids collimation problems of the spent beam and facilitates the generation

Table 1: TESLA linear collider parameters at E_{cm} = 500 GeV and 800 GeV.

		e^+e^-	e^-e^-	$\gamma\gamma$	e^+e^-
Energy (c.m.)	GeV	500			800
Site length	km	32.8			
# of cavities		21024			
Repetition Rate	Hz	5			4
Beam pulse length	μs	950			860
# of bunches		2820			4886
Bunch spacing	ns	337			176
Charge per bunch		2×10^{10}			1.4×10^{10}
Beam size at IP	nm	553/5		157/5	391/2.8
Bunch length at IP	mm	0.3			
Beamstrahlung	%	3.2	2.0	–	4.3
Luminosity	$10^{34}\mathrm{cm}^{-2}\mathrm{s}^{-1}$	3.4	0.47	0.6	5.8
Total beam power	MW	22.6			34
Linac electric power	MW	97			150
Accelerating gradient	MV/m	23.4			35
# of klystrons		584			1240

of polarized positrons, since the emittance requirements for that option are easily met by the incoming electron beam. To achieve the 45% to 60% of e^+ polarization the planar undulator has to be replaced by a helical one.

The damping ring has to accommodate the full TESLA bunch train of 2820 bunches. To keep the ring length reasonable the bunches are stored in a compressed mode in a still 17 km long ring. The "dog-bone" design accommodates 90% of the ring in the linac tunnel with short return loops at either end.

The two main linear accelerators are constructed from roughly ten-thousand 9-cell cavities. Each linac is powered by some 300 klystrons. The cryogenic system is comparable in size to the one currently under construction for the LHC. Seven cryogenic plants are foreseen, each serving a 5 km long section of the linac.

The two times 1.6 km long beam delivery system comprises a beam switch yard to serve the second interaction region, a collimation section to remove the beam halo, which could otherwise cause background at the experiment, and the final focus section. The large bunch spacing allows for head-on collisions in the first interaction region. The second interaction region will have a crossing angle of ≈ 34 mrad and is therefore suitable for the $\gamma\gamma$ option. The luminosity is very sensitive to relative offsets of the colliding beams in vertical position or angle at the IP. A fast orbit feedback system has been designed which, thanks to the large spacing between bunches, can operate on a bunch-to-bunch basis and maintains head-on collision accurate to one tenth of a standard deviation

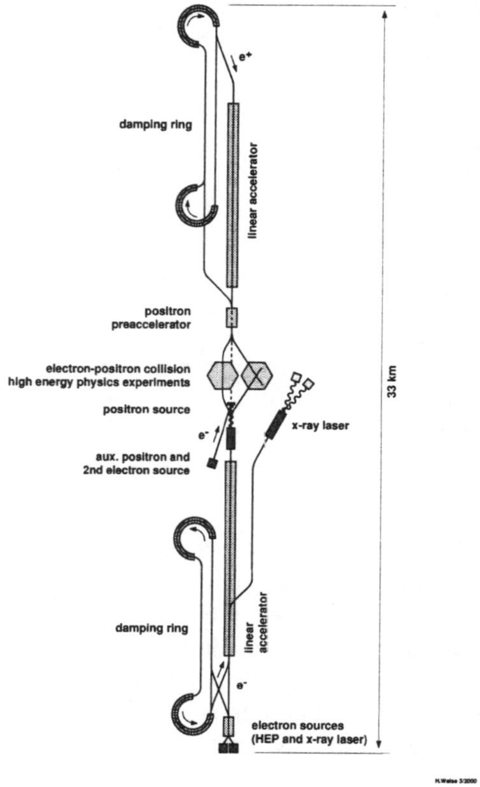

Figure 3: Layout of TESLA linear collider and X-ray FEL.

in orbit and angle.

The linear accelerator and beam delivery system will be installed in an underground tunnel of 5 m diameter. A 2000 m^2 experimental hall is foreseen for the detector. The seven cryogenic plants are located in additional surface halls with a spacing of roughly 5 km.

An integral part of the TESLA proposal is a free electron X-ray laser based on the principle of self amplified spontaneous emission (SASE). Again superconducting accelerating cavities provide the ideal environment for the acceleration of the necessary very small emittance bunches. The first part of the electron linac - operating at a repetition rate of 10 Hz instead of 5 Hz - will be used to alternately accelerate the FEL or collider beam. This operation mode ensures a parallel use of the linac hardware with no interference between the two user groups.

4 The International TESLA Laboratory at DESY in Hamburg

The TESLA collaboration has the strong opinion that the linear collider should be constructed close to an existing laboratory. DESY as the co-ordinating institute within the collaboration is studying a possible site for TESLA with the electron linac extending from the DESY site in Hamburg in north-west direction, tangential to the existing HERA electron-proton collider. The experimental area, hosting the high energy physics experiments and the FEL laboratory, will be located in a rural area close to Hamburg.

5 Conclusions

The TESLA collaboration has worked out a complete design for a high - performance next generation linear collider at 500 – 800 GeV based on superconducting linac technology. The achievements at the test facility demonstrate the availability of the technology for 500 GeV energy. The progress of the ongoing R&D program justifies the optimism that the required higher gradient for 800 GeV is in reach. In spring 2001 the collaboration has published a Technical Design Report which includes project costs, construction schedule and required work force. The costs for the linac and one high-energy physics detector are estimated to be 3344 million Euro. The additional costs for the FEL laboratory, including necessary additional accelerator components, are 531 million Euro. The construction time will be eight years. The Technical Design Report forms the base for the ongoing decision-making within the scientific community and political bodies.

References

1. G. Blair, *Physics with TESLA*, this proceedings, (2001).
2. TESLA Collaboration, *TESLA Technical Design Report*, DESY **2001-11**, ECFA **2001-209**, TESLA **2001-23**, TESLA-FEL **2001-05**, `http://tesla.desy.de/new_pages/TDR_CD/start.html`, (2001).
3. F. Ellinghaus, *TELSA-N: Polarized Electron Nucleon Scattering at TESLA*, this proceedings, (2001).

SEARCH FOR CHARGINOS NEARLY MASS-DEGENERATE WITH THE LIGHTEST NEUTRALINO IN e^+e^- COLLISIONS UP TO $\sqrt{s} = 209$ GeV

N. De FILIPPIS

ALEPH Collaboration

Universitá Degli Studi and INFN, Via Amendola 173 Bari, ITALY

E-mail: nicola.defilippis@ba.infn.it

Search for chargino nearly mass-degenerate with the lightest neutralino is performed with the data collected at a centre-of-mass energy from 189 to 209 GeV by the ALEPH detector at LEP. The analysis uses the detection of an isolated and energetic initial state radiation photon, produced in association with the chargino pair, whose decay products have small visible energy. No evidence of signal is found and lower limits on the chargino mass of 91 and 89 GeV/c^2 at 95% confidence level are derived, for large scalar masses, in gaugino and in higgsino region respectively.

1 Introduction

This paper reports on the search for chargino ($\tilde{\chi}_1^{\pm}$) with a mass difference with respect to the lightest neutralino($\tilde{\chi}_1^0$) of less than 5 GeV/c^2. In the Minimal Supersymmetric extension of the Standard Model (MSSM) [1], $\tilde{\chi}_1^{\pm}$ and $\tilde{\chi}_1^0$ are nearly mass-degenerate for large $M_{1,2}$ (higgsino region); small values of Δm ($\Delta m = m_{\tilde{\chi}_1^{\pm}} - m_{\tilde{\chi}_1^0}$) are possible in gaugino region if the gaugino unification relation between M_1 and M_2 is relaxed. Moreover supersymmetric models with anomaly mediated supersymmetric breaking mechanism [2] foresee the degeneration of chargino and neutralino, since $M_1/M_2 >> 1$.

The analysis is based [3] on the reconstruction of an isolated initial state radiation (ISR) photon together with the low-momentum detectable chargino decay products. The ISR photons are required to have high transverse momentum with respect to the beam line to ensure high trigger efficiency.

Data collected by the ALEPH detector [4] at LEP during 1998–2000, at centre-of-mass energies ranging from 189 to 209 GeV, are used. The total data sample corresponds to an integrated luminosity of about 627 pb^{-1}.

Background events expected from Standard Model processes are generated with a statistics corresponding to at least 20 times the collected luminosity used in the analysis.

Signal events are simulated with the Monte Carlo program `SUSYGEN` [5]. Charginos are produced in pairs with masses ranging from 40 GeV/c^2 up to the kinematic limit, for Δm between 0.1 to 5 GeV/c^2, and with decay length, λ, up to 80 cm. Corrections to the efficiency calculations are applied to take into account the spectral functions of the hadronic decay [6], which are based on the Kühn-Santamaria parametrisation [7].

Parameters are tuned to correctly describe the semileptonic τ decays to pions [8]. The ISR is enabled in the Monte Carlo samples [9]. The photon energy spectrum is different in various regions of the MSSM parameter space; it depends on the relative contribution of the s-channel Z exchange, which, in turn, depends on the $Z\tilde{\chi}_1^+\tilde{\chi}_1^-$ coupling; this effect becomes as important as the chargino is far from the kinematic limit.

2 Event Selection

The signal topology depends on Δm values. Below the threshold of pion production, charginos are long-lived and strongly ionising particles; stable particle analysis [10] efficiently selects this final state. When Δm is larger than 5 GeV/c^2, missing energy chargino analysis [11] are used. In the intermediate range, $m_\pi < \Delta m < 5$ GeV/c^2, the analyses described below is tailored to search for events with few low-momentum particles accompanied by an ISR photon.

Photons are reconstructed by localised energy depositions within groups of neighbouring ECAL towers and by the photon conversion finding [12]. In order to optimise the energy reconstruction of photons, which are not well-contained in ECAL, the sum of the localised energy depositions in ECAL and HCAL is computed inside a cone with half-opening of 11.5°.

Events are selected if at least one photon in the geometrical acceptance ($|\cos\theta_\gamma| < 0.95$) is reconstructed with the transverse momentum with respect to the beam axis , p_t, greater than $\sqrt{s}\sin\theta_{min}$, where $\theta_{min} = 35$ mrad is the lowest polar angle covered by the SICAL detector. The energy deposit is then large enough to ensure the trigger.

Two or more charged tracks, reconstructed with at least 4 hits in the ITC or 4 hits in the TPC, are required. Photons have to be isolated from the closest charged track by 30°.

Good tracks are defined requiring the impact parameter in the transverse plane to the beam axis to be less than 4 cm, if no hits in the TPC are used in the reconstruction. The cut is tightened to 2 cm, and the distance from the interaction point along the beam axis is required to be less than 10 cm, if at least 4 TPC hits are used in the track finder. Most of the two and four fermions events are rejected requiring that the number of good tracks is less than or equal ten.

To prevent from $\gamma(\gamma)\nu\bar{\nu}$ events with one non identified photon conversion in the TPC, at least one track has to be reconstructed with a number of ITC hits greater than zero; secondary vertex reconstruction is enforced for each pair of tracks, the momentum of the tracks is computed at the secondary vertex position and the invariant mass is determined assuming electron mass for each track and it is asked to be greater than 0.1 GeV/c^2. Events with a recoiling mass to the photon smaller than 100 GeV/c^2 are also discarded.

Candidates with a non-vanishing value of the visible energy inside a cone of 14° around the beam axis are rejected. The total visible transverse momentum of the event $p_t^{\rm vis}$ is required to be greater than 3.5% $\sqrt{s}$. Both these cuts improve the signal to $\gamma\gamma$ background discrimination.

The total energy not associated to the reconstructed photons is constrained to the maximum value of the expectation for charginos decaying promptly, at each chargino mass and Δm. Similarly $p_t^{\rm vis}$ is constrained to the maximum value as determined by the Monte Carlo signal simulations. Finally the photon energy is required to be less than the maximum energy for an ISR photon as computed with the following formula: $E_\gamma^{max} = \frac{s-(2m_{\tilde{\chi}_1^\pm})^2}{2\sqrt{s}}$.

The selection criteria have been optimised according to the $\bar{N}_{95}$ [13] prescription over the whole range of Δm.

3 Results

The selection efficiency of pair-produced charginos with an ISR high transverse momentum photon is about 40%. Figure 1a) shows the dependence of the average efficiency on the generated transverse momentum of the photon, p_t^γ. In the range of p_t^γ between 15 and 25 GeV/c, about 20% of the events are lost due to the photon acceptance and reconstruction. Photon isolation criteria reduce the efficiency of about 30%.

The selection efficiency with respect to all chargino pair produced is below 3% and it is mostly due to the small fraction of events with a high p_t^γ photon. The efficiency does not depend on Δm significantly, but on the chargino mass through the ISR photon reconstructed p_t requirements as shown in Figure 1b).

Inefficiency arises when the chargino get a sizeable decay length, due to the cut on the additional reconstructed energy in addition to the photon.

Additional inefficiency due to beam related background leading to depositions in the very forward detectors is estimated using random trigger events. The applied correction to the efficiency amounts to about $(4 \pm 1)\%$. The total systematic uncertainty on the selection efficiency is 15% and it is dominated by the limited Monte Carlo statistics.

The number of observed events is in good agreement with the expectation from standard model processes as shown in Table 1, for $m_{\tilde{\chi}_1^\pm} > 50$, 65, 85 GeV/$c^2$

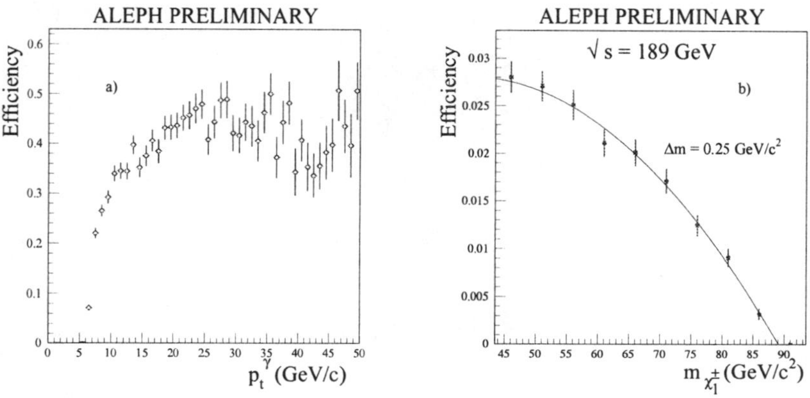

Figure 1: a) Chargino selection efficiency as a function of generated transverse momentum of the ISR photon for chargino masses from 46 to 91 GeV/c^2 and $\sqrt{s} = 189$ GeV; b) Signal efficiency as a function of chargino mass at $\sqrt{s} = 189$ GeV for $\Delta m = 0.25$ GeV/c^2 and $\lambda = 0$.

Table 1: Number of observed events on data and standard model background expected events for $m_{\tilde{\chi}_1^\pm} > 50,\ 65,\ 85$ GeV/c^2 and $\Delta m < 2,\ 1,\ 0.3$ GeV/c^2 respectively.

$\sqrt{s} = 189 - 209\, GeV$		Data	background					
$m_{\tilde{\chi}_1^\pm}$	Δm		$\gamma(\gamma)\nu\bar{\nu}$	$\gamma\gamma \to \tau\tau$	$\gamma\gamma \to ee$	$\tau\tau$	$4f$	Tot Exp
> 50	< 2	13	1.2	2.0	0.3	3.8	1.3	9.0
> 65	< 1	5	0.7	1.8	0.3	1.7	0.6	5.4
> 85	< 0.3	1	0.6	1.4	0	0.5	0	2.9

and $\Delta m < 2,\ 1,\ 0.3$ GeV/c^2 respectively. The main background contributions are also reported.

The main contributions to the background after the selection come from $\tau^+\tau^-$, $\gamma\gamma \to \tau^+\tau^-$ and $\gamma(\gamma)\nu\bar{\nu}$ events.

4 Interpretation in the MSSM

The non-observation of any excess of candidate events is interpreted in terms of upper limits on the chargino production cross section in the MSSM framework. The cross section upper limits are in turn translated into 95% confidence level excluded regions in the $(m_{\tilde{\chi}_1^\pm},\ \Delta m)$ plane.

No background subtraction is performed and systematics are taken into

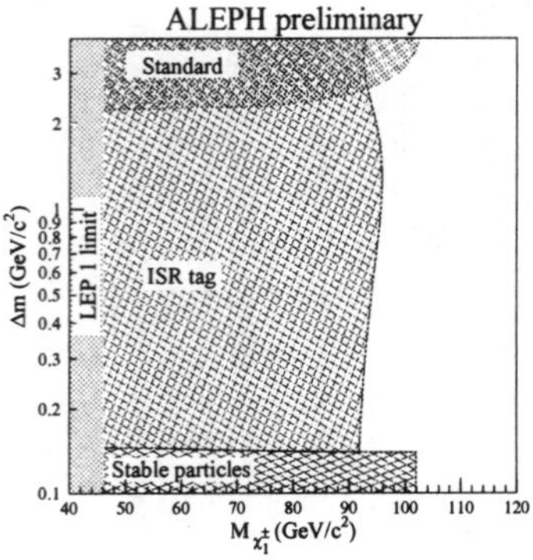

Figure 2: Excluded region in the $(m_{\tilde{\chi}_1^\pm}, \Delta m)$ plane at 95% confidence level, for $m_0 >$ 500 GeV$/c^2$ in the gaugino region. The green region is excluded by the standard missing energy chargino selection [11], the red region is the resulting exclusion of the ISR analysis; the blue region is excluded by the stable particle search [10]. The 95% confidence level lower limit on chargino mass is 91 GeV$/c^2$.

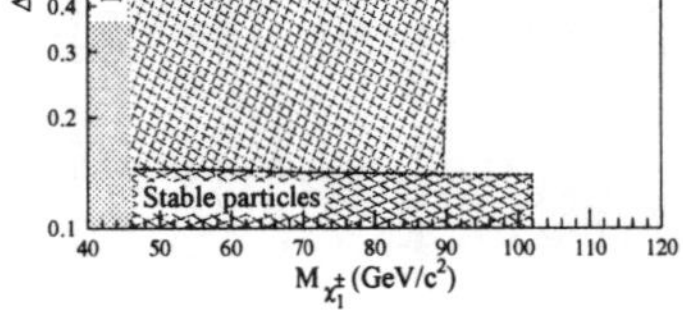

Figure 3: 95% confidence level excluded region for $m_0 >$ 500 GeV$/c^2$ in the higgsino region. The green region is excluded by the results of the standard missing energy chargino results [11], the red region by the ISR analysis; the blue region by the stable particle search [10]. Chargino masses below 89 GeV$/c^2$ are excluded at 95% confidence level.

account according to Ref. [14].

The excluded regions in gaugino and higgsino scenario are shown in Figure 2 and 3 and are valid for all $\tan\beta$ and in the hypothesis of high scalar masses. The results of the standard missing energy selection, the ISR analysis and the stable particles search are distinguished.

Lower limits on chargino mass of 91 GeV/c^2 and 89 GeV/c^2 are set in the gaugino and higgsino region respectively.

These results are improved with respect to those obtained by the DELPHI [15] and L3 [16] collaborations using data collected at energies from $\sqrt{s}=$ 189 GeV up to $\sqrt{s}$ = 202 GeV.

5 Conclusions

Search for charginos nearly mass degenerate with the lightest neutralino is performed using data collected with the ALEPH detector at LEP at a centre-of-mass energies from 189 GeV to 209 GeV. No excess on candidate events with respect to standard model background prediction is found.

In the framework of MSSM, in the high scalar masses scenario and for any $\tan\beta$ mass limits are set at 95% confidence level: $m_{\tilde{\chi}_1^\pm} > 91\,\mathrm{GeV}/c^2$ in gaugino region and $m_{\tilde{\chi}_1^\pm} > 89\,\mathrm{GeV}/c^2$ in higgsino region.

References

1. H.E. Haber and G.L. Kane, *Phys. Rep.* **C117**, 75 (1985);
   R. Barbieri, *Riv. Nuovo Cimento* **11**, 1 (1988).
2. G.F. Giudice, M.A. Luty, H. Murayama and R. Rattazzi, *J. High Energy Phys.* **9812**, 027 (1998);
   L. Randall and R. Sundrum, *Nucl. Phys.* **B557**, 79 (1999);
   T. Gherghetta, G.F. Giudice and J.D. Wells, *Nucl. Phys.* **B559**, 27 (1999).
3. C.H. Chen, M. Drees and J.F. Gunion, *Phys. Rev. Lett.* **76**, 2002 (1996).
4. ALEPH Collaboration, ALEPH: A detector for electron-positron annihilation at LEP. *Nucl. Instrum. and Methods* **A294**, (1990);
   Performance of the ALEPH detector at LEP. *Nucl. Instrum. and Methods* **A360**, 481 (1995).
5. SUSYGEN version 2.2: S. Katsanevas and P. Morawitz, *Comp. Phys. Comm.* **122**, (1998).
6. C.H. Chen, M. Drees and J.F. Gunion, *Phys. Rev.* **D55**, 330 (1997), erratum *Phys. Lett.* **82**, 3192 (1999).
7. J.H. Kühn and A. Santamaria, *Z. Phys.* **C48**, 445 (1990).

8. ALEPH Collaboration, Measurement of the spectral functions of vector current hadronic τ decays, *Z. Phys.* **C76**, 15 (1997);
Measurement of the spectral functions of axial-vector hadronic τ decays and determination of $\alpha_s(M^2_{tau})$, *Eur. Phys. Journal* **C4**, 409 (1998).
9. REMT: F.A. Berends and R. Kleiss, *Nucl. Phys.* **B260**, 32 (1985).
10. ALEPH Collaboration, *Search for gauge mediated SUSY breaking topologies at* $\sqrt{s} \sim 189$ GeV *Eur. Phys. J.* **C16**, 1 (2000);
Single- and multi-photon production and a search for slepton pair production in GMSB topologies in e^+e^- *collisions at a* $\sqrt{s}$ *up to 208 GeV*, ALEPH 2001-010 CONF 2001-007.
11. ALEPH Collaboration, *Search for charginos and neutralinos in* e^+e^- *collisions at centre-of-mass energies near 183 GeV and constraints on the MSSM parameter space*, *Eur. Phys. J.* **C11**, 193 (1999);
Search for charginos and neutralinos in e^+e^- *collisions at* $\sqrt{s}$ *up to 208 GeV and mass limit for the lightest neutralino*, ALEPH 2001-009, CONF 2001-009.
12. ALEPH Collaboration, Single- and multi-photon production in e^+e^- collisions at a centre-of-mass energy of 183 GeV, *Phys. Lett.* **B429**, 201 (1998).
13. ALEPH Collaboration, *Phys. Lett.* **B384**, 427 (1996);
J.-F. Grivaz and F. Le Diberder, *Complementary analyses and acceptance optimization in new particle searches.*, LAL 92-37 (1992).
14. R.D. Cousins and V.L. Highland, *Nucl. Instrum. and Methods* **A320**, 331 (1992).
15. DELPHI Collaboration, Search for charginos nearly mass-degenerate with the lightest neutralino, *Eur. Phys. J.* **C11**, 1 (1999);
Update of the search for charginos nearly mass-degenerate with the lightest neutralino, *Phys. Lett.* **B485**, 95 (2000).
16. L3 Collaboration, Search for charginos with a small mass difference to the lightest supersymmetric particle at $\sqrt{s} = 189$ GeV, *Phys. Lett.* **B482**, 31 (2000);
Search for charginos with a small mass difference to the lightest supersymmetric particle at $189 < \sqrt{s} < 202$ GeV, L3 Note 2582, July 3,2000, Contribution to the *XXXth International Conference on High Energy Physics, Osaka, Japan, July 27-August 2, 2000.*

FERMION PAIR PRODUCTION AT LEP

PATRICK DÉGLON
EP Division, CERN, CH-1211 Genève 23. Switzerland
E-mail: Patrick.Deglon@cern.ch

Fermion pair production $e^+e^- \to f\bar{f}$ is studied with the L3 detector at LEP using data collected at centre–of–mass energies from 130 GeV to 208 GeV. A fit to cross–sections and asymmetries allows the extraction of Standard Model parameters with high precision. No deviation from Standard Model predictions has been found. The measurements are used to derive limits on new physics such as 4–fermion contact interactions, fermion sizes and Quantum Gravity strings.

1 Introduction

The study of fermion pair production, $e^+e^- \to f\bar{f}$, at centre–of–mass energies well above the Z resonance allows the test of the Standard Model (SM), as well as the search for physics beyond the SM.

Cross-sections and asymmetries are measured for τ-pair, μ-pair and electrons. For hadrons, only cross-sections are measured.

Any deviation from the SM prediction could indicate a sign of new physics. We consider, for example, 4–fermion contact interactions, fermion sizes and Quantum Gravity strings.

1.1 LEP and its detectors

The Large Electron Positron collider is located close to Geneva, Switzerland. LEP accelerates electron and positron beams up to the record energy of 105 GeV. The 4 LEP experiments are: L3, ALEPH, OPAL and DELPHI. This report considers results from the L3 experiment[1]. A luminosity of 720 pb^{-1} has been collected by L3 for centre–of–mass energies from 130 to 210 GeV.

1.2 Fermion Pair Production

2 Analysis

Although fermion pair production final states are very different, they all have in common that outgoing fermions are very energetic. Thus, the requirement of large visible energy is used to reject backgrounds.

For hadrons, we use the total energy seen in the detector. To be selected, candidate must have 40 % of the possible energy ($2 \times E_{beam}$).

For τ-pairs, τ decay into one or two neutrinos and a narrow hadronic jet, a muon or an electron. The energy of the most energetic jet or single particle has to be at least 24 % of the beam energy.

For muon pairs, we measure their momentum with the muon chambers. The largest momentum has to be at least 35 GeV for the event to be selected.

For electron pairs, we use the energy deposited in the electromagnetic calorimeter. This energy must be greater than 50% of the beam energy.

Figure 1 shows this energy selection for all channels.

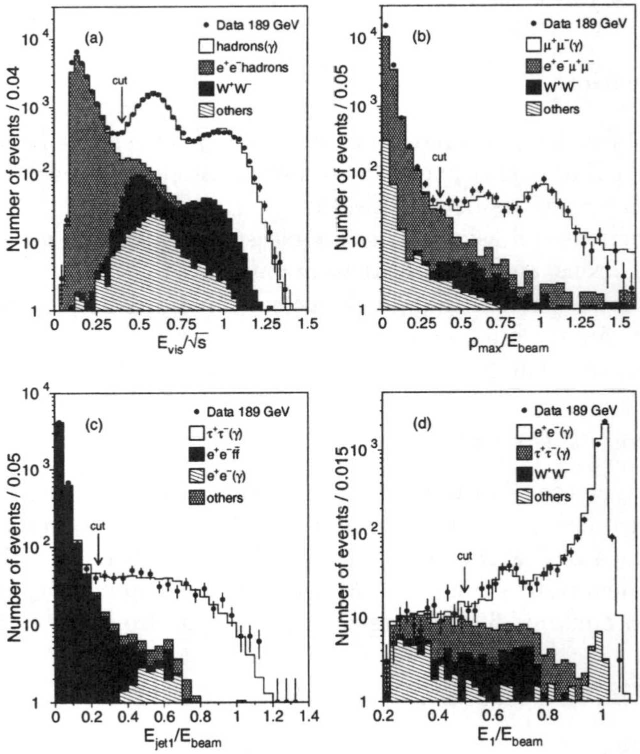

Figure 1: L3 data at $\sqrt{s} = 189$ GeV. (a) The total visible energy normalized to the centre-of-mass energy, $\sqrt{s}$, for the selection $e^+e^- \to hadrons(\gamma)$, (b) highest muon momentum normalized to the beam energy for the selection $e^+e^- \to \mu^+\mu^-(\gamma)$ events, (c) highest tau jet energy normalized to the beam energy for the selection $e^+e^- \to \tau^+\tau^-(\gamma)$ events, and (d) highest electron energy normalized to the beam energy for the selection of $e^+e^- \to e^+e^-(\gamma)$ events.(γ) indicate the presence of additional (mostly ISR) photons.

Other cuts are used to remove other backgrounds, but they are different for the different channels. For example, we require the hadrons to have more than 18 reconstructed objects in the detector, whereas for τ-pairs we require less than 16 objects.

Details of the cuts and the systematics of the selection have been published previously[2].

Due to Initial State Radiation (ISR), some collisions occur at lower energy than twice the beam energy. Many of them even occur at the Z peak, where we already have much data and did not find any sign of new physics. For this reason new physics contributions may be expected only at high effective e^+e^- collision energy $\sqrt{s'}$. To estimate $\sqrt{s'}$, different methods are used, but they all are based on the exact kinematic relation :

$$s' = s - 2\,E_\gamma\sqrt{s} + E_\gamma^2 - \vec{P}_\gamma^2,$$

with E_γ, $\vec{P}_\gamma$ the sum of all ISR photon energies and momenta respectively. To estimate those variables, we use a kinematic fit, or on the assumption that ISR photons are emitted along the beam axis :

$$E_\gamma = \sqrt{s} \cdot \frac{|\sin(\theta_1 + \theta_2)|}{\sin\theta_1 + \sin\theta_2 + |sin\,(\theta_1 + \theta_2)|},$$

with $\theta_{1,2}$ the polar angles of the two jets or particles.

Hence, we consider two samples for each channel : an inclusive sample with almost all events, and a high energy sample with $\sqrt{s'}/\sqrt{s} > 0.85$, where sign of new physics should be more pronouced.

For the electron case where the t-channel, and even more the s/t-interference channel, make the $\sqrt{s'}$ variable less intuitive. Hence, we consider the acollinearity ξ, which refer to as how back-to-back two particles are :

$$\xi = \pi - angle\,between\,electron\,momenta\,in\,space.$$

The high energy sample is selected with $\xi < 25^o$, which means that the two electrons are almost back-to-back, and hence only low energy ISR photon are radiated. Only in very rare kinematical configurations, two or more ISR photons could balance longitudinal momentum to produce a small acollinearity. This effect is corrected for by Monte Carlo when calculating efficiency and background.

The selected events give the total and the differential cross-section for each channel, and, in the case where charge can be determined (all leptonic channels), we also have a measurement of the asymmetry.

Figure 2(left and center) shows the L3 cross-section for both samples for all channels, and the asymmetry for leptons only. The SM prediction is taken from the TopaZ0 program[6].

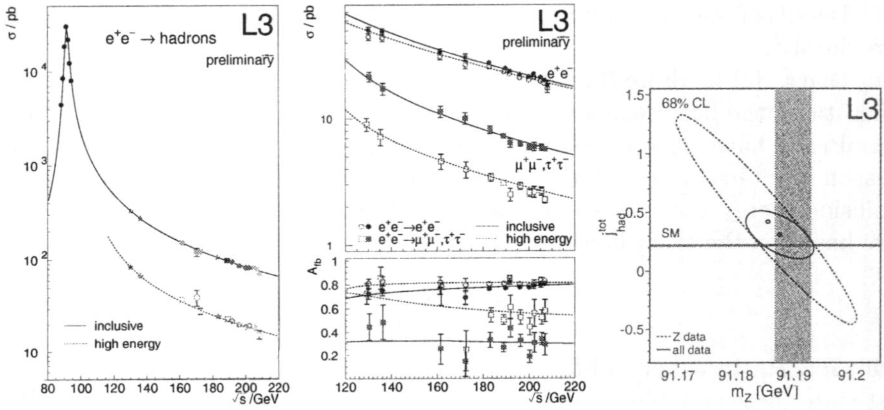

Figure 2: The left part show the L3 hadronic cross-section for the inclusive sample and the high energy sample. – The center part shows the cross-sections and the asymmetries for all three lepton channels for both the inclusive and high energy samples. – The right part shows the contours in the (m_Z, j_h^{tot}) plane at 68% confidence level. The dashed line is obtained from Z data only; the inclusion of 130 GeV to 189 GeV data gives the solid line. The circle (Z data) and the cross (all data) indicate the central values of the fits. The SM prediction for j_h^{tot} is shown as the horizontal band. The vertical band corresponds to the 68% confidence level interval on m_Z in a fit assuming the SM value for γ/Z interference.

3 Interpretation

The measurements can be used both to determine SM parameters[4] and to set limits on new physics beyond the SM[3].

3.1 γ/Z interference

In the *S-matrix* framework[7,8], model independent coupling constants are determined directly from the measured cross sections and asymmetries. For example, at the tree level, we have :

$$\sigma^0_{tot} = \frac{4\pi\alpha^2}{3}\left[\frac{g_f^{tot}}{s} + \frac{j_f^{tot}\,(s-m_Z^2)}{(s-m_Z^2)^2 + m_Z^2\,\Gamma_Z^2} + \frac{r_f^{tot}\,s}{(s-m_Z^2)^2 + m_Z^2\,\Gamma_Z^2}\right],$$

with g_f^{tot}, j_f^{tot} and r_f^{tot} are free parameters for the terms corresponding respectively to photon exchange, photon/Z boson interference and Z boson exchange.

The interference term, being an odd function of $s - m_Z^2$, will displace the position of the maximum of the peak. Hence, a strong correlation is expected between the parameter j_f^{tot} and the mass of the Z boson, m_Z. Figure 2(right) shows measurement of j_{had}^{tot} with the data take at the Z peak only, and all data take at LEP. We can see that measurements performed above from the Z peak clearly improve the extraction of interference term. A fit to all data gives :

$$\begin{aligned} j_{had}^{tot} &= 0.159 \pm 0.082 \\ m_Z &= 91\,190.4 \pm 3.1\,\mathrm{MeV} \end{aligned}$$

3.2 Contact Interactions

In the same manner as the W boson acts in β-decay, a massive particle, much heavier than the interaction energy $\sqrt{s}$, could contribute to the cross section with an additional contact interaction diagram :

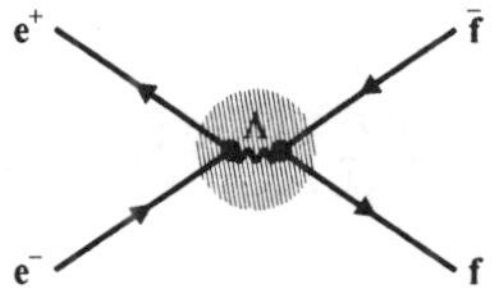

This would yield to an additional term in the Lagrangian[9] :

$$L = \frac{1}{1+\delta_{\mathrm{ef}}} \sum_{i,j=\mathrm{L,R}} \eta_{ij} \frac{g^2}{\Lambda_{ij}^2} (\bar{\mathrm{e}}_i \gamma^\mu \mathrm{e}_i)(\bar{f}_j \gamma_\mu f_j),$$

with $g \equiv 4\pi$ and Λ the energy scale. η_{ij} is $-1, 0, +1$ depending on the fermion helicity states i, j we consider. δ_{ef} is 1 for the electrons and 0 otherwise.

Figure 3(left) shows the upper limits at 95% confident level, using a combination of data from all LEP experiments, for all helicity states considered. Limits on the energy scale Λ of four–fermion contact interactions in the range 3.8 - 14.4 TeV for leptons, and in the range 2.8 - 6.1 TeV for quarks are obtained.

3.3 Fermions Sizes

If fermions are not point-like, the differential cross-sections will be modified. We can consider the Dirac form factor $F(q^2) = 1 + \frac{1}{6} q^2 R^2$, q^2 being t for electrons and s for all other channels. Then, the differential cross-section will

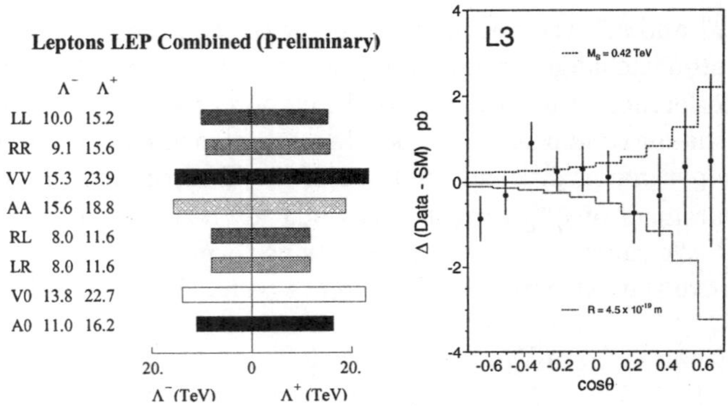

Figure 3: Left, the 95% upper limit confidence level for contact interaction for all helicity states considered (LEP combined). Right, the difference of the L3 $e^+e^- \to e^+e^-(\gamma)$ differential cross-section measurement from the SM expectation and the variations expected for an electron radius of 4.5×10^{-19}, dotted line, and a string scale of 450 GeV, dashed line.

be corrected in the following way[10] :

$$\frac{d\sigma}{dq^2} = \left(\frac{d\sigma}{dq^2}\right)_{SM} F_e^2(q^2) F_f^2(q^2).$$

Figure 3(right) shows the difference between the measured differential cross-section for $e^+e^- \to e^+e^-(\gamma)$ and the expected one for the L3 data at $\sqrt{s} = 189$ GeV. Variation for an electron with radius 4.5×10^{-19} m is shown. No deviation is found, and the following upper limits at 95% CL are obtained :

Channel	R [m]
e^+e^-	$< 3.1 \times 10^{-19}$
$\mu^+\mu^-$	$< 2.4 \times 10^{-19}$
$\tau^+\tau^-$	$< 4.0 \times 10^{-19}$
$\ell^+\ell^-$	$< 2.2 \times 10^{-19}$
$q\bar{q}$	$< 3.0 \times 10^{-19}$

3.4 TeV strings in Quantum Gravity

In Quantum Gravity theories, the Bhabha differential cross-section will be modified as follows[11,12] :

$$\frac{d\sigma}{d\cos\theta} = \left(\frac{d\sigma}{d\cos\theta}\right)_{SM} \left| \frac{\Gamma(1-\frac{s}{M_S^2})\Gamma(1-\frac{t}{M_S^2})}{\Gamma(1-\frac{s}{M_S^2}-\frac{t}{M_S^2})} \right|^2 ,$$

with t and s the Mandestam variables, M_S the string scale, and Γ the usual Riemann Γ function. Figure 3(right) shows the expected deviation for a string scale of $M_S = 420$ GeV. No deviation is found, and a lower limit at 95% CL on the string scale M_S of 0.49 TeV is set.

4 Summary and Conclusions

LEP experiments successfully measured fermion pair production from the Z peak up to 208 GeV to allow high precision SM parameter measurement and fit. In the S-matrix framework, a direct measurement of the γ/Z interference term can be done. No deviation from SM has been observed. Nevertheless, the results obtained are used to derive limits on physics beyond the SM, such as, for example, contact interactions, lepton sizes or strings in Quantum Gravity theories.

Acknowledgements

I would like to thanks all members of the L3 Fermion Pair Production group for their inputs and work. I also thanks the organizers of the Lake Louise Winter Institute 2001 for their wonderful work in such a nice place.

References

1. L3 Collab., *Nucl. Inst. Meth.* **A289**, 35 (1990).
2. L3 Collab., *Phys. Lett.* **B479**, 101 (2000).
3. L3 Collab., *Phys. Lett.* **B489**, 81 (2000).
4. L3 Collab., *Phys. Lett.* **B489**, 93 (2000).
5. ZFITTER v.6.21, D. Bardin *et al.*, preprint HEP-PH/9908433.
6. TOPAZ0 v4.4, G. Montagna *et al.*,
   t NPB **401**, 3 (1993).
7. A. Leike, T. Riemann and J. Rose, *Phys. Lett.* **B273**, 513 (1991).
8. T. Riemann, *Phys. Lett.* **B293**, 451 (1992).
9. E. Eichten, K. Lane, and M. Peskin, *Phys. Rev. Lett.* **50**, 811 (1983).
10. G. Kopp *et al.*, *ZPC*, **65**, 45 (1995).
11. E. Accomando, I. Antoniadis and K. Benakli, preprint HEP-PH/9912287.
12. S. Cullen, M. Perelstein and M. Peskin, preprint HEP-PH/0001166.

SEARCHES AT HERA

NICOLAS DELERUE
(on behalf of the H1 and ZEUS collaborations)
Centre de Physique des Particules de Marseille
163, avenue de Luminy, Case 907,
F-13288 Marseille Cedex 9
France
E-mail: nicolas.delerue@desy.de

Searches for physics beyond the Standard Model have been performed in high-energy $e^{\pm}p$ collisions at HERA. No significant deviation from the Standard Model has been observed while searching for contact interactions, extra dimensions, leptoquarks, R-parity violating squarks and excited fermions. Exclusion limits have been inferred which extend or complement bounds from other colliders. The H1 collaboration has observed a puzzling excess of events with a high P_t isolated lepton and missing transverse momentum, and interpretation as flavour changing neutral currents has been explored.

1 Introduction

1.1 HERA

HERA is the first and only existing lepton-proton collider. Between 1994 and 2000 it has delivered an integrated luminosity of more than 130 pb^{-1} available for physics to each experiment. Leptons (electrons or positrons) have an energy of 27.6 GeV and protons have an energy of 920 GeV (or 820 GeV), this gives an energy in the centre of mass ($\sqrt{s}$) of 318 GeV (or 300 GeV). The HERA collider is now shut down for a luminosity upgrade.

1.2 Physics at HERA

One of the major Standard Model (SM) physics processes at HERA is deep inelastic scattering (DIS) where the incoming lepton exchanges a gauge boson with a quark of the proton. Neutral Current (NC) events result from the exchange of a neutral (γ,Z^0) boson and yield an electron and a recoil jet in the final state, while Charged Current (CC) events result from the exchange of a charged ($W^{\pm}$) boson and yield a neutrino and a recoil jet in the final state.

2 Inclusive NC DIS events at high Q^2

Phenomena characterised by a mass scale Λ larger than HERA's kinematic limit ($\sqrt{s} \simeq 318$ GeV) can be parameterised by four-fermion interactions which could affect the cross-sections for DIS observed at HERA via contact interactions. Comparisons of the cross-section predicted by the SM and those observed at high Q^2 at HERA have been made by both collaborations [1,2]. No deviations have been observed for Q^2 up to 10,000 GeV^2. Thus, exclusion limits in various contact interaction models can be set on Λ up to 9 GeV with the usual coupling strength convention of $\sqrt{4\pi}$. Limits on large extra dimensions have also been set by H1 and exclude scale parameters (M_s) lower than 0.63 TeV and 0.93 TeV for positive and negative interference, respectively.

3 Leptoquarks

Leptoquarks are scalar or vector bosons which couple to a quark and a lepton via a Yukawa coupling λ. They can be resonantly produced at HERA. Their final topology is the same as NC or CC events but their angular distribution can be used to optimise the signal significance.

It had been shown earlier [3,4] that the HERA data collected in 1994-97 showed an excess of electron-quark masses greater than 200 GeV (Figure 1, left). This excess is not confirmed in the new data collected by both experiments[5,6,7] (Figure 1, right).

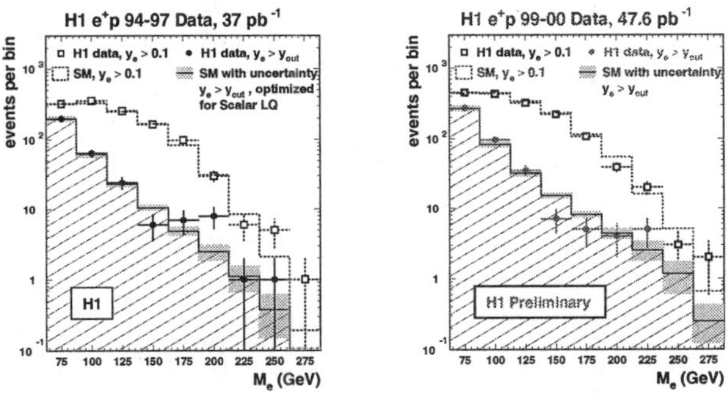

Figure 1: The mass distribution of observed NC-DIS events compared to the expected background for the 1994-1997 data (left) and 1999-2000 data (right).

As no excess is seen, limits have been derived by both collaborations. The results have been interpreted in the model of Buchmüller, Rückl and Wyler (BRW) [8] where the leptoquark branching ratios are fixed. The upper plot of Figure 2 shows the limits set by both collaborations in the BRW framework.

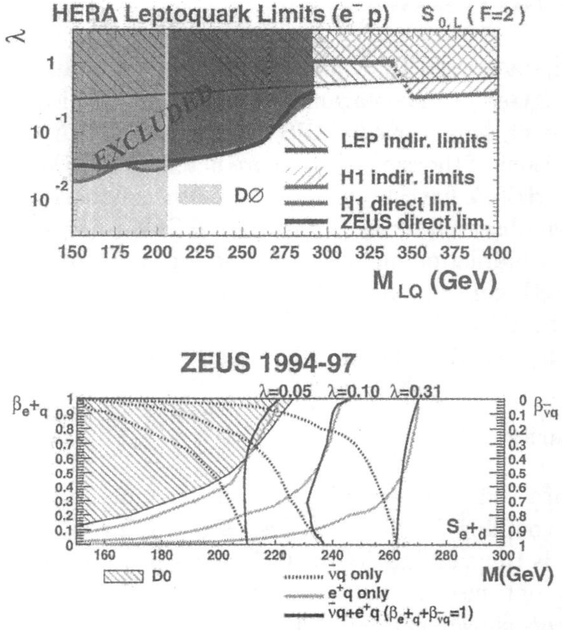

Figure 2: Limits on the Yukawa coupling in the BRW model (upper plot) and limits on the branching ratio $LQ \to eq$ versus mass for fixed values of λ (lower plot).

The lower plot shows the limits derived in a more general model for fixed values of the coupling[9].

4 $\not{R}_p$ supersymmetry

R_p-violating squarks can be produced at HERA up to the kinematic limit in the same way as leptoquarks. They would have many possible decay modes and many different topologies, including some very striking, almost background-free, topologies such as events with a lepton whose charge is opposite to that one of the incoming lepton, or with multileptons.

The signal was looked for in almost all decay modes and no excess was observed[10,11]. Upper limits on the Yukawa coupling λ'_{1j1} have been obtained as a function of the squark mass (see Figure 3, left), by performing a scan in the parameters of the Minimal Supersymmetric Standard Model (MSSM). For a coupling of the electromagnetic strength, squark masses below 258 GeV are ruled out. Constraints on the parameters of the mSUGRA model are also shown in Figure 3 (right) for specific values of the Yukawa coupling.

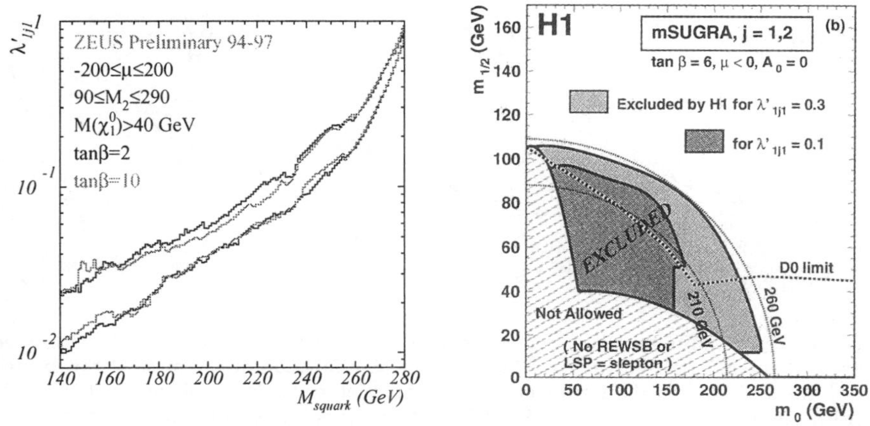

Figure 3: *Left:* Limits on λ'_{1j1} as a function of the squark mass for the unconstrained MSSM. For each $\tan\beta$ hypothesis (dark lines for $\tan\beta = 2$ and light gray lines for $\tan\beta = 10$), the best and worse limit found during the scan are shown *Right:* Domain of the plane $(m_0, m_{1/2})$ excluded by the H1 analysis for $\mu < 0$ and $\tan\beta = 6$ for a $R\!\!\!/_p$ coupling of $\lambda'_{1j1} = 0.3$ (light shaded area) and $\lambda'_{1j1} = 0.1$ (dark grey area). The region below the dashed curve corresponds to the area excluded by the D0 experiment.

5 Excited fermions

Compositeness could manifest itself in excited states of fermions. The single production of excited fermions is considered at HERA. An effective Lagrangian [12,13] describes the production and decay of excited fermions. In this Lagrangian, Λ is the compositeness scale and f, f' and f_s are form factors which determine the couplings of a fermion to its corresponding excited states and to the bosons associated to the three gauge groups (SU(2), U(1) and SU(3)).

The search in the data collected at $\sqrt{s} = 300$ GeV between 1994 and 1997 has been completed [14,15] and no signal was found, thus limits have been inferred (see Figure 4). By assuming $\frac{f}{\Lambda} = \frac{1}{M^*}$ one can exclude excited electrons of mass M^* lower than 229 GeV and excited quarks of mass lower than 203 GeV.

The production cross-section for excited neutrinos (ν^*) is higher (by two orders of magnitude for a ν^* of 200 GeV) in e^-p data than in e^+p data. Preliminary results show that no evidence for ν^* production has been found by either collaboration [15,16] in the recently taken e^-p data. Limits have thus been inferred (see Figure 5). By assuming $\frac{f}{\Lambda} = \frac{1}{M^*}$ one can exclude excited neutrinos of mass below 150 GeV for $f = -f'$ and below 134 GeV for $f = +f'$.

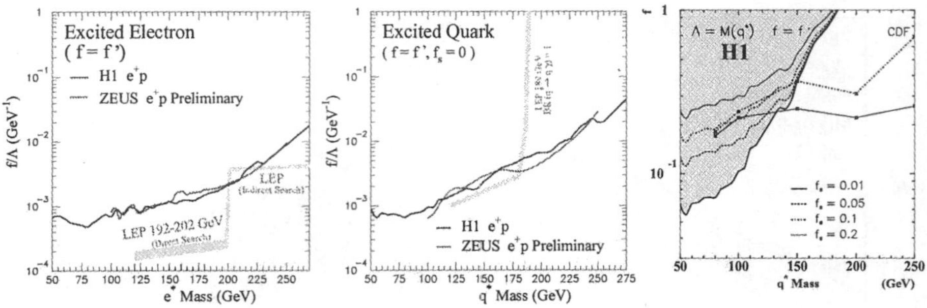

Figure 4: *Left:* Limits on the coupling constant as a function of the mass of the excited electron. *Center:* Limit on the coupling constant as a function of the mass of the excited quark. *Right:* Limit on the form factor f as a function of the mass of the excited quark, with the hypothesis $f = f'$ and $\Lambda = M(q^*)$ for various values of the form factor f_s.

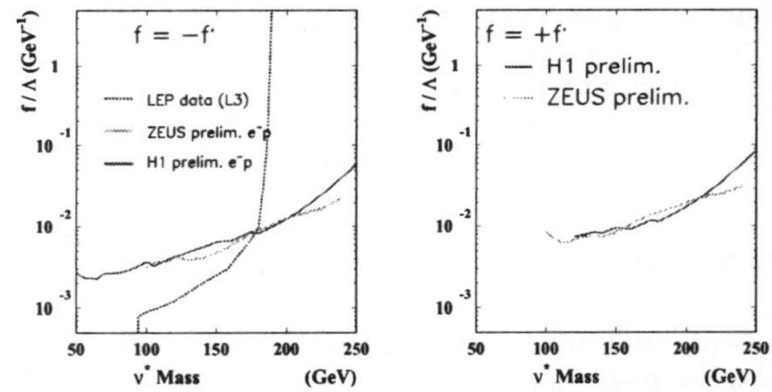

Figure 5: Limits on the coupling constant as a function of the mass of the excited neutrino for the hypothesis $f = -f'$ (left) and $f = +f'$ (right).

As one can see on the Figures 4 (center) and 5, the limits set at HERA reach higher energies than those set at LEP. On Figure 4, the plot on the right shows that for small f_s HERA can extend the limits set at the TeVatron on the excited quark coupling.

6 Events with an isolated lepton and missing transverse momentum, FCNC

Puzzling events with high P_t isolated lepton and missing transverse momentum have been previously observed at HERA. In the SM, such final states mainly come from W production. Both collaborations have a comparable analysis [17,18] (see Table 1). H1 still sees a deviation from the SM prediction whereas ZEUS does not and this needs to be resolved with more statistics. Beyond the SM, an explanation of these events could be flavour changing neutral currents (FCNC) leading to single top production which is highly suppressed in the SM. Limits

	H1 data preliminary	SM expectation H1	SM expectation ZEUS	ZEUS data preliminary
$P_T^X(e) > 25$ GeV	3	0.84	0.78	1
$P_T^X(\mu) > 25$ GeV	6	0.94	0.82	0

Table 1: Comparison of the numbers of events with high P_t isolated leptons compared to the Standard Model expectation. The results presented here correspond, for both collaborations, to $82pb^{-1}$ analysed before July 2000. In the H1 numbers, further cuts were applied after the nominal selection cuts such that both analyses correspond to similar acceptances for W events.

on this phenomena have been set by both collaborations[18,19] (see Figure 6).

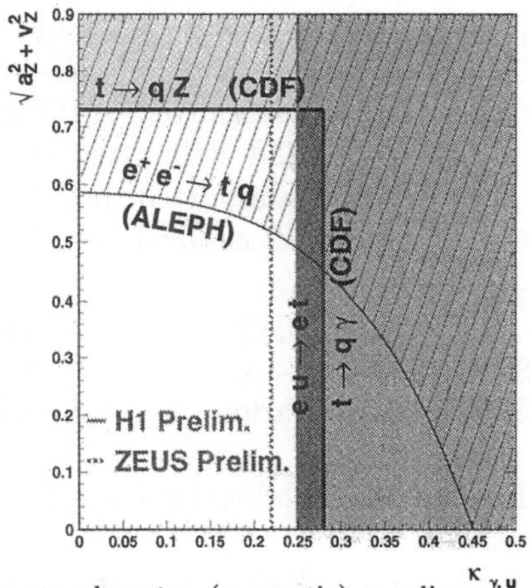

Figure 6: Limits on the anomalous $tu\gamma$ (magnetic) coupling $\kappa_{\gamma,u}$ and on the anomalous tuZ (vector) coupling $\sqrt{a_Z^2 + v_Z^2}$ obtained at the TeVatron (CDF), LEP (ALEPH) and HERA (H1 and ZEUS) colliders.

7 Conclusion

Searches have been performed for contact interactions, extra dimensions, leptoquarks, $R\!\!\!/_p$ supersymmetry and excited fermions and no evidence was found for any of these processes. Limits have thus been derived and are comparable to those set at LEP or the TeVatron. The isolated lepton events observed by H1 are still puzzling. HERA is currently being upgraded to produce 5 times more luminosity starting this autumn (2001), thus searches have an exciting future at HERA.

Acknowledgments

I wish to thank all my ZEUS and H1 colleagues who have contributed to the results discussed here.

References

1. C. Adloff *et al.* [H1 Collaboration], *Phys. Lett.* **B479**, 358 (2000) [hep-ex/0003002]. H1 Collaboration, Abstract **951**, submitted to ICHEP 2000, Osaka.
2. J. Breitweg *et al.* [ZEUS Collaboration], *Eur. Phys. J.* **C14**, 239 (2000) [hep-ex/9905039].
3. C. Adloff *et al.* [H1 Collaboration], *Eur. Phys. J.* **C11**, 447 (1999) [hep-ex/9907002].
4. J. Breitweg *et al.* [ZEUS Collaboration], *Eur. Phys. J.* **C16**, 253 (2000) [hep-ex/0002038].
5. H1 Collaboration, Abstract **953**, submitted to ICHEP 2000, Osaka.
6. ZEUS Collaboration, Abstract **1038**, submitted to ICHEP 2000, Osaka.
7. H1 Collaboration, Abstract **954**, submitted to ICHEP 2000, Osaka. ZEUS Collaboration, Abstract **552** submitted to EPS HEP 1999, Tampere.
8. W. Buchmuller, R. Ruckl and D. Wyler, *Phys. Lett.* **B191**, 442 (1987). Erratum-ibid.B448:320,1999
9. J. Breitweg *et al.* [ZEUS Collaboration], *Phys. Rev.* **D63**, 052002 (2001) [hep-ex/0009059].
10. C. Adloff *et al.* [H1 Collaboration], hep-ex/0102050. DESY-01-021, Submitted to *Eur. Phys. J.* **C**.
11. ZEUS Collaboration, Abstract **1042**, submitted to ICHEP 2000, Osaka.
12. J. Kuhn and P. Zerwas, *Phys. Lett.* **B147**, 189 (1984).
13. K. Hagiwara, D. Zeppenfeld and S. Komamiya, *Z. Phys.* **C29**, 115 (1985).
14. C. Adloff *et al.* [H1 Collaboration], *Eur. Phys. J.* **C17**, 567 (2000) [hep-ex/0007035].
15. ZEUS Collaboration, Abstract **1040**, submitted to ICHEP 2000, Osaka.
16. H1 Collaboration, Abstract **956**, submitted to ICHEP 2000, Osaka.
17. H1 Collaboration, Abstract **974**, submitted to ICHEP 2000, Osaka.
18. H1 Collaboration, Abstract **1041**, submitted to ICHEP 2000, Osaka.
19. H1 Collaboration, Abstract **961**, submitted to ICHEP 2000, Osaka.

DETECTION OF A HYPERCHARGE AXION IN ATLAS

ERIK ELFGREN, GEORGES AZUELOS
Université de Montréal, C.P. 6128, succ. Centre-ville, Montréal (Québec) H3C 3J7, Canada
E-mail: elf@lps.umontreal.ca, azuelos@lps.umontreal.ca

GILLES COUTURE
Université du Québec à Montréal, C.P. 8888, succ. Centre-ville, Montréal (Québec) H3C 3P8, Canada
E-mail: couture.gilles@uqam.ca

SVERKER FREDRIKSSON
Luleå University of Technology, 971 87 Luleå, Sweden
E-mail: sverker@eol.mt.luth.se

The Hypercharge Axion is a pseudo-scalar particle introduced to explain the dominance of matter over antimatter in the universe, and the existence of large scale magnetic fields. The search for this particle is possible at the LHC. The distinguishing characteristics are the decay channels into pairs of photons and Z^0s with high transverse momenta and invariant mass equal to that of the axion. For a Hypercharge Axion mass and an interaction scale of the order of a TeV, the Hypercharge Axion could be detected in ATLAS.

1 Introduction

It is a mystery that the universe is mainly constituted of matter and not equal amounts of matter and antimatter. Equal amounts would have been expected from the symmetry between matter and antimatter. The CP-violation of the weak force is far too low to explain the present dominance of matter. But how do we know that the universe is made of matter?

The deduction that the universe is largely dominated by matter is based on the facts[1] that (1) when matter meets antimatter, they annihilate, creating γ-rays. (2) The amount of antimatter in our solar-system must be negligible, otherwise we (or our satellites) would have been annihilated. (3) Clusters of antimatter in or close to our galaxy would produce a surplus of antimatter in the cosmic radiation, which is not observed. (4) For large, distant bodies of antimatter, the γ-ray emission from the interface between matter and antimatter would have been observed, which it isn't. Hence, any significant amounts of antimatter would have been detected by now.

In 1967 Sakharov[2] proposed three general conditions which have been commonly accepted to be necessary for the appearance of the asymmetry:

1. **Non-conservation of Baryon number**
   Can be obtained in the Standard Model.

2. **Violation of C and CP symmetries**
   Are too weak in the Standard Model.

3. **Thermodynamic nonequilibrium**
   Is probably too weak in the Standard Model.

More details can be found in the M.Sc. thesis by Elfgren[3].

2 The Hypercharge Axion

It can be shown that the divergence of the baryon[a] current due to the anomaly equation is proportional to:

$$\vec{B}_Y \cdot \vec{E}_Y, \tag{1}$$

where B_Y and E_Y are respectively the hypermagnetic and the hyperelectric fields. In other words, the baryon asymmetry could be driven by the hyperelectromagnetic fields, which existed before electroweak phase transition, if they are strong enough during the critical moment of the phase transformation[4] (the third of Sakharov's conditions). Otherwise they will be eradicated by so called sphaleron processes[1].

The hyperelectromagnetic fields behave similarly to ordinary electromagnetic fields and if we introduce the Hypercharge Axion, Maxwell's equations yield a resonance equation of the form [5]:

$$\frac{\partial^2 \vec{\mathcal{B}}_Y}{\partial t^2} + \sigma \frac{\partial \vec{\mathcal{B}}_Y}{\partial t} = a(X(t))\mathcal{B}_Y, \tag{2}$$

where $\mathcal{B}_Y$ is the cosmic hypermagnetic field, σ is the conductivity of the plasma, t is cosmic time and $a(X(t))$ is a time dependent function resulting from the introduction of the Hypercharge Axion, X. We see that for periodic functions $a(X(t))$ a resonance can be achieved with certain modes in the plasma.

If we suppose that some (weak) primordial homogeneous large-scale hypermagnetic fields existed, then these could have been amplified by the oscillating Hypercharge Axion. According to Brustein and Oaknin[5] this amplification

[a] The Baryon asymmetry is considered to be a good measure of the matter-antimatter asymmetry, partly because most of the universe mass is made of baryons, partly because there is a more fundamental symmetry which says that there are always as many baryons as there are leptons.

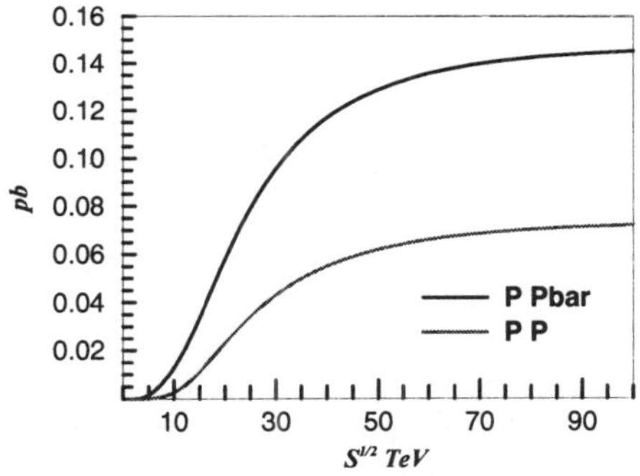

Figure 1: Cross section for $f+\bar{f}\rightarrow Z^*/\gamma^*\rightarrow Z/\gamma+X$ from Brustein and Oaknin[8] for $m_X=800$ GeV and $M=1$ TeV.

could be 10^{12} or even higher. If this amplification happens at the right moment, this could result in the required field-strengths producing the observed asymmetry.

The second of Sakharov's conditions, the violation of C and CP, is expected to be explained elsewhere and according to Brustein and Oaknin[5] the third one, thermodynamic nonequilibrium, could be fulfilled by the hyperelectromagnetic fields.

More details can be found in the M.Sc thesis by Elfgren[3].

3 Production of a Hypercharge Axion at LHC

The Hypercharge Axion is a pseudoscalar particle with electroweak interactions and thus it couples only to photons and Z^0. The Lagrangian can be written as:

$$L=\frac{1}{8M_Y}XY_{\mu\nu}Y_{\alpha\beta}\epsilon^{\alpha\beta\mu\nu} \tag{3}$$

where X is the Hypercharge Axion field, $Y_{\mu\nu}$ is the hypercharge field (consisting of γ and Z^0) and M_Y is a mass scale, which we set to 1 TeV.

3.1 The Process

Brustein and Oaknin proposed the process $f+\bar{f}\rightarrow Z^*/\gamma^*\rightarrow Z/\gamma+X$ which has a cross section of about 15 fb at a center of mass energy of 14 TeV (see Fig. 1).

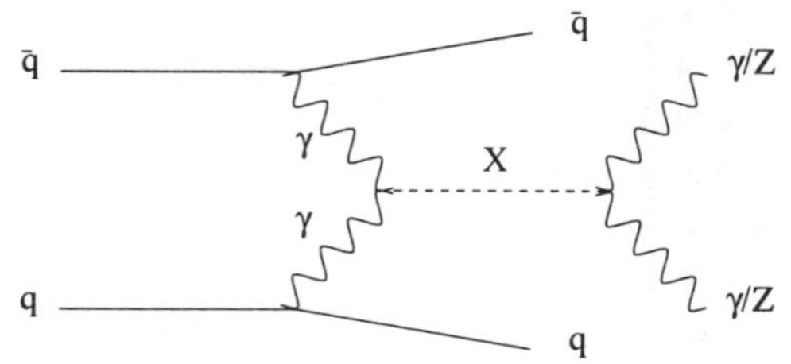

Figure 2: The production process that is the object of this study. Two quarks within the proton emit two photons, which collide and produce the axion, which disintegrates to photons and/or Z^0s.

Here, an alternative production process for the Hypercharge Axion at LHC is considered which is the object of this presentation (see Fig. 2).

The luminosity of the $\gamma\gamma$-production as a function of their invariant mass with a $\mathcal{L}(pp) = 10^{-33}\text{cm}^{-2}\text{s}^{-1}$ was calculated, in the Weizsäcker-Williams approximation, to be as in Figure 3.

3.2 Characteristics of Decay Products

The search for the Hypercharge Axion will exploit the following signal characteristics:

- The decay products are photons and Z^0s which consequently decay to leptons or jets.
- The decay products are back-to-back in the transversal plane.
- The decay products have high p_T because of the heaviness of the Hypercharge Axion.
- The forward jets are not useful because of the Weizsäcker-Williams approximation for which the jets do not obtain a transverse momentum. This approximation was used to calculate the number of events produced for the three processes ($X \to \gamma\gamma$, $X \to \gamma Z^0$, $X \to Z^0 Z^0$) for an integrated luminosity of 300 fb^{-1} (see Fig. 4).

3.3 Signal and Backgrounds

The simulation was implemented in PYTHIA[6] where the backgrounds also were calculated and ATLFAST[7] was used to simulate the detector.

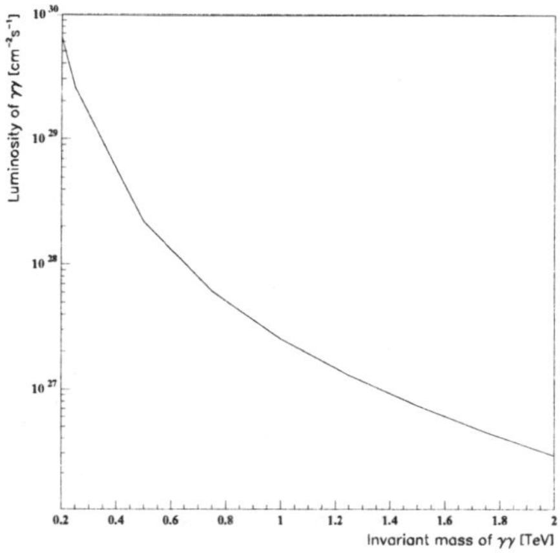

Figure 3: Luminosity of $\gamma\gamma$-production as a function of their invariant mass.

The different final states that we are looking for are:

$$\begin{array}{l} \gamma\gamma, \\ \gamma\ell\bar{\ell}, \\ \gamma jj,\ jjjj,\ \ell\bar{\ell}jj, \\ \ell\bar{\ell}\ell\bar{\ell}, \end{array} \tag{4}$$

where γ is the photon, ℓ is a lepton (e or μ) and j is a jet which is the remnant of an emitted quark (or gluon). The most useful signal will be from $\gamma\gamma$ where the background is low (basically from $qq \to \gamma\gamma$) and the signal is strong. The signal from $\gamma\ell\bar{\ell}$ can also be useful because of its very low background (principally from $qq \to \gamma Z^0$). However, it also has the drawback of very low cross section. All the signals containing jets are almost useless because of the enormous jet-background produced in LHC. The only remaining possibility, $\ell\bar{\ell}\ell\bar{\ell}$, is ruled out by its tiny cross section.

4 Detection of the Hypercharge Axion

We require a signal of at least 5σ and more than 10 events to claim a discovery. The integrated luminosity is taken to be 300 fb^{-1}, which corresponds to three years of high luminosity. The simulation was done for two different masses of the Hypercharge Axion, 800 GeV and 1000 GeV. Analyses were also performed

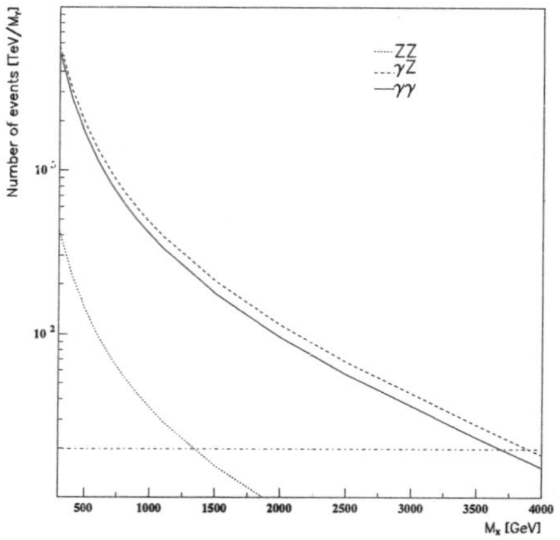

Figure 4: Number of events produced for the different processes for three years of high-luminosity LHC operation.

Channel	m_X [GeV]	$n = S/\sqrt{B}$	# of events
$\gamma\gamma$	1000	14.7σ	210
	800	13.0σ	255
$\gamma\ell\bar{\ell}$	1000	2.0σ	12
	800	2.4σ	15

Table 1: Signal over background for axion masses $m_X = 800, 1000$ GeV for the two principal final states.

for different masses but for higher masses the signal died out and for lower masses it drowned in the background. For the two principal channels, $\gamma\gamma$ and $\gamma\ell\bar{\ell}$, the Monte Carlo simulation gives the results in Table 1 (see also Figures 5 and 6).

5 Conclusions

As can be seen in Eq. 3 the cross section will be proportional to M_Y^{-2} which means that the Hypercharge Axion could be discovered if the mass scale is below about 1600 GeV for $m_X = 1000$ GeV in the $X \to \gamma\gamma$ channel. Before the background, our results are comparable to those of Brustein and Oaknin[8] (who did not calculate the background). The principal channel of discovery

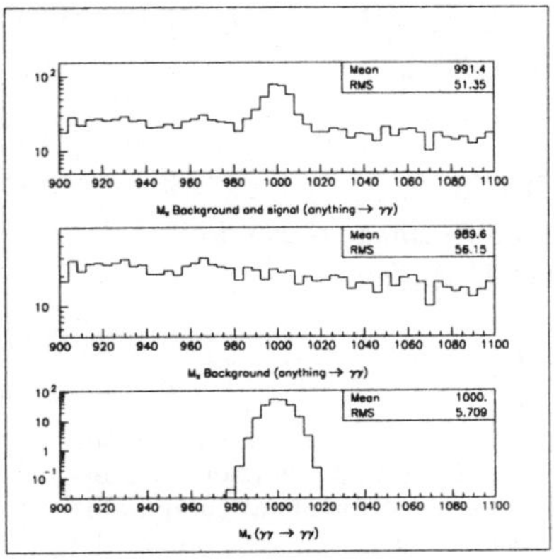

Figure 5: Histograms for signal and background for $X \to \gamma\gamma$ at $m_X = 1000$ GeV.

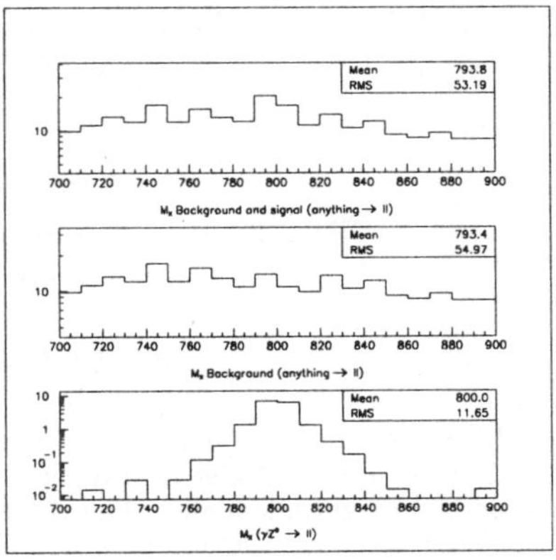

Figure 6: Histograms for signal and background for $X \to \gamma\ell\bar{\ell}$ at $m_X = 800$ GeV.

would be $X \to \gamma\gamma$ and $X \to \gamma\ell\bar{\ell}$ could be used for confirmation.

References

1. M. Trodden, *Rev. Mod. Phys.* **71**, 1463 (1999).
2. A.D. Sakharov, *Zh. Eksp. Teor. Fiz. Pis'ma Red.* **5**, 32 (1967); *JETP Lett.* **5**, 24 (1967) (from Trodden[1]).
3. E. Elfgren and G. Azuelos, hep-ph/0105290 (2001).
4. M. Giovanni and M.E. Shaposhnikov, *Phys. Rev. Lett.* **80**, 22 (1998).
5. R. Brustein and D.H. Oaknin, hep-ph/9901242, *Phys. Rev.* **D60**, 023508 (1999).
6. T. Sjöstrand, *Computer Physics Commun.* **82**, 74 (1994).
7. E. Richter-Was, D. Froidevaux and L. Poggioli, ATLFAST, a package for fast simulation of ATLAS detector, ATL-PHYS-98-131.
8. R. Brustein and D.H. Oaknin, hep-ph/9906344, *Phys. Rev.* **D62**, 015001 (2000).

TESLA-N: FUTURE POLARIZED ELECTRON-NUCLEON SCATTERING AT DESY

FRANK ELLINGHAUS
DESY Zeuthen, Platanenallee 6, 15738 Zeuthen, Germany
E-mail: Frank.Ellinghaus@desy.de

for the TESLA-N Study Group[a]

Measurements of polarized electron-nucleon scattering can be realized at the TESLA linear collider facility with projected luminosities that are about two orders of magnitude higher than those expected of other experiments at comparable energies. Longitudinally polarized electrons, accelerated as a small fraction of the total current in the e^+ arm of TESLA, can be directed onto a polarized solid state target. A large variety of polarized parton distribution and fragmentation functions can be determined with unprecedented accuracy, many of them for the first time. A main goal is the precise measurement of the x- and Q^2-dependence of the unknown transversity distributions. The additional possibilities of using unpolarized targets and of experiments with a real photon beam turn TESLA-N into a versatile next-generation facility at the intersection of particle and nuclear physics.

1 Introduction

Measurements of polarized deep-inelastic scattering (DIS) were up to now mostly performed with longitudinally polarized nucleons. This is characterized through the helicity distributions $\Delta q(x, Q^2)$, where q is the quark flavor, x is the 'Bjorken-variable' that is the fraction of the nucleon momentum carried by the interacting parton and Q^2 is the virtuality of the exchanged photon. However, of equal importance for a complete understanding of the spin structure of the nucleon are the hitherto unmeasured transversity distributions $\delta q(x, Q^2)$, which can only be measured with transversely polarized nucleons.

First results on transversity distributions can be expected from HERMES [1] and COMPASS [2] within 3-5 years from now, while a complete high precision mapping of their Q^2- and x-dependence requires high statistics measurements that are beyond the scope of presently or soon running experiments. Therefore, there exists an increasing conviction in the necessity of a new facility to study polarized lepton-nucleon/nucleus scattering with very high luminosity and a high enough center-of-mass energy to cover a sufficient kinematic domain. This document outlines the TESLA-N project [3], which would use one arm of the TESLA linear collider for this purpose.

[a] http://www-zeuthen.desy.de/hermes/future

2 Physics Prospects

The HERMES results that have emerged over the recent past are demonstrating the richness of polarized electron-nucleon physics. The higher energy and the much higher luminosity of TESLA-N are expected to again significantly enlarge the number and variety of observable effects as well as the precision with which they can be studied. Only a few examples can be discussed in the following.

2.1 Transversity Distributions

The transversity distributions $\delta q(x, Q^2)$ are not accessible in inclusive measurements, because they are chirally odd and only occur in combinations with other chirally odd objects. In semi-inclusive DIS of unpolarized leptons off transversely polarized nucleons several methods have been proposed to access $\delta q(x, Q^2)$ via specific single target-spin asymmetries. The experimentally most direct access is to extract an asymmetry that involves $\delta q(x, Q^2)$ in combination with the chirally odd polarized fragmentation function $H_1^{\perp(1)}(z)$. This fragmentation function correlates the transverse spin of the struck quark with the azimuthal distribution of the produced single hadron [4,5,6].

The projected statistical accuracy for the measurement of the (x, Q^2)-dependence of the u_v-quark transversity distribution at TESLA-N is shown in Figure 1. A broad range of $0.003 < x < 0.7$ can be accessed in conjunction with $1 < Q^2 < 100$ GeV^2, with an impressive statistical accuracy over almost the full range.

2.2 Polarized Gluon Distribution Function

The polarized gluon distribution $\Delta G(x, Q^2)$ in the nucleon is essentially unknown as of today. For polarized DIS at TESLA-N energies the most promising approaches to determine $\Delta G(x, Q^2)$ are the analysis of pairs of high-$p_\perp$ hadrons [7], and open charm production [8], in semi-inclusive measurements.

A first indication for the sign and approximate size of $\Delta G(x)$ has already been provided by HERMES [9]. However, due to the limited c.m. energy the systematic 'theoretical' errors are large. In contrast, for the considerably higher energies of COMPASS and TESLA-N these problems should be tractable. The anticipated COMPASS results will provide very valuable information, but a high precision experiment like TESLA-N with at least 50 times more statistics (cf. section 3.4) over a wide range of x is eventually needed for a reliable result. At RHIC, the determination of $\Delta G(x)$ via direct photon and heavy quark

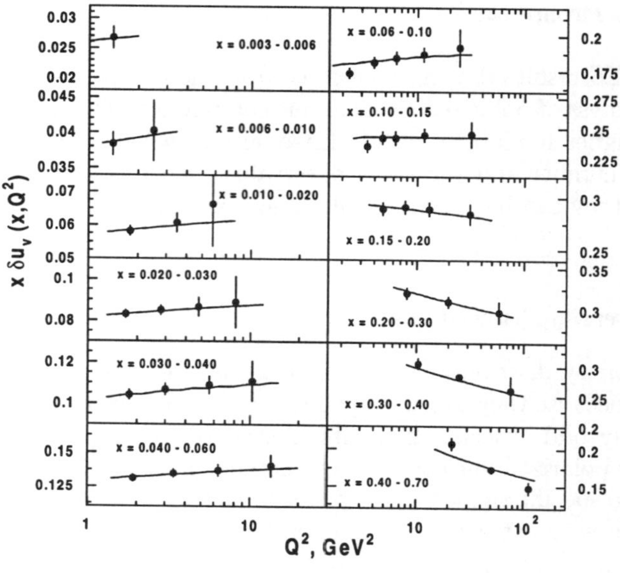

Figure 1: The valence u-quark transversity distribution as a function of x and Q^2 as it would be measured at TESLA-N, based on an integrated luminosity of 100 fb^{-1}. This represents a conservative estimate for one year of data taking (cf. section 3.4). The curves show the LO Q^2-evolution of the u_v-quark transversity distribution obtained with a fit to the simulated asymmetries.

pair production is also not without systematic problems. An independent determination in lepton-nucleon scattering is required.

2.3 Precision Measurements of $g_1(x, Q^2)$ and $g_2(x, Q^2)$

The luminosity and kinematic range available at TESLA-N will allow the determination of the polarized structure functions $g_1(x, Q^2)$ and $g_2(x, Q^2)$ through inclusive measurements with unprecedented accuracy. The structure function $g_1(x, Q^2)$ represents a particular combination of the longitudinal quark spin distributions Δu, Δd and Δs and the corresponding antiquark distributions. The anticipated precision in conjunction with the wide kinematic range will allow studies that so far have not been possible. One example is the determination of $\Delta G(x, Q^2)$ from NLO fits which provides an alternative (indirect) method to the above described measurement (cf. section 2.2). The measurements of $g_2(x, Q^2)$ obtained so far[10] will be improved by TESLA-N, extending the x range down to values of $5{\cdot}10^{-3}$.

3 Layout of the Experiment

3.1 Polarized Target

One of the main ingredients of the TESLA-N apparatus is the polarized target. To reach the required high luminosity with a small fraction (20 nA) of the total TESLA current (45 μA) a polarized solid state target of about 1 g/cm^2 areal density will be appropriate.

The system will consist of a ^{4}He evaporator cryostat, a 5 T Helmholtz-type magnet and a 140 GHz microwave system for permanent Dynamic Nuclear Polarization. The polarization is measured by Nuclear Magnetic Resonance. NH_3 and ^{6}LiD have been chosen as target materials because of their low dilution and their good resistance against radiation damage.

3.2 Polarized Electron Beam

The electrons for TESLA-N will be accelerated together with the positrons in the north arm of the TESLA main accelerator. This 'opposite charge option' was chosen in order to realize a separation between the beam for eN physics and the main beam for e^+e^- physics by a static magnet system with a length of about 150 m, located upstream of the separation building. A separate electron gun system is required at the north end of the TESLA machine. It is envisaged to use a laser driven strained GaAs SLAC-type gun that must be made capable to deliver 20k highly polarized electrons per 0.77 ns. It must be followed by a separate preaccelerator. The beam energy initially will be 250 GeV; energies up to 500 GeV may be possible in a later phase of TESLA. In addition, lower beam energies (25 – 50 GeV) are possible with a second magnetic separation system. Together with a recoil detector [11] this would open up the field of exclusive measurements.

3.3 Overview of the Apparatus

In a fixed-target electron-nucleon scattering experiment at 250 GeV, acceptable resolutions in particle momentum and scattering angle may only be achieved by using a multi-stage spectrometer. A schematic sketch of a possible TESLA-N apparatus is shown in Figure 2. All three stages of the spectrometer will use large dipole magnets (SM1-3) for momentum analysis. The electron-hadron separation will be provided by combinations of transition radiation detectors (TRD1/2) with electromagnetic calorimeters (ECAL1/2/3), complemented by ring-imaging Cerenkov detectors for hadron identification (RICH1/2). Both ECAL1 and ECAL2 must not cover the entrance cone to the next spectrometer

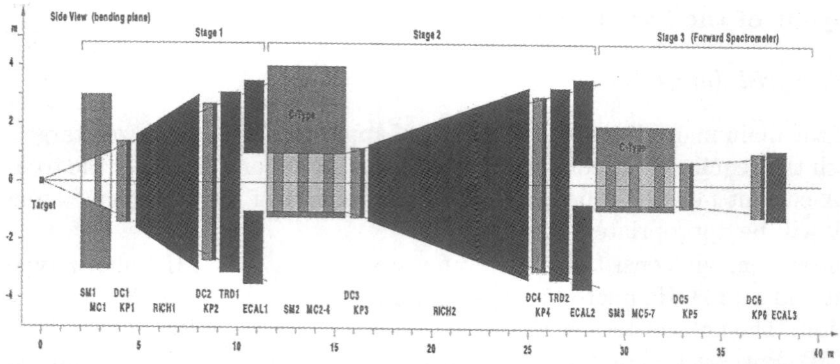

Figure 2: Schematic side view of the envisaged TESLA-N apparatus. For an explanation of the acronyms see the text.

stage. Due to the very high fluxes it is planned to combine fast tracking detectors (e.g. scintillating fibres), so-called 'key planes' (KP1-6), with precise tracking detectors, e.g. drift chambers (DC1-6).

About 20% of the incoming electrons will cause bremsstrahlung at a target thickness of 1 g/cm^3. Due to the magnetic deflection, the resulting lower momentum electrons form a 'sheet of flame' on their way down the spectrometer. The electrons and the bremsstrahlung photons must not meet any material in their way to avoid background showers. The safest way to ensure this is to provide an extended vacuum vessel. This vessel forms a narrow 'slit' whose height (in the bending plane) is increasing along the spectrometer, while its width can be as low as $\pm$2 cm. The dipole fields in SM1-3 should be oriented horizontally and the vacuum vessel extends vertically down from the beamline. As a further consequence, SM2 and SM3 will be C-type magnets. The integrated magnetic fields will be 2 Tm for SM1 and 5 Tm each for SM2 and SM3.

The acceptance of the TESLA-N spectrometer for leading hadrons as a function of $z = E_h/\nu$ is shown in Figure 3. More than 80% of all leading hadrons are detected in the spectrometer and about 40% of them (for $z \geq 0.15$) are detected both in Stage 1 and 2. As a result, these hadrons are detected with good momentum resolution independently of the vertex reconstruction. Scattered electrons with high Q^2 (>10 GeV^2) are predominantly detected in Stage 2, while low-Q^2 electrons (<10 GeV^2) are detected in Stage 3. The opening of SM1 limits the acceptance to $\theta_x \leq 225$ mrad and $\theta_y \leq 280$ mrad. The lowest possible detection angle θ_{min} is about 5 mrad for momentum analysis within Stage 2 including the detection plane in front of SM 2. If Stage 3 is

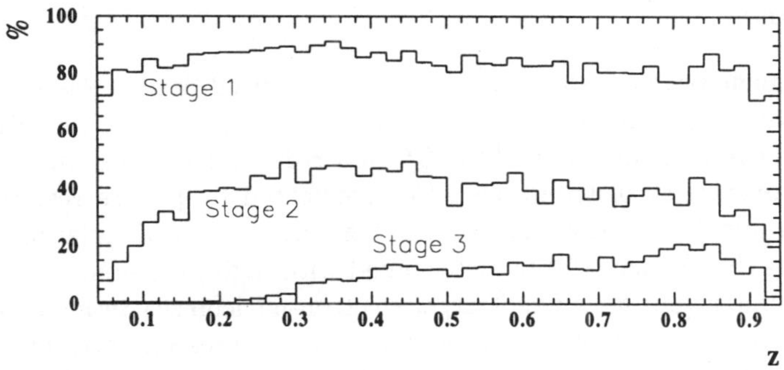

Figure 3: Acceptance of the TESLA-N spectrometer for the leading hadron.

used including the detection plane in front of SM 3, θ_{min} can be reduced to values as low as 2-3 mrad. These figures are based on a width of the vacuum vessel of $\pm$ 2 cm and a width of the DC frame next to it of 3 cm. For electrons, θ_{min} directly determines the lowest reachable Bjorken-x.

3.4 Luminosity

An areal target density of 1 g/cm^2 of polarizable material that is hit every 0.77 ns by bunches carrying 20k electrons each, leads to a maximum possible luminosity of 12 mb^{-1} per bunch or $7.5 \cdot 10^{34}$ cm^{-2} s^{-1}.

Adopting a conservative ansatz for efficiencies, namely a combined up-time of accelerator and experiment of 0.33 in conjunction with an efficiency of the experiment of 0.75, results in the conservative overall efficiency factor of 0.25. This factor leads to maximum achievable integrated luminosities for TESLA-N of 600 fb^{-1} per effective year.

In certain areas more work has to be invested to solidify the assumptions made above:

i) It is presently assumed that a time resolution of 0.77 ns can be realized in the future for the fast tracking detectors at TESLA-N. As it was proven recently, today's technology already allows to reach 2 ns [12]. In a conservative approach a beam current lower by a factor of 3 would have to be assumed.

ii) Reducing the beam current by another factor of 2 would have to be assumed to keep the multiple event fraction in a DIS event small.

Altogether, 100 fb^{-1} per effective year is the conservative integrated luminosity of TESLA-N. This is still a factor of 50 above the maximum achievable integrated luminosity of 2 fb^{-1}, presently planned for one effective year of COMPASS running with the same overall efficiency factor of 0.25.

4 Summary

This document presents the prospects for a polarized deep-inelastic electron-nucleon scattering experiment at the TESLA facility. For the first time, this will allow a complete mapping of the Q^2- and x-dependence of both the helicity and the transversity distributions. Complemented by precise results on the polarized gluon distribution most of the components of the angular momentum structure of the nucleon will be determined with high precision.

A possible layout for a fixed-target electron-nucleon scattering experiment at TESLA is presented as well. A separate hall is foreseen north of the main e^+e^--interaction point. First design considerations for a polarized target and a three-stage spectrometer are discussed. It is concluded that the experiment is technically feasible, highly competitive and very cost-effective, although many aspects of the design require further study. It will be unique as it combines high luminosity with large center-of-mass energies, using highly polarized electron beams and targets. The possibilities of using unpolarized targets and of experiments with a real photon beam turn TESLA-N into a versatile next-generation facility at the intersection of particle and nuclear physics.

References

1. HERMES Collaboration, Hermes internal note 00–003, DESY, 2000.
2. COMPASS Collaboration, CERN/SPSLC 96–14 (1996).
3. M. Anselmino et al., *Electron Scattering with Polarized Targets at TESLA*, hep–ph/0011299, DESY 00–160, 2000.
4. J.C. Collins, *Nucl. Phys.* **B396**, 161 (1993);
   J.C. Collins, S.F. Heppelmann, G.A. Ladinsky, *Nucl. Phys.* **B420**, 565 (1994).
5. A.M. Kotzinian, P.J. Mulders, *Phys. Lett.* **B406**, 373 (1997).
6. P.J. Mulders, R.D. Tangermann, *Nucl. Phys.* **B461**, 197 (1996);
   D. Boer and P.J. Mulders, *Phys. Rev.* **D57**, 5780 (1998).
7. A. Bravar, D. v.Harrach, A. Kotzinian, *Phys. Lett.* **B421**, 349 (1998).
8. M. Glück, E. Reya, W. Vogelsang, *Nucl. Phys.* **B351**, 579 (1991).
9. HERMES Coll., A. Airapetian et al., *Phys. Rev. Lett.* **84**, 2584 (2000).
10. E155x Coll., P. Bosted et al., *Nucl. Phys.* **A663–664** 297 (2000).
11. J.J.M. Steijger, G. v.d. Steenhoven, NIKHEF 2000-013.
12. D. von Harrach, private communication.

PARTICLE PRODUCTION AND MULTIPLICITY IN HEAVY ION COLLISIONS AT PHENIX

LARS EWELL(BNL)
for the PHENIX Collaboration
E-mail: ewell@bnl.gov

In June of 2000, the PHENIX detector at the Relativistic Heavy Ion Collider (RHIC) accelerator began to acquire data from $\sqrt{s_{NN}} = 130GeV$ Au on Au collisions. These data will facilitate the investigation of hot dense matter similar to that thought to have existed ~ 10 μs after the Big Bang. An important observable of these collisions is the charged particle multiplicity. The PHENIX collaboration has measured the charged particle multiplicity as a function of centrality for central to peripheral collisions. This dependency, as well as comparisons to earlier measurements, will be discussed.

1 Introduction

1.1 Detector

PHENIX[1] is one of the two large detectors at the RHIC accelerator at the Brookhaven National Laboratory. It was designed with the goal of detecting photons and leptons from the collisions. It consists of eleven different subsystems. The subsystems that are used in this analysis are the trigger detectors, the Zero Degree Calorimeters and the Beam Beam Counters, and the Pad Chambers. The way in which these detectors were used to measure the particle multiplicity is explained below.

1.2 Particle Multiplicity

The total particle multiplicity in a typical $\sqrt{s_{NN}} = 130GeV$ Au on Au central collision is on the order of 5,000[2] with the bulk ($> 80\%$) of the particles being pions. How the charged particle multiplicity varies with pseudorapidity η ($\eta = -ln(tan(\theta/2))$, with θ the polar angle measured from the beam axis) is expressed in the global variable, $\frac{dN_{ch}}{d\eta}$. This variable is a favorite of study since it reveals information regarding the energy density of the collision as well as the initial conditions of the collision environment. In addition, the well known heavy ion models such as HIJING[3] provide predictions against which the data can be judged. Finally, other models such as EKRT[4] which invoke gluon saturation to describe the centrality dependence also provide a prediction for comparison.

2 Heavy Ion Collisions vs Particle Collisions

2.1 e^+e^- *vs.* $p\,\overline{p}$

It is instructive to review particle collisions before discussing heavy ion collisions. In e^+e^- annihilation ($\sqrt{s} \gtrsim 5\text{GeV}$), the collision is dominated by hard processes which involve large momentum transfers. In $p\,\overline{p}$ collisions, there is a mixture of hard and soft with the total cross section being dominated by soft processes. As the collision energy increases, hard processes become more important.

2.2 Heavy Ions, Au on Au

In heavy ion collisions, the colliding projectiles have enough extent that the collision can be described in terms of 'spectators' and 'participants'. This description is illustrated in Figure 1.

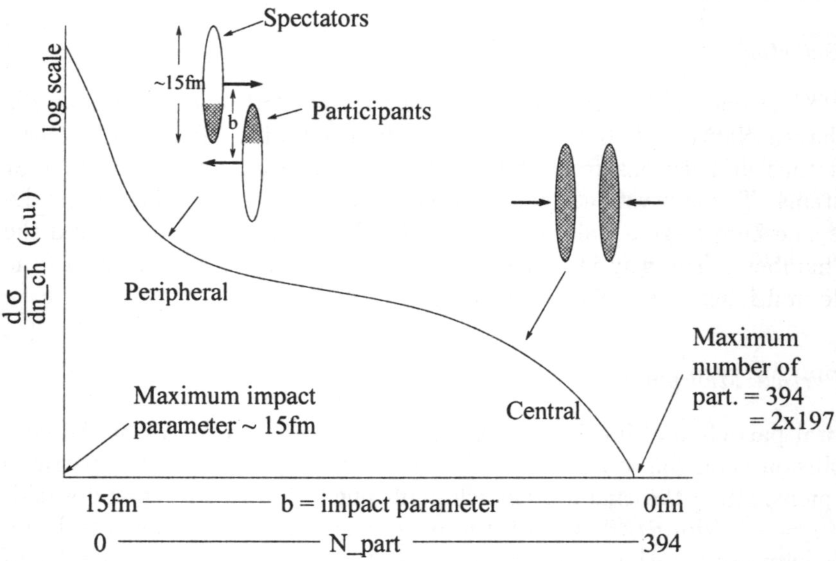

Figure 1: Diagram illustrating difference between peripheral (left) and central (right) heavy ion collisions.

In this figure, a semi-peripheral collision can be seen in the upper left-hand part. We can see that a large fraction of the nucleons within the nuclei are just 'spectators' (clear/white) and only a small fraction of the nucleons are 'participants' (cross hatched). Conversely, we see a much more central collision on the right side of the diagram. Here almost all of the nucleons within the nuclei are 'participants'.

From the horizontal axis, we see that the number of participants is inversely proportional to the impact parameter. From the vertical axis, we see that the cross section for a peripheral collision with few charged particles produced is much larger than the cross section for a central collision with the production of many charged particles.

It is important to remember that the mixture of hard and soft processes discussed briefly above needs to be folded into this (spectator - participant) picture when one makes predictions regarding global observables. Due in part to the unknown interplay between these processes, a wide range of theoretical estimates emerges from various different models as will be discussed below.

2.3 Glauber Model

To describe the dynamics of nuclear collisions, the Glauber model[5] is used. In this model a number of simplifying assumptions are made, among them:

1) The nucleons within colliding nuclei are assumed to travel in a straight line trajectory (no curvature).
2) The nucleon - nucleon cross section is assumed to be constant (independent of prior collisions), $\sigma_{nn} = 40 \pm 5mb@\sqrt{s_{NN}} = 130GeV$.

In addition to these Glauber Model assumptions, the Woods-Saxon distribution is used to describe the nucleon density within the nucleus.

3 Prior Measurements

3.1 $p - \overline{p}$

Charged particle multiplicity has a long history and was measured in $p\,\overline{p}$ collisions by, e.g., the UA5 collaboration in the early 1980's at CERN[6]. $p\,\overline{p}$ results are of primary importance when making model comparisons since many of the most used heavy ion multiplicity predictors (e.g. HIJING) use $p\,\overline{p}$ collision data as input and/or constraints. They found that the central ($\eta = 0$) charged particle rapidity density, $\rho_n(0)$, scaled with energy as a power law. i.e.

$$\rho_n(0) = \frac{1}{\sigma}\frac{d\sigma}{d\eta}|_{\eta=0} = (0.74 \pm 0.04)s^{0.105\pm0.006}.$$

3.2 Pb-Pb

In addition to $p\,\overline{p}$ data, charged particle multiplicity has been measured in Pb-Pb collisions by, e.g., the NA49 collaboration again at CERN[7]. The beam energy was 158 · A GeV corresponding to $\sqrt{s_{NN}}$=17GeV for this fixed target configuration. They found that the charged particle pseudorapidity density scaled as $\frac{dN_{ch}}{d\eta} \propto N_{part}^{\alpha}$ with $\alpha = 1.07 \pm 0.04$.

4 Current Measurements

4.1 PHENIX Trigger Detectors

As mentioned above, the PHENIX detector uses a combination of the Zero Degree Calorimeters (ZDCs) and Beam Beam Counters (BBCs) to trigger on events. In addition to triggering, these detectors allow for a measure of centrality. In Figure 2, a schematic of these detectors and the data derived from them is shown.

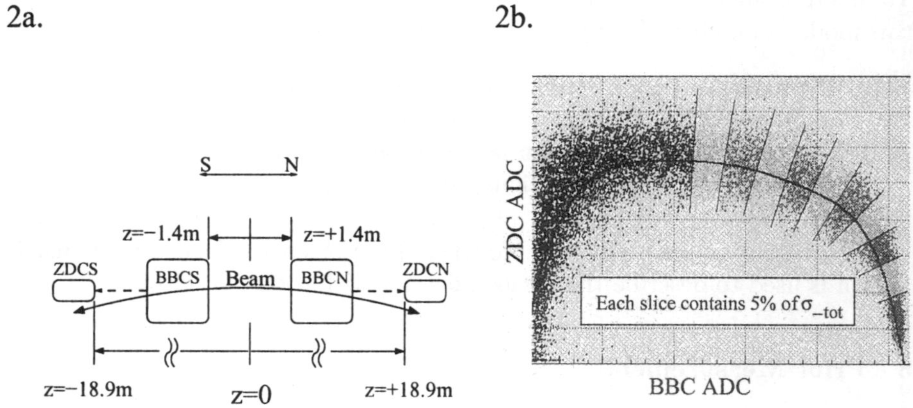

Figure 2: PHENIX Trigger Detectors - 2a. Centrality Data - 2b.

In Figure 2a, we see the ZDCs located at ±18.9m (south,north) from the collision vertex and the BBCs located at ±1.4m. If a Au Au collision occurs at z=0, the charged particles produced trigger a coincident signal between the north and south BBCs and the north and south ZDCs. If the collision is very central, many charged particles are produced which results in a large BBC signal, and there are few spectators. Since the ZDCs are sensitive only to neutral particles (neutrons) not swept away by the Dx sweeping magnets (not shown in Figure 2a, but located between the ZDCs and BBCs), the signal they see , which is proportional to the number of spectators, is small in such an event.

Therefore, the most central events correspond to the bottom right part of Figure 2b. As an event gets more peripheral, the signal in the BBCs gets smaller while the ZDCs see a maximum signal around mid centrality. Finally, in very peripheral events both the ZDCs and the BBCs see small signals so that these events correspond to the bottom left part of the figure. In this way, the centrality of an event is defined in traversing Figure 2b in a counterclockwise fashion from lower right to lower left.

4.2 PHENIX Pad Chambers

The PHENIX detector subsystems used in this analysis of charged particle multiplicity are the Pad Chambers (PCs). The PCs are gas detectors sensitive to charged particles. If a charged particle traverses the PC, a hit is registered in the two dimensional array of cells that make up the PC. By combining all of the hits from PC1 (r=2.5m) and PC3 (r=5.0m), a number of tracks (two points each) are reconstructed. When these tracks are projected back to the collision area, a peak is extracted which defines the collision vertex. This is depicted schematically in Figure 3a.

3a. 3b.

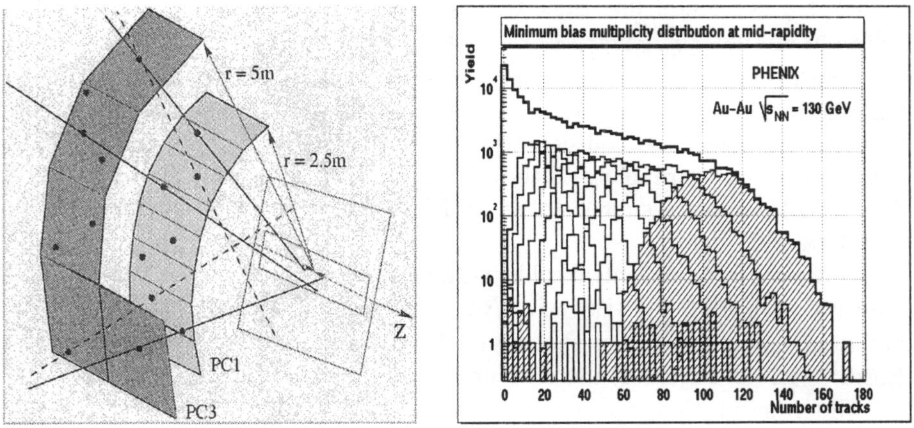

Figure 3: PHENIX Multiplicity Detectors PC1 and PC3 - 3a, Multiplicity Data - 3b

In that figure, we see three real tracks (solid lines) and two random tracks (dotted lines). To determine the total multiplicity, the total number of reconstructed tracks are counted. To determine the combinatorial background the

PC sectors are interchanged, thereby generating a sample of completely random tracks, and the track reconstruction is repeated. After the background is subtracted and other corrections, like limited detector acceptance, are applied[8], the total charged particle multiplicity is determined. This quantity is plotted for different centralities in Figure 3b.

In this figure, we see the number of tracks (charged particles) plotted. The shape of this curve should be compared to the shape of the curve shown in Figure 1. As can be seen in this comparison, the central collisions consist of those events with the largest number of tracks while the peripheral events generate fewer tracks. The cross hatched area represents the 5% most central events (see Figure 2b, bottom right) with the remaining 5% slices of centralities (traversing Figure 2b counter-clockwise from lower right) represented by sucessive regions to the left.

4.3 Multiplicity vs. $N_{participants}$ vs. Different Models

In Figure 4, we see plotted the quantity $\frac{dN_{ch}}{d\eta}$, normalized to the number of participant pairs ($0.5N_{part}$).

When comparing the different models to the data, a good place to start out is with the 'Wounded Nucleon' (WN) model. As can be seen, the WN model goes right through the UA5 $p\overline{p}$ point as it must. Furthermore, the WN model is a horizontal line since it represents a straight superposition of p-p collisions so that $\frac{dN_{ch}}{d\eta(0.5N_{part})}$ stays constant with N_{part}.

At the most central collisions, we see the single PHOBOS data point[9], followed by the PHENIX data at different centralities. It is interesting to note that in very peripheral collisions (low number of participants), the PHENIX data appears to extrapolate to the UA5 data point indicating that peripheral Au Au collisions are similar to $p\overline{p}$ collisions with regard to charged particle pseudorapidity density.

However, in more central collisions a clear deviation from the WN model is observed as $\frac{dN_{ch}}{d\eta(0.5N_{part})}$ rises with N_{part}. This trend is predicted by HIJING, but at a lower value than the data. A recent prediction by Kharzeev and Nardi[10] (KN in Figure 4) more accurately describes the data.

Gluon saturation models [4] predict a flat or falling relationship between $\frac{dN_{ch}}{d\eta(0.5N_{part})}$ and N_{part} which is not observed.

4.4 Comparison to Previous Measurements

If a comparison of the most central bin (0-5%) in the PHENIX data is made with the fixed target experiment NA49 discussed above, the values are $\frac{dN_{ch}}{d\eta}|_{\eta=0}$

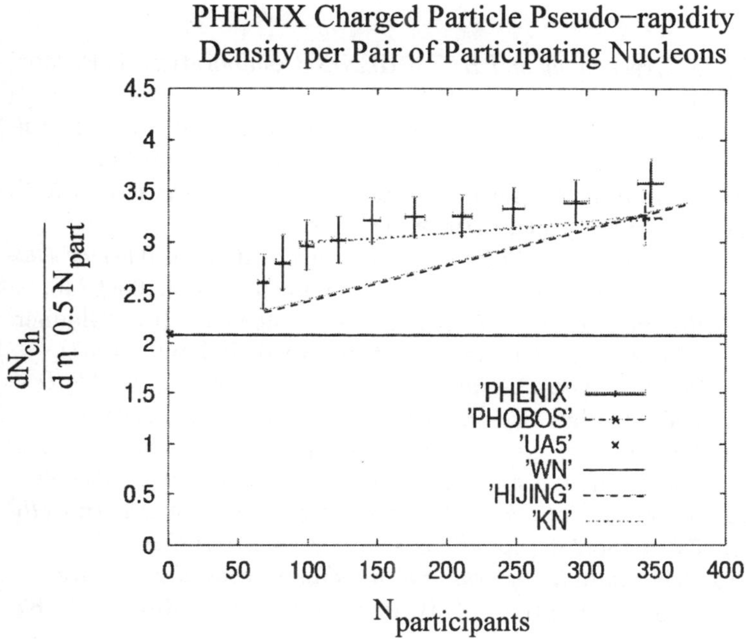

Figure 4: $\frac{dN_{ch}}{d\eta(0.5N_{part})}$ vs. N_{part} vs. different models

$= 622$ vs $\frac{dN_{ch}}{d\eta}|_{\eta=0} = 369$ for NA49 or an increase of $\sim 68\%$.

Although the increase in the most central bin is substantial, a more modest, but significant, increase in the *scaling* with N_{part} is observed. As mentioned above, NA49 found $\frac{dN_{ch}}{d\eta} \propto N_{part}^{\alpha}$ with $\alpha = 1.07 \pm 0.04$. If the same fit is attempted with the PHENIX data, a value of $\alpha = 1.16 \pm 0.04$ is obtained. In particular, the new PHENIX scale value is clearly larger than unity.

5 Conclusion

The PHENIX experiment at the RHIC accelerator has measured the charged particle multiplicity as a function of centrality for $\sqrt{s_{NN}} = 130GeV$ Au on Au collisions. Compared with the previous highest energy measurement, a substantial increase ($\sim 68\%$) of the most central bin is observed. The increase in scaling with N_{part} is more modest, but significant. A clear deviation from the 'Wounded Nucleon' model is observed as $\frac{dN_{ch}}{d\eta(0.5N_{part})}$ is found to increase with N_{part}.

References

1. *The PHENIX Experiment at RHIC*, PHENIX Collaboration, D.P. Morrison, Nucl. Phys. **A638**, 565c (1998).
2. *Charged-particle pseudorapidity density distributions from Au + Au collisions at $\sqrt{s_{NN}}$ = 130 GeV*, B.B. Back et al., nucl-ex/0106006.
3. *Energy and Centrality Dependence of Rapidity Densities at RHIC*, X.N. Wang and M. Gyulassy, nucl-th/0008014.
4. *Centrality Dependence of Multiplicities in Ultrarelativistic Nuclear Collisions*, K.J. Eskola, K. Kajantie and K. Tuominen, hep-ph/0009246.
5. *Multiple Diffraction Theory of High-Energy Collisions*, R.J. Glauber, High Energy Physics and Nuclear Structure, North-Holland, (1967).
6. *Scaling of Pseudorapidity Distributions at c.m. Energies up to 0.9 TeV*, C.J. Alner et al., Z. Phys. C - Particles and Fields **33**, 1-6 (1986).
7. *Hadron production in nuclear collisions from the NA49 experiment at 158GeV/c·A*, J. Bachler et al., Nuclear Physics **A661**, (1999) 45c-54c.
8. *Centrality Dependence of Charged Particle Multiplicity in Au-Au Collisions at* $\sqrt{s_{NN}} = 130GeV$, K. Adcox et al., nucl-ex/0012008.
9. *Charged particle multiplicity near mid-rapidity in central Au + Au collisions $\sqrt{s}$ = 56 and 130AGeV*, B.B. Back et al., Phys. Rev. Lett. **87**, 3100 (2000).
10. *Hadron production in nuclear collisions at RHIC and high density QCD*, D. Kharzeev and M. Nardi, nucl-th/0012025.

RECENT QCD RESULTS FROM DØ AT THE TEVATRON

E.J. GALLAS
for the DØ Collaboration
Fermi National Accelerator Laboratory, Batavia, Illinois 60510, USA
E-mail: eggs@fnal.gov

The DØ experiment at the Tevatron proton-antiproton collider accumulated a large sample of high energy jet production data during Run I (1992-1996). Presented here are measurements of the inclusive jet cross section at a center-of-mass energy $\sqrt{s} = 1800$ GeV, and a ratio of central inclusive jet cross sections at two different center-of-mass energies (1800 and 630 GeV). Also included is a measure of the ratio of multijet production cross sections at the higher $\sqrt{s}$. All measurements are compared to next-to-leading order QCD predictions with recent parton distribution functions (PDFs). Due to decreased statistical and systematic errors in the measurement, comparison with the theory shows that the prediction may benefit from an increased order in the calculation and from inclusion of proton-antiproton jet data in new global PDF fits.

1 Introduction

The Tevatron collider program completed its first run of data taking in early 1996. The large sample collected by the DØ experiment enables a number of jet measurements which can be compared to predictions from Standard Model QCD. Measurements included in this presentation are:

- The inclusive jet cross section at a proton-antiproton center-of-mass energy $\sqrt{s} = 1800$ GeV, extending previous measurements into the forward region of pseudorapidity ($|\eta| < 3.0$).
- The ratio of this cross section to that at a lower energy ($\sqrt{s} = 630 GeV$) in the central region of pseudorapidity ($|\eta| < 0.5$).
- The ratio of inclusive three-jet to inclusive two-jet production cross sections (R_{32}) at the higher $\sqrt{s} = 1800$ GeV.

These measurements are compared to next-to-leading order (NLO) QCD predictions of JETRAD [1] using a number of recent parton distribution functions [2,3,4]. Further details of these results may be found in the References [5,6,7].

2 The Event Samples

All jet measurements described here use primarily calorimetry to identify jets and measure their energy and orientation. Jets are reconstructed using an iterative algorithm with a fixed cone size of radius $R = 0.7$ in $\eta - \phi$ space.

The pseudorapidity is defined as $\eta = -\ln[\tan\frac{\theta}{2}]$, where θ is the angle of the jet relative to the incoming proton beam. The angle ϕ is the azimuthal angle about the beam axis. Vertex, jet and event quality criteria reduce backgrounds caused by electrons, photons, noise and cosmic rays. Additional corrections are applied to obtain jet energies corrected for calorimeter response, noise, showering outside of the cone radius, and energy deposits from spectator interactions. For the inclusive cross sections, unsmearing corrections are applied to remove the effect of a finite jet energy resolution.

The measurements of inclusive cross section at the higher energy ($\sqrt{s} = 1800$ GeV) used an integrated luminosity of nearly 100 pb^{-1} of data. An integrated luminosity of 537 nb^{-1} was available for the inclusive cross section at the lower energy ($\sqrt{s} = 630$ GeV). About 10 pb^{-1} was selected to measure the multijet ratio to reduce systematic uncertainties in the energy scale at higher luminosities. Details of the jet finding algorithm, jet reconstruction, event selection, and measurements can be found through the References [5,6,7].

Jets are the manifestation of partons produced in the primary parton-parton collision. Partons in the NLO calculation may have any angular separation, while the cone of fixed size used to experimentally reconstruct jets may envelop more than one NLO parton. Therefore, partons in the NLO calculations are required to be separated by an angular difference of $\Delta R_{sep} = 1.3 \times R$ in order to be counted as distinct jets.

A χ^2 statistic including both statistical and systematic uncertainties in the theory and data is used in all cases to quantitatively assess the agreement of the predictions with the measurements.

3 Inclusive Jet Cross Section

Data recorded by the DØ detector in Run I yields a high statistics measurement of the inclusive jet cross section over a wide range in jet transverse energy (E_T), where the cross section spans seven orders of magnitude (Figure 1). This new measurement represents the lowest systematic uncertainties to date and includes new data reaching far into the forward regions of pseudorapidity.

An increased cross section at high jet E_T could indicate quark compositeness and/or new physics beyond the Standard Model. Comparison of these cross sections to QCD predictions indicates how well we understand the proton structure: the adequacy of NLO predictions and sensitivity of the predictions to current PDF's (parton distribution functions).

A χ^2 comparison of the measurements to the theory shows that the theory is consistent with the measurements for many of the modern PDF parameterizations, such as CTEQ4M and MRSTgU, but finds other PDF's such as

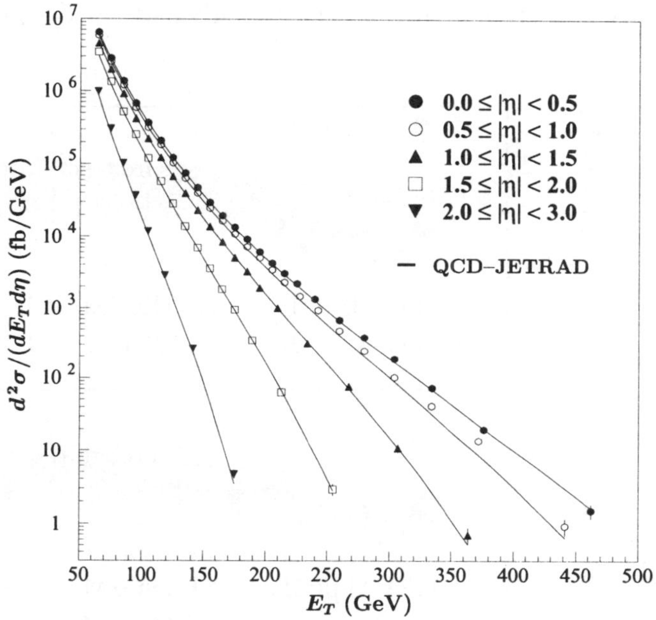

Figure 1: The single inclusive jet production cross section as a function of jet E_T, in five pseudorapidity intervals, showing only statistical uncertainties, along with theoretical predictions.

CTEQ3M, MRST, and MRSTgD unlikely. A good agreement is seen with the newer CTEQ4HJ PDF, which was contrived using high E_T jet Tevatron collider data, suggesting that the theory should incorporate a higher gluon content at high x than previous PDF's have included. The new jet data is being incorporated in global PDF fits by MRST and CTEQ groups.

4 Ratio of Cross Sections

DØ also recorded data at a center-of-mass energy of 630 GeV during Run I. A dimensionless ratio provides a more stringent test of NLO QCD because of the substantial reduction in both experimental and theoretical uncertainties when taking a ratio. More specifically, we measure the ratio of $R = \sigma_s^{630}/\sigma_s^{1800}$ as a function of x_T, where σ_s is the scale invariant cross section: $\sigma_s = (E_T^3/2\pi)(d^2\sigma/dE_Td\eta)$ and x_T is the jet transverse momentum fraction: $x_T = 2E_T/\sqrt{s}$.

Shown in Figure 2 is the ratio of dimensionless cross sections, R, compared to NLO QCD as given by JETRAD. In the naive parton model, this ratio is flat

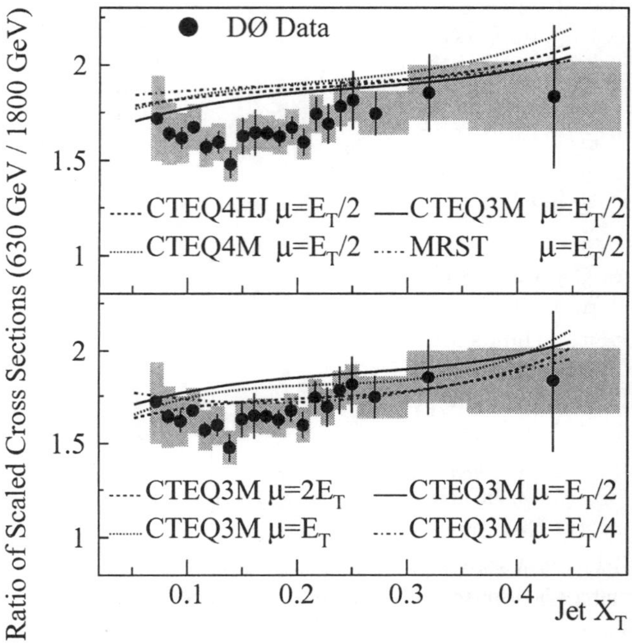

Figure 2: Ratio of dimensionless cross sections (numerator $\sqrt{s}$ = 630 GeV, denominator $\sqrt{s}$ = 1800 GeV) compared to NLO QCD as given by JETRAD.

in x_T. Deviations from this model result from scaling evolution (of both PDF's and α_s) and terms beyond LO. It is interesting to note that as predicted by the theory and verified by experiment, the ratio is greater than 1, expressing that the probability of jet emission is higher for the lower center-of-mass energy at the same x_T (intuitively, emission of lower E_T jets is more likely).

In the χ^2 comparison, NLO predictions for the ratio exhibit satisfactory agreement with the shape of the observed ratio. The choice of PDF has little effect on the result. In terms of only the normalization however, the absolute values of the predictions lie significantly higher than the data, especially for the standard scale ($\mu_R = E_T^{max}/2$). To summarize, further corrections to the NLO calculation seem to be required to reduce the scale dependence and improve the predictive power of the calculation.

5 R_{32}

The ratio of inclusive three-jet to inclusive two-jet production reflects the rate of gluon emission in QCD jet production processes. A three jet cross section explicitly offers the opportunity to investigate scale differences at a secondary vertex (a subsequent gluon emission). Taking the ratio reduces systematic uncertainties.

We measure the ratio

$$R_{32} = \frac{\sigma_3}{\sigma_2} = \frac{\sigma(p\bar{p} \to n \text{ jets} + X; n \geq 3)}{\sigma(p\bar{p} \to m \text{ jets} + X; m \geq 2)}$$

as a function of the scalar sum of jet transverse energies ($H_T = \sum E_T^{\text{jet}}$). The measurement is performed with four distinct sets of selection criteria for all jets in the event: E_T thresholds of 20, 30, or 40 GeV for $|\eta_{\text{jet}}| < 3$, and $E_T > 20$ GeV for $|\eta_{\text{jet}}| < 2$. Three thresholds were chosen to study threshold dependence, and the minimum threshold was chosen to maximize statistics for which jet reconstruction efficiency was nearly 100%. For each threshold, the ratio R_{32} (or roughly the gluon emission rate) rises rapidly below $H_T = 200$ GeV and levels off to about 70%, 60% and 50% as the minimum jet E_T threshold rises from 20 to 30 to 40 GeV, respectively ($|\eta_{\text{jet}}| < 3$).

To study the effect of a different μ_R scale on the production of additional jets, the renormalization scale of the third jet is varied from $\mu_R^{(3)} = \lambda H_T$ (same as for the leading jets) to a scale proportional to the E_T of the third jet $\mu_R^{(3)} \propto E_T^{(3)}$. Also, a scale proportional to the maximum jet transverse energy (E_T^{max}) is studied, as this is the usual form used for comparisons of JETRAD to measured jet cross sections.

Figure 3 shows the measured R_{32} as a function of H_T for jet $E_T > 20$ GeV and $|\eta_{\text{jet}}| < 2$. The 20 GeV threshold has good sensitivity to scale in the JETRAD prediction and has reduced statistical uncertainty. The central rapidity region ($|\eta_{\text{jet}}| < 2$) has the best understood jet energy uncertainties and correlations. Although some of the predictions do not visually overlap with the data, acceptable agreement is found for some scales because of the strong point-to-point correlations of the data uncertainties which are taken into account in the χ^2 comparison of the various predictions to the data.

Generalizing for all criteria: The scale prescriptions using a softer scale for lower E_T jets did not provide a good fit to the data for any value of λ. The λ independent scale $\mu_R = 0.6E_T^{\text{max}}$ for all jets yielded good agreement with measurement (probability $p > 57\%$) for the $E_T > 20$ GeV criteria, but the χ^2 rises (and the corresponding probabilities decrease) for the higher E_T thresholds (not shown). The JETRAD prediction assuming a scale $\mu_R = \lambda H_T$

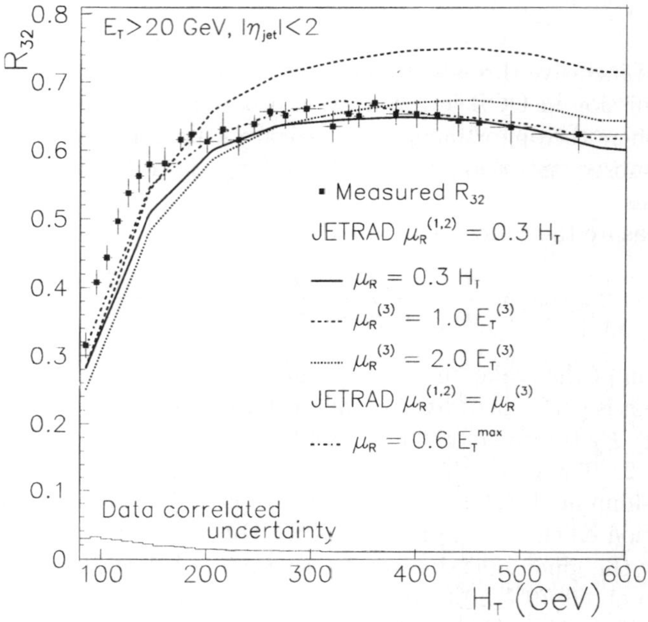

Figure 3: The ratio R_{32} as a function of H_T, requiring jet $E_T > 20$ GeV and $|\eta_{\text{jet}}| < 2$. Error bars indicate statistical and uncorrelated systematic uncertainties, while the histogram at the bottom shows the correlated systematic uncertainty. The four smoothed distributions show the JETRAD prediction for the renormalization scales indicated in the legend.

provides the best description of the data for λ between 0.30 and 0.35 ($p > 80\%$). Moreover, the χ^2 is also minimized in the $\lambda \approx 0.30$ region for the other selection criteria (not shown) making this scale choice the most robust of all the μ_R scales studied.

Although a few scale prescriptions did provide predictions which yielded good fits to this particular measurement, the applicability of those scales in predicting other jet production characteristics is unclear. Reduced scale dependence would improve the predictive power of the theory.

6 Conclusions

It is interesting to note that for Run I jets, overall uncertainties in the measurement are now less than the theoretical uncertainties. In Run II analysis and future calculations, it is therefore important to consider in more detail how comparisons can be improved. In particular, DØ is studying the use of new jet reconstruction algorithms (which are more readily compared to theory) and refining techniques for the further reduction of systematic uncertainties in the analysis of the incoming data of Tevatron Run II. We look forward to improved predictions coordinated with these improved techniques to facilitate a deeper understanding of the data/theory comparisons.

Acknowledgments

We thank David Summers, Dieter Zeppenfeld, and Walter Giele for stimulating and helpful discussions. We thank the staffs at Fermilab and collaborating institutions, and acknowledge support from the Department of Energy and National Science Foundation (USA), Commissariat à L'Energie Atomique and CNRS/Institut National de Physique Nucléaire et de Physique des Particules (France), Ministry for Science and Technology and Ministry for Atomic Energy (Russia), CAPES and CNPq (Brazil), Departments of Atomic Energy and Science and Education (India), Colciencias (Colombia), CONACyT (Mexico), Ministry of Education and KOSEF (Korea), CONICET and UBACyT (Argentina), The Foundation for Fundamental Research on Matter (The Netherlands), PPARC (United Kingdom), Ministry of Education (Czech Republic), and the A.P. Sloan Foundation.

References

1. W.T. Giele, E.W.N. Glover, David A. Kosower, *Nucl. Phys.* **B403**, 633 (1993).
2. H.L. Lai *et al.*, *Phys. Rev.* **D51**, 4763 (1995).
3. A.D. Martin *et al.*, *Eur. Phys. J.* **C4**, 463 (1998).
4. H.L. Lai *et al.*, *Phys. Rev.* **D55**, 1280 (1997).
5. B. Abbott *et al.*, *Phys. Rev. Lett.* **86**, 1707 (2001).
6. B. Abbott *et al.*, *Phys. Rev. Lett.* **86**, 2523 (2001).
7. B. Abbott *et al.*, *Phys. Rev. Lett.* **86**, 1955 (2001).

TWO PHOTON PHYSICS AT LEP

DANIEL HAAS
*University of Basel, Klingelbergstrasse 82,
CH-4056 Basel, Switzerland
E-mail:Daniel.Haas@cern.ch*

LEP offers an excellent opportunity to measure two photon processes over a large kinematical range and thus study the complex nature of the photon. This article reviews the experimental status of "Two Photon Physics" at LEP. The recent results on resonances, multi-hadron and heavy quark production are discussed.

1 Introduction

Over the past decade two photon physics has proven to be a very productive source of information about QED, QCD and hadron spectroscopy. The Feynman diagram responsible for a two photon collision process at LEP is shown in Figure 1, where the high energy incident electrons and positrons split off virtual photons and the scattered electrons take most of the beam energy. These two photons can then interact to form a state X with mass $W_{\gamma\gamma}$. The four-momentum transfer q_i to the photons depends on the angle and energy of the scattered electrons[a].

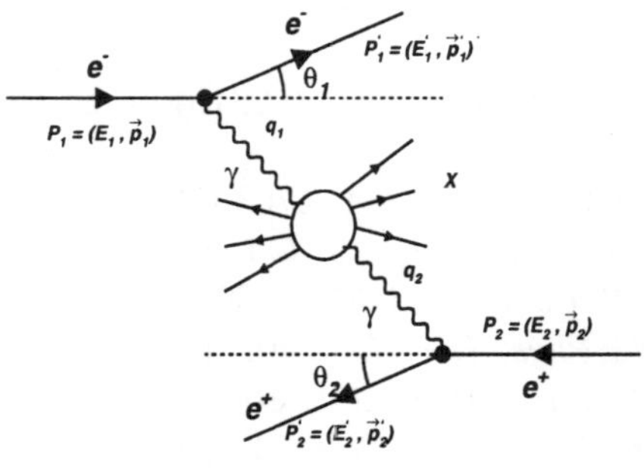

Figure 1: $\gamma\gamma$ collision in e^+e^- scattering

When neither of the scattered electrons is detected (untagged events), the virtual photons are referred to as nearly real i.e. $q_1^2 \approx q_2^2 \approx 0$. This class of events allows several tests of QCD by studying hadronic resonances, the inclusive hadron cross section and jet production rates. If there is detection of

[a]Electron stands for electron and positron throughout this article

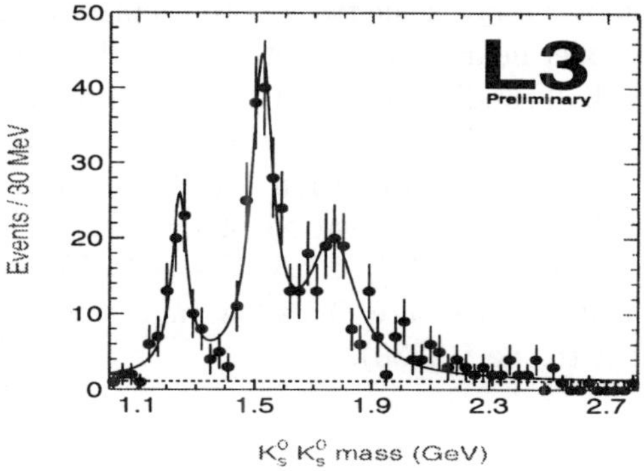

Figure 2: The $K_s^0 K_s^0$ mass spectrum

one of the scattered electrons $Q^2 = -q_1^2$ (single tagged events), it is possible to probe the other photon $q_2^2 \approx 0$ regarded as a "target" and study its structure. Finally, if both the scattered electrons are detected $Q_i^2 = -q_i^2, (i = 1, 2)$ (double tagged events), the structure of the reaction of highly virtual photons is probed. In the following sections, a review is given of the $\gamma\gamma$ results obtained at LEP, with special attention to recent results.

2 Resonance production

Two photon formation of C-even meson resonances provides valuable information on the internal structure of mesons. In particular it is interesting to look for resonances whose $\gamma\gamma$ couplings are much smaller than quark-model predictions; e.g. glueball or hybrid quark-gluon states. At LEP, many exclusive channels are studied. As an example the study of resonances for $K^0 K^0$ final states will be discussed

2.1 $K_s^0 K_s^0$ Resonances and GlueBall Search

The resonance formation process $\gamma\gamma \to R \to \mathrm{K}_s^0 \mathrm{K}_s^0 \to \pi^+\pi^-\pi^+\pi^-$ has been studied[1] with the L3 detector. The $K_s^0 K_s^0$ mass spectrum Figure 2, shows clear evidence for the formation of the $f_2'(1525)$ tensor meson. Around 1300 MeV, $f_2(1270) - a_2(1320)$ destructive interference is observed consistent with theoretical predictions[3]. In addition, there is an enhancement of $\approx$6 standard deviations around 1750 MeV which is possibly due to the formation of a radially excited state of the f_2', according to theoretical predictions[2]. A study of the

angular distribution of the $f_2^{'}$ in the two-photon centre-of-mass system favours helicity-2 formation over helicity-0, consistent with theoretical predictions[6].
A search for the glueball candidate ξ (2230) has been performed at LEP in the $K_s^0 K_s^0$ decay channel. The search is motivated due to the previous observation of ξ (2230) by the Mark III Collaboration[5] which has been confirmed by BES Collaboration[4]. At LEP, non observation of signal gives an upper limit for $\Gamma_{\gamma\gamma}(\xi(2230)) \times \mathrm{Br}(\xi(2230) \to \mathrm{K_s^0 K_s^0}) < 1.5$ eV at 95% CL under the hypothesis it is a pure spin 2, helicity two state. This low value is most likely inconsistent with a $\mathrm{q\bar{q}}$ assignment to the $\xi(2230)$.

3 The Two Photon Total Cross-section

At LEP II energies, the two photon process $e^+e^- \to e^+e^-\gamma^*\gamma^* \to e^+e^- hadrons$ is a copious source of hadron production. In this reaction the photons either

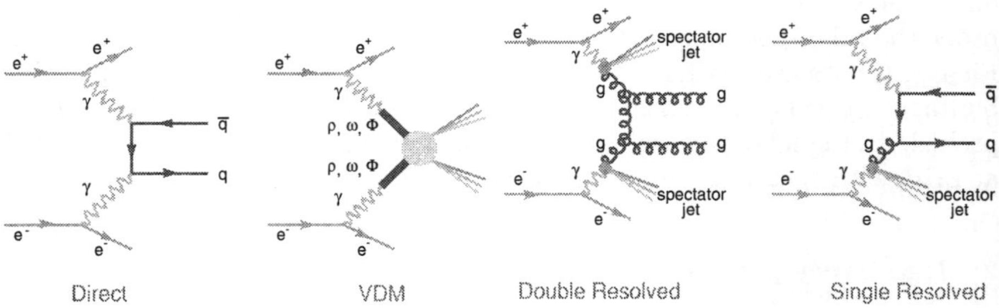

Figure 3: Some diagram contributing to hadron production in $\gamma\gamma$ collisions at LEP.

interact as a point-like particle or undergo quantum fluctuation (resolved photon) into a resonant (VMD) or non-resonant virtual states opening up all the possibilities of hadronic interactions as shown in Figure 3.

These interactions can be described in terms of Regge poles[7,8], (Pomeron or Reggeon exchange). A measurement of the total hadronic cross section as a function of $\sqrt{s}$, improves our understanding of the hadronic nature of the photon. At LEP, using the high energy runs above the Z peak, L3 and OPAL have measured the cross section[9,10] $\sigma(\gamma\gamma \to hadrons)$ in the range $5 \leq W_{\gamma\gamma} \leq 145$ GeV as shown in the Figure 4. The cross-section measurement of the two experiments show a clear rise at high energies, described by a "Soft Pomeron" and the data of the L3 experiment show a fast decrease at low energies due to "Reggeon exchange". The rise of $\sigma_{\gamma\gamma}$ is faster than the one observed in hadron-hadron or γp collisions; a simple factorization ansatz[11]

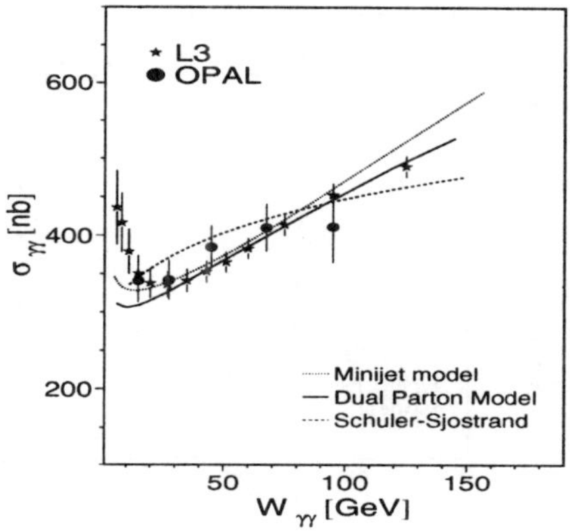

Figure 4: The measured cross-section $\sigma(\gamma\gamma \to hadrons)$ as a function of $W_{\gamma\gamma}$.

$\sigma_{\gamma\gamma} = \sigma^2_{\gamma \rm p}/\sigma_{\rm pp}$ is excluded as can be seen in Figure 4 from the predictions of Schuler and Sjostrand[14]. The data are rather well described by the dual parton model of Engel and Ranft[15] or by analytical calculations which take into account the importance of QCD effects at high transverse momentum. In Figure 4, the minijet model of Godbole and Pancheri[16] is also represented. One has to notice that all models has some dependence which can change the cross section predictions by 10-30%. The Monte Carlo models PYTHIA and PHOJET which are used to correct the data, differ by ≈20% in the absolute normalization. In future, improvements in the theoretical predictions especially the description of diffractive processes are desirable.

4 Heavy Quark Production

The study of heavy quark (c,b) production in two photon collisions at LEP provides not only an excellent test of perturbative QCD but also gives an estimate of the gluon density in the photon. At LEP energies, the direct and resolved photon processes are predicted to give comparable contributions to the charm and beauty quark production cross-sections[17]. The resolved process is dominantly quark-gluon fusion: $\gamma {\rm g} \to q\bar{q}$.

The cross-section of the processes $\rm e^+e^- \to e^+e^- c\bar{c}, b\bar{b}$ X has been measured by the L3[18] and OPAL[19] experiments. At L3, the charm and beauty quark are identified by tagging leptons $({\rm e}, \mu)$ from semileptonic charm and beauty decays. Charm quark were also identified by the reconstruction of $\rm D^{\pm *}$ meson decays. The total inclusive cross-sections are plotted in Figure 5 to-

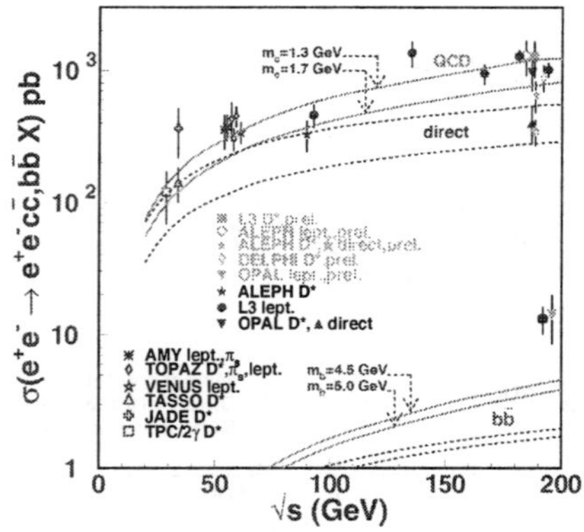

Figure 5: The cross-section for heavy quark production as measured at LEP and at previous e^+e^- colliders

gether with previous measurements. The data are compared to NLO QCD calculations[17]. The direct process $\gamma\gamma \to c\bar{c}, b\bar{b}$, shown with dotted line, is insufficient to describe the data, even if real and virtual gluon corrections are included. The cross sections requires contributions from the resolved processes which are dominantly $\gamma g \to c\bar{c}, b\bar{b}$. The data therefore requires a significant gluon content in the photon.

The $b\bar{b}$ cross section is measured for the first time in two photon collisions. The preliminary value of b cross section lie somewhat above QCD predictions.

5 $\gamma^*\gamma^*$ Collisions

The cross-section of $\gamma^*\gamma^*$ collisions has been measured at LEP with L3[20] and OPAL[21] experiments in the range of 3 GeV^2 < $Q^2_{1,2}$ < 37 GeV^2. Since the two photons are highly virtual and unlike the proton, they do not contain constituent quarks with an unknown density distribution, so one may hope to have a complete perturbative QCD calculation under particular kinematical conditions. An alternative QCD approach is based on the BFKL equation[22]. Here the highly virtual two-photon process, with $Q_1^2 \simeq Q_2^2$, is considered as the "golden" process where the calculation can be verified without phenomenological inputs[23,24]. The $\gamma^*\gamma^*$ interaction can be seen as the interaction of two $q\bar{q}$ pairs scattering off each other via multiple gluon exchange. In this scheme the cross-section for the collision of two virtual photons [23,24] depends upon the "hard Pomeron" intercept $\alpha_P - 1 = 0.53$[23,24] in the LO, whereas in the next-to-leading order the BFKL contribution is calculated to be smaller,

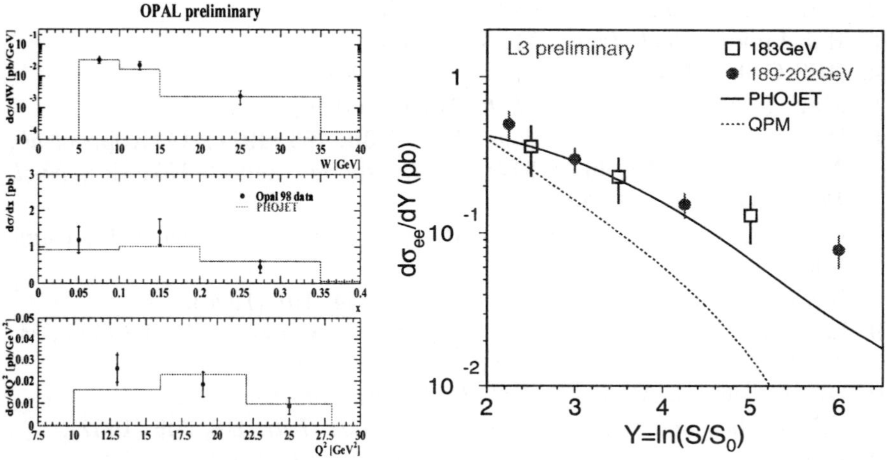

Figure 6: The differential cross-section of double tag events compared to PHOJET Monte Carlo predictions.

$\alpha_P - 1 \simeq 0.17$[25]. The results from L3 and OPAL (Figure 6) show that the events are well described by the PHOJET Monte Carlo model which uses the GRV-LO parton density in the photon and leading order perturbative QCD. By leaving α_P as a free parameter in the BFKL LO calculations, a combined fit to the L3 data obtained at $\sqrt{s} \approx 91$, 183 and 189 GeV gives a value of $\alpha_P - 1 = 0.29 \pm 0.025$ with χ^2/d.o.f =7/9.

Outlook

Progress in the field of the two photon physics at LEP is significant, most notable are multi-hadron production and the measurements on heavy quarks as well as our refined understanding in the study of resonances.

Acknowledgements

I would like to thank all the LEP colleagues for their contribution. I am grateful to J.H. Field, M.N. Kienzle-Focacci and M. Wadhwa for encouraging discussions. This work is supported by the Swiss National Science Foundation.

References

1. L3 Coll., M. Acciarri *et al.*, *Phys. Lett.* **B363**, 118 (1995).

2. C.R. Münz, *Nucl. Phys.* **A609**, 364 (1996).
3. H.J. Lipkin, *Nucl. Phys.* **B7**, 321 (1968).
4. BES Coll., *Phys. Rev. Lett.* **76**, 3502 (1996).
5. Mark III Coll., *Phys. Rev. Lett.* **56**, 107 (1986).
6. B. Schrempp-Otto *et al.*, *Phys. Lett.* **B36**, 463 (1971); G. Köpp *et al.*, *Nucl. Phys.* **B70**, 461 (1974); P. Grassberger and R. Kögerler, *Nucl. Phys.* **B106**, 451 (1976).
7. P.D.B. Collins, Introduction to Regge theory (Cambridge U.P., Cambridge, 1977).
8. Review of Particle Physics, *Phys. Rev.* **D54**, 191 (1996).
9. L3 Coll., M. Acciarri *et al.*, *Phys. Lett.* **B408**, 450 (1997); L3 Coll., M. Acciarri *et al.*, Submitted to EPS, Tampere, 1999.
10. OPAL Coll., G. Abbiendi *et al.*, CERN-EP/99-076.
11. V.N. Gribov and L. Ya. Pomeranchuk, *Phys. Lett.* **B8**, 343 (1962).
12. A. Donnachie and P.V. Landshoff, *Phys. Lett.* **B296**, 227 (1992).
13. Review of Particle Physics, *Eur. Phys. J.* **C3**, 205 (1998).
14. G.A. Schuler and T. Sjöstrand, *Nucl. Phys.* **B407**, 539 (1993); G.A. Schuler and T. Sjöstrand, *Z. Phys.* **C73**, 677 (1997).
15. R. Engel, *Z. Phys.* **C66**, 203 (1995); R. Engel and J. Ranft, *Phys. Rev.* **D54**, 4246 (1996).
16. R.M. Godbole and G.Pancheri, hep-ph/9903331 (1999).
17. M. Drees *et al.*, *Phys. Lett.* **B306**, 371 (1993).
18. L3 Coll., M. Acciarri *et al.*, Submitted to EPS, Tampere, 1999.
19. OPAL Coll., G. Abbiendi *et al.* , Submitted to EPS, Tampere, 1999.
20. L3 Coll., M. Acciarri *et al.*, *Phys. Lett.* **B453**, 333 (1999); L3 Coll., M. Acciarri *et al.*, Submitted to EPS, Tampere, 1999.
21. OPAL Coll., G. Abbiendi et al. , Submitted to EPS, Tampere, 1999.
22. E.A. Kuraev, L.N. Lipatov and V.S. Fadin, *Sov. Phys. JETP* **45**, 199 (1977); Ya.Ya. Balitski and L.N. Lipatov, *Sov. J. Nucl. Phys.* **28**, 822 (1978).
23. S.J. Brodsky, F. Hautmann and D.E. Soper, *Phys. Rev.* **D56**, 6957 (1997).
24. J. Bartels, A. De Roeck and H. Lotter, *Phys. Lett.* **B389**, 742 (1996); J. Bartels, A. De Roeck, C. Ewerz and H. Lotter, hep-ph/9710500.
25. S.J. Brodsky, V.S. Fadin, V.T. Kim, L.N. Lipatov and G.B. Pivovarov, hep-ph/99101229.

LOW x AND DIFFRACTION AT HERA

R.J. HALL-WILTON
On Behalf of the H1 and ZEUS Collaborations
Petrie Science Building, York University,
4700 Keele Street, Toronto, Ontario, M3J 1P3, Canada
E-mail: r.hall-wilton@desy.de

Recent results from H1 and ZEUS on low x and diffraction are reviewed. The rôle of gluons and the consistency of the results with predictions using DGLAP evolution are explored.

1 Introduction

HERA collides electrons (positrons) with protons at a centre of mass energy $\sqrt{s} \sim 300$ GeV. In the low x regime, only the neutral current exchange of a photon is important. Two kinematic variables used to describe the exchange are Q^2, the virtuality of the photon, and x, the fraction of the proton's momentum carried by the struck parton. The kinematic region accessed at HERA spans six orders of magnitude in x and Q^2.

The cross section factorises into a non-perturbative part, the parton densities in the proton ($f_p(Q^2, x)$), and a partonic cross section ($\hat{\sigma}$), calculable in perturbative QCD (pQCD). The cross section can be expressed as

$$\frac{d^2\sigma^{e^\pm p}}{dxdQ^2} = \frac{2\pi\alpha^2}{xQ^4}(Y_+F_2 - y^2F_L \pm Y_-xF_3) \text{ with } Y_\pm = 1 \pm (1-y)^2 \tag{1}$$

where F_2 is the proton structure function, F_L is the longitudinal structure function and xF_3 is the γ-Z interference term. At leading order (LO), F_2 can be related to the individual quark densities

$$F_2(x, Q^2) = \Sigma_i e_i^2 (xq_i(x, Q^2) + x\overline{q}_i(x, Q^2)). \tag{2}$$

Structure functions are not calculable perturbatively, so they must be measured, and perturbative QCD can be used to predict how they vary with Q^2. In particular, the DGLAP equations describe parton splitting (P):

$$\frac{\delta f(x, Q^2)}{\delta \ln(Q^2)} = f(x, Q^2) \otimes P. \tag{3}$$

The fact that DGLAP evolution uses an expansion in $\ln(Q^2)$, indicates two areas where evidence for non-DGLAP evolution can be sought in the data.

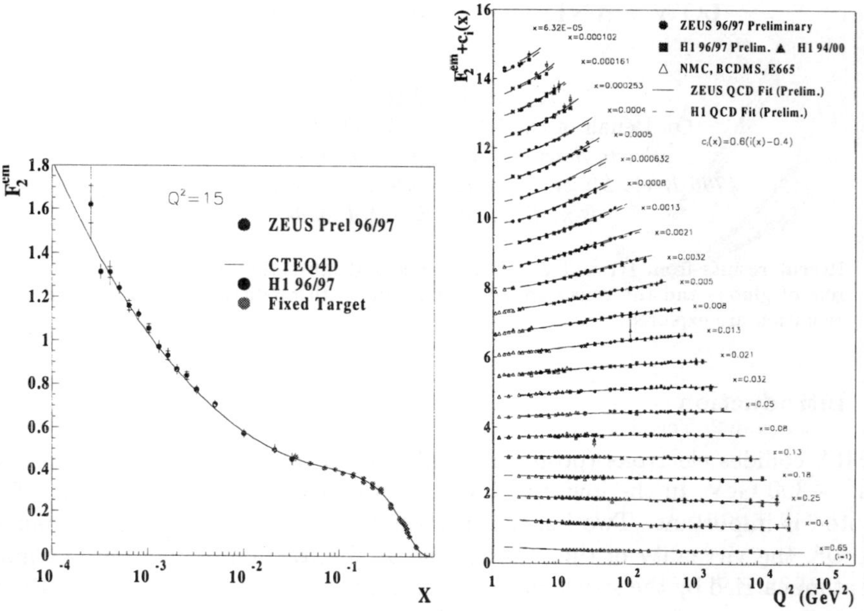

Figure 1: Data from HERA and fixed target experiments on F_2. The left hand plot shows F_2 versus x at Q^2=15 GeV^2. The right hand plot shows F_2 versus Q^2 at fixed x.

Firstly, how low in Q^2 does DGLAP continue to describe the measured data? Secondly at low x, where $ln\frac{1}{x}$ terms are large (neglected in DGLAP evolution), does DGLAP describe the data? Note that by kinematics low x always implies low Q^2 at HERA.

Diffractive scattering is often thought of as a *"soft"* process. However, results will be presented where DGLAP evolution has been applied to diffraction, where a *"hard"* scale (Q^2) is present.

2 Results at Low x

2.1 *Inclusive Measurements*

Recent data[1,2] on the structure function F_2 are shown in Figure 1. The steep rise of F_2 observed at small x indicates rising parton densities. The line is a NLO DGLAP fit by the CTEQ group[3]. The DGLAP fit shows good agreement with the data down to Q^2 of a few GeV^2.

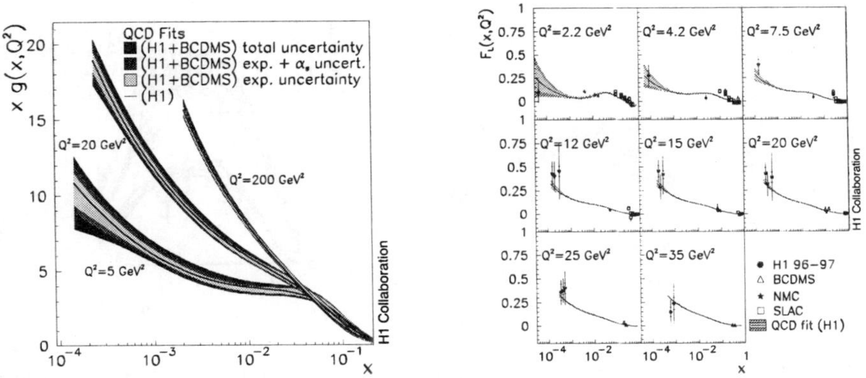

Figure 2: The gluon distribution (left) and F_L (right) extracted by H1.

From Figure 1, when F_2 is plotted at constant x, F_2 is independent of Q^2 for $x \sim 0.1$. As x decreases, the deviations from this scaling increase. At leading order, these scaling violations can be directly related to the gluon density,

$$\frac{\delta f(x, Q^2)}{\delta \ln(Q^2)} \sim xg(x, Q^2). \tag{4}$$

The NLO picture is less trivial, but both H1 and ZEUS have done NLO fits to the data to extract the gluon density, as shown from H1[2] in Figure 2. Here the gluon density rises strongly at small x. However the QCD fits by both H1 and ZEUS lead at $Q^2 \sim 1$ to a gluon density close to 0. Thus, the meaning of a gluon density here is unclear, and it is also questionable whether the DGLAP evolution used in the fit to extract the gluon density is applicable at such low values of Q^2.

H1 have extracted F_L from cross section data, by comparing data with a DGLAP fit, with F_L set to 0. The deviations of the measured data from this fit give the value of F_L. These values of F_L (shown in Figure 2) give vastly improved knowledge of the behaviour of F_L from the previous fixed target data, and are consistent with the H1 DGLAP fit. A rise in F_L is also seen at low x, which is again indicative of the rise in the gluon density at low x.

2.2 Charm Production

One final state which is particularly sensitive to the gluon distribution is charm production, where the leading order contribution is directly sensitive to the

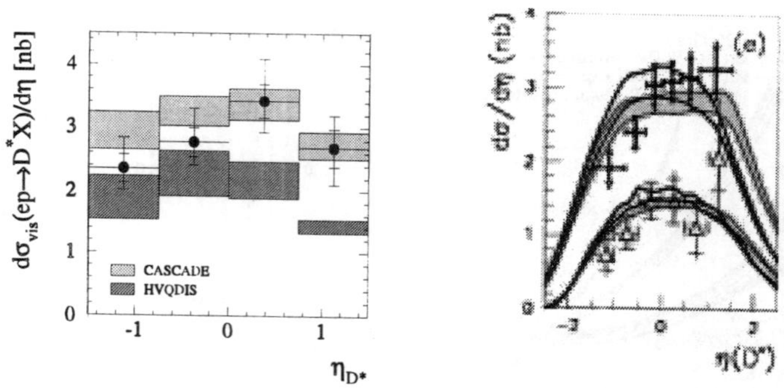

Figure 3: H1 (left) and ZEUS (right) D* differential cross sections versus the pseudorapidity of the D*, $\eta(D^{\star})$.

gluon density. Figure 3 shows results from H1[5] on D* production, compared to two models - HVQDIS, which uses DGLAP evolution, and CASCADE, which uses non-DGLAP (CCFM) evolution, including $\ln \frac{1}{x}$ terms. It appears that DGLAP evolution is disfavoured for charm production, as the predicted cross sections undershoot the data in the forward region. ZEUS, however, found agreement between data and HVQDIS after adding a more realistic model of the charm hadronisation (which affected only the shape of the η distribution and not the normalisation). Therefore it is too early to say that the behaviour observed here is indicative of non-DGLAP evolution. However, results on forward jet production[7,8] have found similar indications of non-DGLAP evolution.

2.3 Very Low x and Q^2

ZEUS[9] has also measured F_2 at very low values of Q^2. Very low Q^2 implies very small x, due to kinematics. At low Q^2 there is no *"hard"* scale for pQCD calculation, and therefore DGLAP evolution is expected to fail here. This transition to *"soft"* physics can clearly be seen from Figure 4. At $Q^2 \sim 1$ GeV^2 the DGLAP fit gives a good description of data . For $Q^2 < 1$ GeV^2 a Regge-type phenomenological fit gives a good description of the data. As $Q^2 \to 0$, $F_2 \to k\, Q^2$ from current conservation.

3 Diffraction

One of the early surprises from HERA data was that about 10% of the DIS events at HERA have little or no activity in the forward (proton) direction -

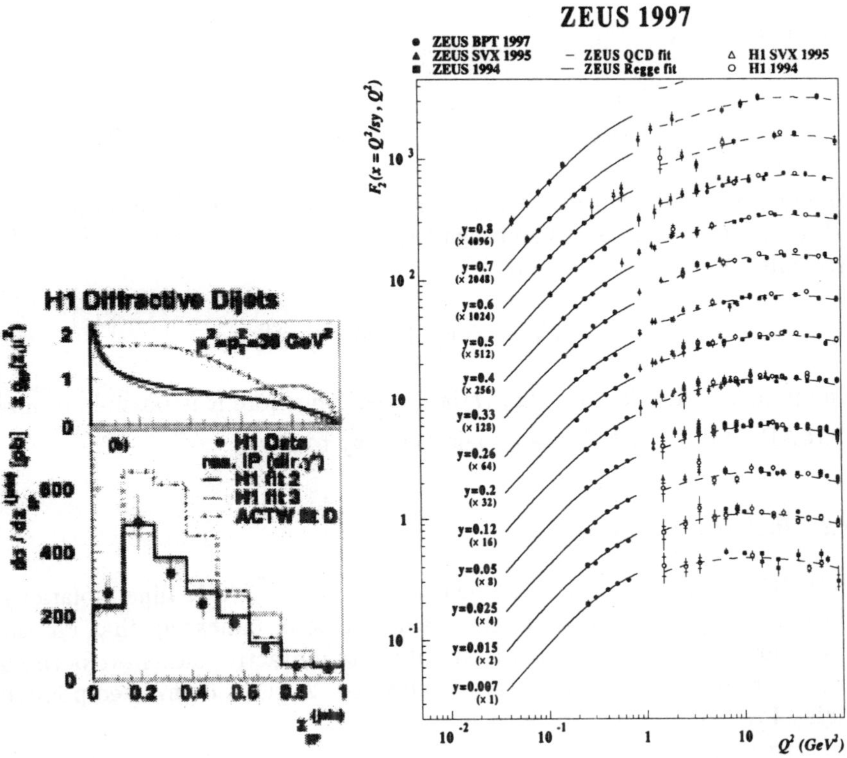

Figure 4: Diffractive dijets differential cross sections against $z_{I\!P}$ from H1 (left). $z_{I\!P}$ is the fraction of energy of system X contained within the dijets. F_2 as a function of Q^2 at low Q^2 from ZEUS (right).

these are termed diffractive events. To describe these events, the exchange of a colourless object, called the pomeron ($I\!P$), is often used.

3.1 Resolved Pomeron Approach

The resolved pomeron approach describes the diffractive scatter in terms of a partonic pomeron, where a parton within the pomeron scatters off the photon. To describe this extra vertex, two new variables are needed - $x_{I\!P}$, which is the fraction of the proton's momentum carried by the pomeron, and β, which is the fraction of the pomeron's momentum which the struck parton carries. The diffractive cross section can be written in terms of a diffractive structure function, $F_2^{D(3)}$,

$$\frac{d^3\sigma^D}{d\beta dQ^2 dx_{I\!P}} = \frac{4\pi\alpha^2}{\beta Q^2}(1 - y + \frac{y^2}{2})F_2^{D(3)}(\beta, Q^2, x_{I\!P}). \tag{5}$$

Analogous to the inclusive case, the cross section factorises[10] into the universal diffractive parton densities and the pQCD partonic cross section, $F_2^D \sim f^D \otimes \hat{\sigma}$. Assuming Regge factorisation,

$$F_2^D \sim f(x_{I\!P}, t) \cdot F_2^{I\!P}(\beta, Q^2). \tag{6}$$

The pomeron flux factor, $f(x_{I\!P}, t) is \sim \frac{1}{x_{I\!P}}$. Therefore it is possible to make a DGLAP analysis of $F_2^{I\!P}$ in an analogous way to F_2.

3.2 Inclusive Diffraction

Both H1 and ZEUS have published results on $F_2^{D(3)}$[12]. Scaling violations are observed in the data also at large values of beta, indicating that gluons are important in diffraction. NLO DGLAP fits to diffractive data have been made, and the diffractive parton densities extracted. A gluon-dominated pomeron is highly favoured by the fits.

ZEUS has also released diffractive data at very low Q^2, for both $F_2^{D(3)}$ and $F_2^{D(4)}$[11]. The behaviour is again analogous to the inclusive data, with a clear transition at low Q^2 ($Q^2 < 1$ GeV^2).

3.3 Diffractive Final States

It is interesting to see whether the data for diffractive final states form a consistent picture with the resolved pomeron model. The parton densities extracted from the DGLAP fits to diffractive data can be fed into calculations for particular diffractive final states, and compared to the measured data.

One final state for which this has been done is the diffractive dijet production - this process is sensitive to the gluon distribution at LO. Results from H1[13] are shown in Figure 4. The diffractive gluon content of the pomeron is tightly constrained by the dijet data. The data favour a large gluon component in the pomeron. Overall, the agreement between the predictions from the resolved pomeron model and the data indicate consistency in the resolved pomeron approach.

4 Summary

There are many new precise results in low x and diffraction at HERA of which only a small selection is shown here. The results emphasise the importance of the gluon distribution in the low x regime. DGLAP evolution seems to describe the measured data across most of the kinematic range down to $Q^2 \sim 1$ GeV^2 There are some indications of non-DGLAP parton dynamics in charm and forward jet production, although neither of these is conclusive yet. For $Q^2 < 1$ GeV^2, where DGLAP evolution is not applicable, a Regge-type phenomenological approach fits the data.

Diffractive data can be fitted to give diffractive parton densities, using DGLAP evolution, giving a consistent picture for a range of results. A gluon-dominated pomeron is favoured. The transition to the low Q^2 regime in diffraction is qualitatively similar to the inclusive case.

Overall, these results show that HERA is an ideal place to study evolution schemes and the transition between *"soft"* and *"hard"* physics.

Acknowledgments

The author would like to thank the ZEUS and H1 colleagues who helped during the preparation of this talk.

References

1. ZEUS Collaboration, Abstract 1048, ICHEP, Osaka (2000).
2. H1 Collaboration, C. Adloff *et al*, submitted to *Eur. Phys. J.* **C** (2000).
3. H.L. Lai *et al*, *Phys. Rev.* **D55**, 1280 (1997).
4. ZEUS Collaboration, J. Breitweg *et al*, *Eur. Phys. J* **C7**, 609 (1999).
5. H1 Collaboration, Abstract 984, ICHEP, Osaka (2000).
6. ZEUS Collaboration, J. Breitweg *et al*, *Eur. Phys. J* **C12**, 35 (2000).
7. ZEUS Collaboration, J. Breitweg *et al*, *Eur. Phys. J* **C6**, 239 (1999).
8. H1 Collaboration, C. Adloff *et al*, *Nucl. Phys.* **B538**, 3 (1999).
9. ZEUS Collaboration, J. Breitweg *et al*, *Phys. Lett.* **B487**, 53 (2000).
10. J.C. Collins, *Phys. Rev.* **D57**, 3051 (1998).
11. ZEUS Collaboration, Abstract 875, ICHEP, Osaka (2000).
12. ZEUS Collaboration, J. Breitweg *et al*, *Eur. Phys. J* **C6**, 43 (1999), H1 Collaboration, C. Adloff *et al*, *Z. Phys.* **C76**, 613 (1997).
13. H1 Collaboration, C. Adloff *et al*, accepted by *Eur. Phys. J.* **C** (2000).

HIGGS SEARCHES WITH THE DELPHI DETECTOR

JØRGEN HANSEN

Department of Physics, University of Oslo, P.O. Box 1048 Blindern, 0316 Oslo, Norway

E-mail: jorgen.hansen@fys.uio.no

Preliminary results for the search for the Higgs boson(s) are presented in the framework of the Standard Model (SM) and the Minimal Supersymmetric Standard Model (MSSM). The results from the year 2000 running are presented, and show no sign of a Higgs signal. Lower limits on the Higgs boson mass(es) are obtained using the data from the entire LEP-*II* running period (presented in full in Ref.[1]), which in a few cases are updates of what was shown at the conference.

1 Introduction

In the Standard Model, a single complex scalar Higgs doublet is sufficient in order to accomodate the Electroweak symmetry breaking and the assignment of fermionic masses. After the Higgs mechanism has absorbed three of the four parameters of the complex field, a single free parameter, the Higgs boson mass m_{H^0}, remains. The tree level production of the SM Higgs boson is at LEP-*II* the s-channel Z^0 exchange Higgs strahlung process $e^+e^- \to H^0 Z^0$.

In a two doublet Higgs model, which is needed for a supersymmetric theory, five Higgs bosons remain: two neutral scalars h^0 and H^0(by definition, $m_{h^0} < m_{H^0}$), one neutral pseudoscalar A^0 and a charged particle/anti-particle pair H^+H^-. In addition to the Higgs Strahlung process, Higgs bosons can now be produced through s-channel Z^0 exchange Higgs pair production, either as a pair of charged Higgs bosons $e^+e^- \to H^+H^-$, or as a pair of neutrals, one of which must be the A^0, $e^+e^- \to h^0A^0$ or H^0A^0.

During the running of LEP in the year 2000, DELPHI has collected a total of 224.3 pb^{-1} at an efficiency of 95.4%, distributed at centre-of-mass energies ranging from 200 GeV to 209 GeV. Details of the simulation of background, and an overview of the DELPHI detector in the year 2000, can be found in the first paper listed in this Ref. [1].

2 The SM search

2.1 The leptonic channels

The leptonic channels, where the Z^0 decays to a pair of leptons (e, μ or τ) and the H^0 decays to a $b\bar{b}$ quark pair, each make up about 2.5% of the total decay width of the H^0Z^0 channel (for a Higgs signal of around m_{h^0}=115 GeV/c^2). The $H^0e^+e^-$ and $H^0\mu^+\mu^-$ channels are both sequential cut analyses using the b-tagging (i.e. the sum of the b-tag for the two hadronic jets) as the final discriminant variable, and the Higgs mass estimator being the invariant dijet mass of the two hadronic jets after a 5C fit, imposing energy and momentum conservation, and in addition forcing the invariant mass of the e^+e^- or $\mu^+\mu^-$ system to the mass of the Z^0.

The $\tau^+\tau^-q\overline{q}$ analysis covers three different channels, two in the SM H^0Z^0 (either the Z^0 going to a $q\overline{q}$ quark pair and the H^0 decaying into $\tau^+\tau^-$, or the Z^0 decaying into $\tau^+\tau^-$ and the H^0 going to a $b\overline{b}$ quark pair) and one in the MSSM h^0A^0 (as the analysis has no way of distinguishing the h^0 from the A^0, the signal consists of a $b\overline{b}$ quark pair recoiling against a $\tau^+\tau^-$ system). In all three cases, the analysis starts with a search for τ candidates in one- or three-prong decays. After rescaling the energies and masses of the τ jets, imposing energy and momentum conservation to give a better estimate of the masses of both di-jets ($\tau^+\tau^-$ and $q\overline{q}$), the final discriminant variable is built from these rescaling factors and the b-tag information. The Higgs mass estimator is constructed by calculating the sum of the invariant masses of the $\tau^+\tau^-$ and the $q\overline{q}$ system, and in the two SM channels subtracting the nominal mass of the Z^0.

2.2 The neutrino channel

The neutrino channel, where $Z^0 \to \nu\overline{\nu}$ and $H^0 \to b\overline{b}$, makes up about 15% of the total decay width of the H^0Z^0 channel. The analysis consists of a series of stringent cuts[2] before a final discriminant variable is constructed, taking b-tagging information and various event shape characteristics into account. The Higgs mass estimator is calculated as the visible mass after a 1C fit is performed, the constraint being that the invisible recoil system is an on-shell Z^0 boson.

2.3 The hadronic four-jet channel

The hadronic four-jet channel, where $Z^0 \to q\overline{q}$ and $H^0 \to b\overline{b}$ (about 52% of the H^0Z^0 final states) starts with a common four-jet preselection[3] and forces the event into a four-jet topology using the DURHAM algorithm. The final discriminant variable is defined as the output of an artificial neural network combining 13 variables: the global b-tag of the event, four variables which test the compatibility of the event kinematics with the hypotheses of W^+W^- and Z^0Z^0 production to either four or five jets, and eight variables intended to distinguish the $q\overline{q}(\gamma)$ background from the true four-jet signal.

Table 1: Expected background and signal together with the observed data for the different SM Higgs boson channels at two different cut levels of the final discriminant variable of the analyses. The signal efficiency quoted corresponds to a signal of m_{H^0}=114 GeV/c^2.

Channel	Data	Background	$q\bar{q}(\gamma)$	4-fermion	Other	Signal Eff.
			Candidates selection level			
$H^0e^+e^-$	7	11.6	0.5	10.4	0.7	57%
$H^0\mu^+\mu^-$	7	10.6	0.2	10.4	—	67%
$\tau^+\tau^-q\bar{q}$	5	6.0	0.4	5.6	—	22%
$H^0\nu\bar{\nu}$	90	99.7	50.4	49.3	—	60%
$H^0q\bar{q}$	398	423.7	154.9	268.8	—	79%
			Tight selection			
$H^0e^+e^-$	1	3.5	0.1	3.2	0.2	49%
$H^0\mu^+\mu^-$	2	3.6	0.1	3.5	—	56%
$\tau^+\tau^-q\bar{q}$	2	4.1	0.1	4.0	—	19%
$H^0\nu\bar{\nu}$	3	4.9	1.4	3.5	—	30%
$H^0q\bar{q}$	8	7.4	2.8	4.6	—	36%
Total	16	23.5	4.5	18.8	0.2	27%

The Higgs mass estimator is determined from the likelihood pairing function,

$$P_b^{j_1} \cdot P_b^{j_2} \cdot \Big(\big(1 - R_b^Z - R_c^Z\big) \cdot P_q^{j_3} \cdot P_q^{j_4} + R_b^Z \cdot P_b^{j_3} \cdot P_b^{j_4} + R_c^Z \cdot P_c^{j_3} \cdot P_c^{j_4} \Big) \cdot P_{j_3,j_4}^{5C}$$

which is calculated for each of the six possibilities to combine the jets j_1, j_2, j_3 and j_4. $P_b^{j_i}, P_c^{j_i}, P_q^{j_i}$ are the probability densities of getting the observed b-tagging value for the jet j_i when originating from b, c or light (u, d or s) quarks, respectively, R_b^Z and R_c^Z are the hadronic branching fractions of the Z^0 into b or c quarks, respectively, and P_{j_3,j_4}^{5C} is the probability corresponding to the kinematical 5C-fit with the jets j_3 and j_4 assigned to the Z^0. The pairing which minimizes this function is now chosen, and the Higgs mass estimator is set to the invariant mass of the jet pair j_1, j_2.

2.4 Results and limits for the SM search

The agreement between data and expected background at different cut levels for the different SM Higgs search channels is shown in Table 1. The plot of the Higgs mass estimator at the tight selection level for the sum of the SM Higgs search channels is shown in Figure 1.

In the SM Higgs boson search, the only free parameter of the model is the Higgs boson mass, and the final statistical analysis is therefore concerned with

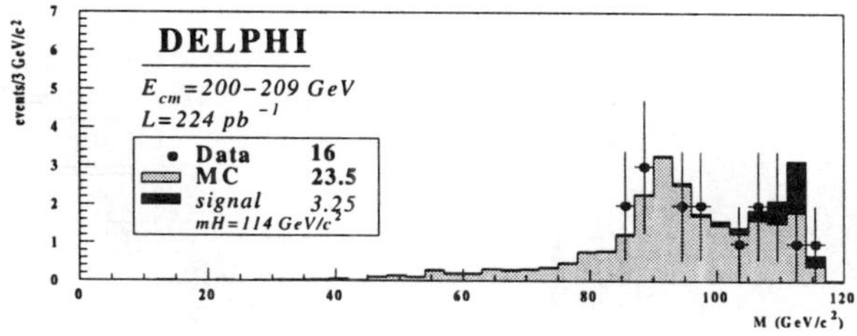

Figure 1: Distribution of the Higgs mass estimator at the tight selection level for the sum of all SM search channels in the year 2000. Data (dots) are compared to the SM background expectation and the expectation from a Higgs boson of mass 114 GeV/c^2.

the confidence of a specific Higgs boson mass hypothesis. A likelihood method[4] based on the ratio of the confidence level for the signal plus background (CL_{s+b}) to the background (CL_b) is used to extract the 95% CL on the SM Higgs boson mass. This result for the SM search is shown in Figure 2.

3 The MSSM search

3.1 The h^0A^0 channel

The h^0A^0 channel is rather similar to the SM $H^0q\bar{q}$ channel, the main difference being a higher content of B-hadrons in the h^0A^0 signal events, leading to a better discrimination of signal and background due to the b-tagging. After the common four-jet preselection (as in the $H^0q\bar{q}$ channel), a further preselection is performed before the final discriminant variable is constructed by a likelihood method using eight input variables accounting for the b-tag and event kinematics properties. At the final cut selection level, 6.0 events, of which the majority (5.6 events) originates from the 4-fermion processes, is expected in background, whereas the observed number of events is 5. At this level, the signal efficiency is 25% for a Higgs boson signal of m_{A^0}=90 GeV/c^2, $\tan\beta$=20.

As the h^0A^0 production mode dominates over the h^0Z^0 production at large values of $\tan\beta$, where the mass difference between the h^0 and A^0 is small at LEP-*II* energies, the pairing defining the Higgs boson candidates is chosen as that which minimizes the mass difference between the two di-jets after a 4C fit is performed. The Higgs mass estimator is now chosen to be the sum of the invariant masses of these two di-jet objects.

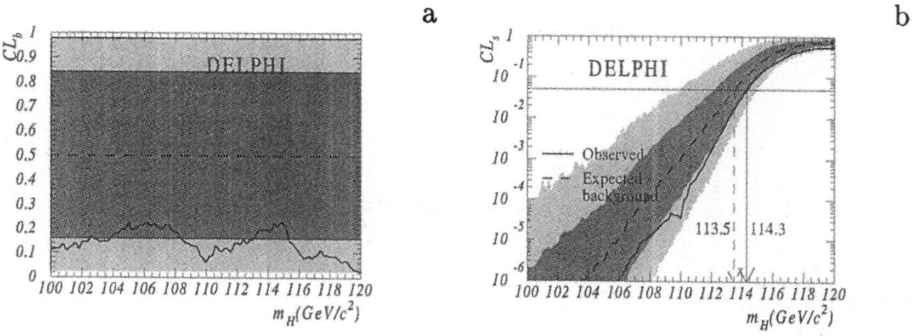

Figure 2: The results from the statistical treatment of the SM search results. In part a, the CL for the background is shown, where the observed result can be seen to lie about one sigma away from the expectation on the non-signal like side, whereas part b shows the CL for the signal, which gives the lower limits for the Higgs boson mass.

3.2 The H^+H^- channel

The decays considered in the H^+H^- analysis include one hadronic and one leptonic decay mode for the charged Higgs bosons. Thus, there are three search channels for the H^+H^- analysis: the leptonic channel $(\tau^-\bar{\nu}_\tau\tau^+\nu_\tau)$, the hadronic channel $(\bar{c}sc\bar{s})$, and the semileptonic channel $(\tau^-\bar{\nu}_\tau c\bar{s}$ or $\tau^+\nu_\tau\bar{c}s)$. After a channel-specific preselection, the main background of all three channels comes from W^+W^- events. Therefore, all channels construct a channel-specific anti-WW likelihood, containing information from several of the following characteristics and variables: the boson polar angle, τ polarization and identification, di-jet momentum polar angle, hadronic jet flavour tagging information and variables containing information on pairing. Also, the hadronic and semileptonic channels use an anti-QCD likelihood based on event kinematics information.

After the final likelihoods, the number of events remaining in the three channels amounts to 585 for an expected background of 627.6. The same statistical procedure as in the SM search [4] was used to extract 95% CL lower limits on the charged Higgs mass. This was done keeping the branching ratio of $H \to \tau\nu$ as a free parameter, and the results are shown in part a of Figure 3, which gives an overall lower limit on the charged Higgs mass of 73.8 GeV/c^2.

3.3 The invisible Higgs channel

For this channel, where the Higgs boson decays into stable, non-interacting particles, rendering it invisible, the analysis covers four different channels,

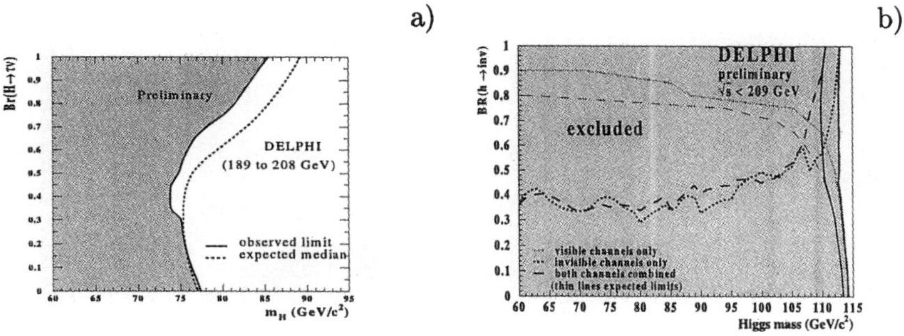

Figure 3: Excluded Higgs mass regions for the H^+H^- and invisible Higgs search. Part a shows, for the charged Higgs search, the regions excluded at 95% CL in the plane of the branching ratio of $H \to \tau\nu$ (the branching ratio of $H \to cs$ is assumed to be 1-BR($H \to \tau\nu$)) vs. the charged Higgs mass. Part b shows, for the invisible Higgs search, the regions excluded at 95% CL in the plane of the branching ratio of $h \to$invisible vs. the Higgs mass.

according to the decay of the Z^0: one hadronic ($Z^0 \to q\bar{q}$) and three leptonic ($Z^0 \to e^+e^-$, $\mu^+\mu^-$ or $\tau^+\tau^-$). After a preselection, the hadronic channel uses an Iterative Discriminant Analysis Program (IDA) to combine information on event energy and momentum deposits as well as event shape in a discriminant variable. The leptonic channels apply cuts on lepton identification and event kinematics to reduce the background. When adding the remaining candidates, the number of observed events is 41 for an expected background of 58.4 events. 95% CL lower limits on the Higgs mass was found as a function of the branching ratio into invisible decays BR_{inv}, assuming visible branchnig ratios BR_{vis}=1-BR_{inv} and SM production cross sections. Part b of Figure 3 shows the result, which yields a global lower Higgs boson mass limit of 112.7 GeV/c^2.

3.4 Results and limits for the MSSM search

In the MSSM, there are at tree level only two free parameters of the Higgs sector, usually chosen as two of the three parameters $\tan\beta$, m_{A^0} and m_{h^0}. Radiative corrections introduce additional parameters[2]. The search translates into regions in the MSSM parameter space excluded at the 95% CL or more, and are given for two different benchmarks, called the m_h^{max} scenario (which gives the maximal value of m_{h^0} as a function of $\tan\beta$) and the no mixing scenario[2]. The results are shown in Figure 4.

Finally, the excluded regions in the different Higgs boson masses and, for the MSSM, $\tan\beta$ values are summarized in Table 2.[5]

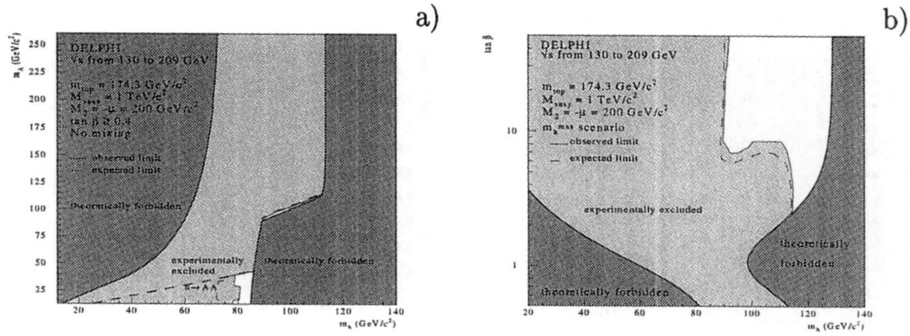

Figure 4: MSSM regions in the $(m_{h^0},\, m_{A^0})$ plane for the no mixing scenario (part a) and in the $(m_{h^0},\, \tan\beta)$ plane for the m_h^{max} scenario (part b) excluded at 95% CL. All SM and MSSM channels up to $\sqrt{s}$=209 GeV are included.

Table 2: The search results, given in terms of excluded ranges of Higgs boson masses and, for the MSSM case, values of $\tan\beta$. For MSSM exclusion of Higgs boson masses, the assumptions m_{A^0}>12 GeV/c^2 and $\tan\beta$<0.6 is made.

Framework	Observed limit	Median expected limit	Scenario
SM	m_{H^0}<114.3 GeV/c^2	m_{H^0}<113.5 GeV/c^2	—
MSSM	m_{A^0}<90.9 GeV/c^2	m_{A^0}<90.1 GeV/c^2	m_h^{max}
	m_{h^0}<89.8 GeV/c^2	m_{h^0}<89.0 GeV/c^2	
	$\tan\beta \in$ [0.49,2.36]	$\tan\beta \in$ [0.54,2.36]	
MSSM	m_{A^0}<90.8 GeV/c^2	m_{A^0}<90.0 GeV/c^2	No mixing
	m_{h^0}<90.0 GeV/c^2	m_{h^0}<89.1 GeV/c^2	
	$\tan\beta \in$ [0.59,9.36]	$\tan\beta \in$ [0.72,8.49]	

References

1. DELPHI Collaboration, *Phys. Lett.* **B499**, 23 (2001);
   DELPHI Higgs Research Line Teams, DELPHI 2001-017 CONF 458;
   M. Battaglia, M. Ellert, T. Ekelof, G. Gómez-Ceballos, A. Kiiskinen, P. Lutz and F. Matorras, DELPHI 2001-030 CONF 471;
   A. Sopczak and M. Stanitzki, DELPHI 2001-025 CONF 466.
2. DELPHI Collaboration, DELPHI 2000-077 CONF 376.
3. DELPHI Collaboration, *Eur. Phus. J.* **C17**, 187 (2000).
4. A.L. Read, in CERN Report 2000-005, p. 81 (2000).
5. http://delphiwww.cern.ch/teams/searches/delwww/higgs/inputs/results/results.html#010227

INDIRECT CP VIOLATION RESULTS FROM BELLE

T. HARA for the Belle collaboration

Graduate School of Science, Osaka University, 1-1 Machikaneyama, Toyonaka, Osaka 560-0043, Japan

E-mail: hara@oskjcc.hep.sci.osaka-u.ac.jp

We present a measurement of the Standard Model CP violation parameter $\sin 2\phi_1$ based on a 10.5 fb^{-1} data sample collected at the $\Upsilon(4S)$ resonance with the Belle detector at the KEKB asymmetric e^+e^- collider. One neutral B meson is reconstructed in the $J/\psi K_S$, $\psi(2S)K_S$, $\chi_{c1}K_S$, $\eta_c K_S$, $J/\psi K_L$ or $J/\psi\pi^0$ CP-eigenstate decay channel and the flavor of the accompanying B meson is identified from its charged particle decay products. ¿From the asymmetry in the distribution of the time interval between the two B-meson decay points, we determine $\sin 2\phi_1 = 0.58^{+0.32}_{-0.34}(\mathrm{stat})^{+0.09}_{-0.10}(\mathrm{syst})$.

1 Introduction

In the Standard Model (SM), CP violation arises from a complex phase in the Cabibbo-Kobayashi-Maskawa (CKM) quark mixing matrix[1]. In particular, the SM predicts a CP violating asymmetry in the time-dependent rates for B_d^0 and $\overline{B_d}^0$ decays to a common CP eigenstate, f_{CP}, without theoretical ambiguity due to strong interactions[2]:

$$A(t) \equiv \frac{\Gamma(\overline{B_d}^0 \to f_{CP}) - \Gamma(B_d^0 \to f_{CP})}{\Gamma(\overline{B_d}^0 \to f_{CP}) + \Gamma(B_d^0 \to f_{CP})} = -\xi_f \sin 2\phi_1 \sin \Delta m_d t,$$

where $\Gamma(\overline{B_d}^0\ (B_d^0) \to f_{CP})$ is the decay rate for a $\overline{B_d}^0 (B_d^0)$ to f_{CP} at a proper time t after production, ξ_f is the CP-eigenvalue of f_{CP}, Δm_d is the mass difference between the two B_d^0 mass eigenstates, and ϕ_1 is one of the three internal angles of the CKM Unitarity Triangle, defined as $\phi_1 \equiv \pi - \arg\left(\frac{-V_{tb}^* V_{td}}{-V_{cb}^* V_{cd}}\right)$ [3].

In this Letter, we report a measurement of $\sin 2\phi_1$ using $B_d^0\overline{B_d}^0$ meson pairs produced at the $\Upsilon(4S)$ resonance, where the two mesons remain in a coherent p-wave state until one of them decays. The decay of one of the B mesons to a self-tagging state, f_{tag}, i.e. a final state that distinguishes between B_d^0 and $\overline{B_d}^0$, at time t_{tag} projects the accompanying meson onto the opposite b-flavor at that time; this meson decays to f_{CP} at time t_{CP}. The CP violation manifests itself as an asymmetry $A(\Delta t)$, where Δt is the proper time interval $\Delta t \equiv t_{CP} - t_{tag}$.

The data sample corresponds to an integrated luminosity of 10.5 fb^{-1} collected with the Belle detector [4] at the KEKB asymmetric e^+e^- (3.5 on

8 GeV) collider [5]. At KEKB, the $\Upsilon(4S)$ is produced with a Lorentz boost of $\beta\gamma = 0.425$ along the electron beam direction (z direction). Because the B_d^0 and $\overline{B_d}^0$ mesons are nearly at rest in the $\Upsilon(4S)$ center of mass system (cms), Δt can be determined from the z distance between the f_{CP} and f_{tag} decay vertices, $\Delta z \equiv z_{CP} - z_{tag}$, as $\Delta t \simeq \Delta z/\beta\gamma c$.

2 Data analysis and Experimental results

We reconstruct B_d^0 decays to the following CP eigenstates: $J/\psi K_S$, $\psi(2S)K_S$, $\chi_{c1}K_S$, $\eta_c K_S$ for $\xi_f = -1$ and $J/\psi\pi^0$, $J/\psi K_L$ for $\xi_f = +1$. The J/ψ and $\psi(2S)$ mesons are reconstructed via their decays to $\ell^+\ell^-$ ($\ell = \mu, e$). The $\psi(2S)$ is also reconstructed via its $J/\psi\pi^+\pi^-$ decay, the χ_{c1} via its $J/\psi\gamma$ decay, and the η_c via its $K^+K^-\pi^0$ and $K_S(\pi^+\pi^-)K^-\pi^+$ [a] decays. For J/ψ and $\psi(2S) \to \ell^+\ell^-$ decays, we use oppositely charged track pairs where both tracks are positively identified as leptons. Candidate $K_S \to \pi^+\pi^-$ decays are oppositely charged track pairs that have an invariant mass within $\pm 4\sigma$ of the K^0 mass ($\sigma \simeq 4$ MeV$/c^2$).

We isolate reconstructed B meson decays using the energy difference $\Delta E \equiv E_B^{cms} - E_{beam}^{cms}$ and the beam-energy constrained mass $M_{bc} \equiv \sqrt{(E_{beam}^{cms})^2 - (p_B^{cms})^2}$, where E_{beam}^{cms} is the cms beam energy, and E_B^{cms} and p_B^{cms} are the cms energy and momentum of the B candidate. After a ΔE selection [b], the B meson signal region (other than $J/\psi K_L$) is defined as $5.270 < M_{bc} < 5.290$ GeV$/c^2$; the M_{bc} resolution is 3.0 MeV$/c^2$. Table 1 lists the numbers of observed events (N_{ev}) and the background (N_{bkgd}) determined by extrapolating the event rate in the non-signal ΔE *vs.* M_{bc} region into the signal region.

Candidate $B_d^0 \to J/\psi K_L$ decays are selected by requiring the observed K_L direction to be within 45° from the direction expected for a two-body decay (ignoring the B_d^0 cms motion). We reduce the background by means of a likelihood quantity that depends on the J/ψ cms momentum, the angle between the K_L and its nearest-neighbor charged track, the charged track multiplicity, and the kinematics that obtain when the event is reconstructed assuming a $B^+ \to J/\psi K^{*+}(K_L\pi^+)$ hypothesis. In addition, we remove events that are reconstructed as $B_d^0 \to J/\psi K_S$, $J/\psi K^{*0}(K^+\pi^-, K_S\pi^0)$, $B^+ \to J/\psi K^+$, or $J/\psi K^{*+}(K^+\pi^0\ K_S\pi^+)$ decays. Among the total of 131 entries in the $0.2 \le p_B^{cms} \le 0.45$ GeV$/c$ signal region, the fit finds 77 $J/\psi K_L$ events.

The leptons and charged pions and kaons among the tracks that are not

---

[a] Throughout this Letter, when a mode is quoted the inclusion of the charge conjugate mode is implied.

[b] a ΔE selection that varies from ± 25 MeV to ± 100 MeV (corresponding to $\sim \pm 3\sigma$), depending on the mode.

Table 1: The numbers of CP eigenstate events

Mode	N_{ev}	N_{bkgd}
$J/\psi(\ell^+\ell^-)K_S(\pi^+\pi^-)$	123	3.7
$J/\psi(\ell^+\ell^-)K_S(\pi^0\pi^0)$	19	2.5
$\psi(2S)(\ell^+\ell^-)K_S(\pi^+\pi^-)$	13	0.3
$\psi(2S)(J/\psi\pi^+\pi^-)K_S(\pi^+\pi^-)$	11	0.3
$\chi_{c1}(\gamma J/\psi)K_S(\pi^+\pi^-)$	3	0.5
$\eta_c(K^+K^-\pi^0)K_S(\pi^+\pi^-)$	10	2.4
$\eta_c(K_SK^+\pi^-)K_S(\pi^+\pi^-)$	5	0.4
$J/\psi(\ell^+\ell^-)\pi^0$	10	0.9
Sub-total	194	11
$J/\psi(\ell^+\ell^-)K_L$	131	54

associated with f_{CP} are used to identify the flavor of the accompanying B meson. Tracks are selected in several categories that distinguish the b-flavor by the track's charge: high momentum leptons from $b \to c\ell^-\overline{\nu}$, lower momentum leptons from $c \to s\ell^+\nu$, charged kaons from $b \to c \to s$, high momentum pions from decays of the type $B_d^0 \to D^{(*)-}(\pi^+, \rho^+, a_1^+, \text{etc.})$, and slow pions from $D^{*-} \to \overline{D}^0\pi^-$. For each track in one of these categories, we use the MC to determine the relative probability that it originates from a B_d^0 or $\overline{B_d}^0$ as a function of its charge, cms momentum and polar angle, particle-identification probability, and other kinematic and event shape quantities. We combine the results from the different track categories (taking into account correlations for the case of multiple inputs) to determine a b-flavor q, where $q = +1$ when f_{tag} is more likely to be a B_d^0 and -1 for a $\overline{B_d}^0$. We use the MC to evaluate an event-by-event flavor-tagging dilution factor, r, which ranges from $r = 0$ for no flavor discrimination to $r = 1$ for perfect flavor assignment. We only use r to categorize the event. For the CP asymmetry analysis, we use the data to correct for wrong-flavor assignments.

The probabilities for an incorrect flavor assignment, w_l ($l = 1, 6$), are measured directly from the data for six r intervals using a sample of exclusively reconstructed, self-tagged $B_d^0 \to D^{*-}\ell^+\nu$, $D^{(*)-}\pi^+$, and $D^{*-}\rho^+$ decays. The b-flavor of the accompanying B meson is assigned according to the above-described flavor-tagging algorithm, and values of w_l are determined from the amplitudes of the time-dependent B_d^0-$\overline{B_d}^0$ mixing oscillations[6]: $(N_{OF} - N_{SF})/(N_{OF} + N_{SF}) = (1 - 2w_l)\cos(\Delta m_d \Delta t)$. Here N_{OF} and N_{SF} are the numbers of opposite and same flavor events. Table 2 lists the resulting w_l values together with the fraction of the events (f_l) in each r interval. All events in Table 1 fall in one of the six r intervals. The total effective tagging efficiency is $\sum_l f_l(1 - 2w_l)^2 = 0.270^{+0.021}_{-0.022}$, where the error includes both statistical and

Table 2: Experimentally determined event fractions (f_l) and incorrect flavor assignment probabilities (w_l) for each r interval.

l	r	f_l	w_l
1	$0.000 - 0.250$	0.393 ± 0.014	$0.470^{+0.031}_{-0.035}$
2	$0.250 - 0.500$	0.154 ± 0.007	$0.336^{+0.039}_{-0.042}$
3	$0.500 - 0.625$	0.092 ± 0.005	$0.286^{+0.037}_{-0.035}$
4	$0.625 - 0.750$	0.100 ± 0.005	$0.210^{+0.033}_{-0.031}$
5	$0.750 - 0.875$	0.121 ± 0.006	$0.098^{+0.028}_{-0.026}$
6	$0.875 - 1.000$	0.134 ± 0.006	$0.020^{+0.023}_{-0.019}$

systematic uncertainties, in good agreement with the MC result of 0.274.

The vertex positions for the f_{CP} and f_{tag} decays are reconstructed using tracks that have at least one 3-dimensional coordinate determined from associated $r\phi$ and z hits in the same SVD layer plus one or more additional z hits in other SVD layers. Each vertex position is required to be consistent with the IP profile smeared in the $r\phi$ plane by the B meson decay length. [c] The f_{CP} vertex is determined using lepton tracks from the J/ψ or $\psi(2S)$ decays, or prompt tracks from η_c decays. The f_{tag} vertex is determined from tracks not assigned to f_{CP} with additional requirements of $\delta r < 0.5$ mm, $\delta z < 1.8$ mm and $\sigma_{\delta z} < 0.5$ mm, where δr and δz are the distances of the closest approach to the f_{CP} vertex in the $r\phi$ plane and the z direction, respectively, and $\sigma_{\delta z}$ is the calculated error of δz. Tracks that form a K_S are removed. The MC indicates that the average z_{CP} resolution is 75 μm (rms); the z_{tag} resolution is worse (140 μm) because of the lower average momentum of the f_{tag} decay products and the smearing caused by secondary tracks from charmed meson decays. The overall efficiency of the vertex reconstruction is estimated to be 87%.

The resolution function $R(\Delta t)$ for the proper time interval is parameterized as a sum of two Gaussian components: a *main* component due to the SVD vertex resolution, charmed meson lifetimes and the effect of the cms motion of the B mesons, plus a *tail* component caused by poorly reconstructed tracks. The relative fraction of the main Gaussian is determined to be 0.982 ± 0.013 from a study of $B_d^0 \to D^{*-}\ell^+\nu$ events. The reliability of the Δt determination and $R(\Delta t)$ parameterization is confirmed by lifetime measurements of the neutral and charged B mesons that use the same procedures and are in good agreement with the world average values [7].

We determine $\sin 2\phi_1$ from an unbinned maximum-likelihood fit to the

[c] The IP size, determined run-by-run, is typically $\sigma_x \simeq 100$ μm, $\sigma_y \simeq 5$ μm and $\sigma_z \simeq$ 3 mm.

observed Δt distributions. The probability density function (pdf) expected for the signal distribution is given by

$$\mathcal{P}_{sig}(\Delta t, q, w_l, \xi_f) = \frac{e^{-|\Delta t|/\tau_{B_d^0}}}{2\tau_{B_d^0}}\{1 - \xi_f q(1-2w_l)\sin 2\phi_1 \sin(\Delta m_d \Delta t)\},$$

where we fix the B_d^0 lifetime and mass difference at their world average values[7]. The pdf used for background events is $\mathcal{P}_{bkg}(\Delta t) = f_\tau e^{-|\Delta t|/\tau_{bkg}}/2\tau_{bkg} + (1-f_\tau)\delta(\Delta t)$, where f_τ is the fraction of the background component with an effective lifetime τ_{bkg} and $\delta(\Delta t)$ is the Dirac delta function.

The pdfs are convolved with $R(\Delta t)$ to determine the likelihood value for each event as a function of $\sin 2\phi_1$:

$$\mathcal{L}_i = \int \{f_{sig}\mathcal{P}_{sig}(\Delta t', q, w_l, \xi_f) + (1-f_{sig})\mathcal{P}_{bkg}(\Delta t')\}R(\Delta t - \Delta t')d\Delta t',$$

where f_{sig} is the probability that the event is signal, calculated as a function of p_B^{cms} for $J/\psi K_L$ and of ΔE and M_{bc} for other modes. The most probable $\sin 2\phi_1$ is the value that maximizes the likelihood function $L = \prod_i \mathcal{L}_i$, where the product is over all events. The application of the fit to all the events listed in Table 1 produces the result $\sin 2\phi_1 = 0.58^{+0.32+0.09}_{-0.34-0.10}$, where the first error is statistical and the second systematic. The systematic errors are dominated by the uncertainties in w_l ($^{+0.05}_{-0.07}$) and the $J/\psi K_L$ background (± 0.05). Separate fits to the $\xi_f = -1$ and $\xi_f = +1$ event samples give $0.82^{+0.36}_{-0.41}$ and $0.10^{+0.57}_{-0.60}$, respectively[d]. Figure 1(a) shows $-2\ln(L/L_{max})$ as a function of $\sin 2\phi_1$ for the $\xi_f = -1$ and $\xi_f = +1$ modes separately and for both modes combined. Figure 1(b) shows the asymmetry obtained by performing the fit to events in Δt bins separately, together with a curve that represents $\sin 2\phi_1 \sin(\Delta m_d \Delta t)$ for $\sin 2\phi_1 = 0.58$.

We check for a possible fit bias by applying the same fit to non-CP eigenstate modes, where "$\sin 2\phi_1$" should be zero, and the charged mode $B^+ \to J/\psi K^+$. For all the modes combined we find 0.065 ± 0.075, consistent with a null asymmetry.

3 Conclusion

We have presented a measurement of the Standard Model CP violation parameter $\sin 2\phi_1$ based on a 10.5 fb^{-1} data sample collected at the $\Upsilon(4S)$:

$$\sin 2\phi_1 = 0.58^{+0.32}_{-0.34}(\text{stat})^{+0.09}_{-0.10}(\text{syst}).$$

[d] A fit to only the $B_d^0 \to J/\psi K_L$ events yields -0.04 ± 0.60. Separate fits to the $q = +1$ and $q = -1$ event samples give sin2ϕ_1 values of $0.40^{+0.47}_{-0.49}$ and $0.73^{+0.41}_{-0.46}$, respectively.

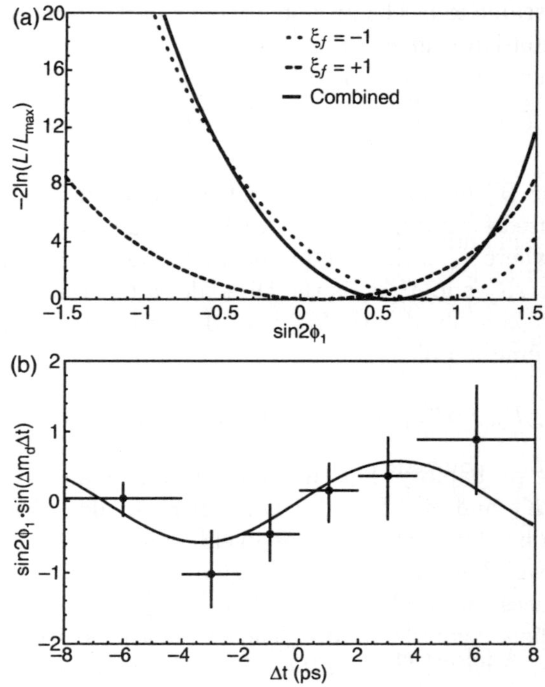

Figure 1: (a) Values of $-2\ln(L/L_{max})$ *vs.* $\sin 2\phi_1$ for the $\xi_f = -1$ and $+1$ modes separately and for both modes combined. (b) The asymmetry obtained from separate fits to each Δt bin; the curve is the result of the global fit ($\sin 2\phi_1 = 0.58$).

The probability of observing $\sin 2\phi_1 > 0.58$ if the true value is zero is 4.9%. Our measurement is more precise than the previous measurements [8] and consistent with SM constraints [9].

Acknowledgments

We wish to thank the KEKB accelerator group for the excellent operation. We acknowledge support from the Ministry of Education, Culture, Sports, Science and Technology of Japan and the Japan Society for the Promotion of Science; the Australian Research Council and the Australian Department of Industry, Science and Resources; the Department of Science and Technology of India; the BK21 program of the Ministry of Education of Korea and the SRC program of the Korea Science and Engineering Foundation; the Polish State Committee for Scientific Research under contract No.2P03B 17017; the Ministry of Science and Technology of Russian Federation; the National Science Council and the Ministry of Education of Taiwan; the Japan-Taiwan Cooperative Program of the Interchange Association; and the U.S. Department of Energy.

References

1. M. Kobayashi and T. Maskawa, *Prog. Theor. Phys.* **49**, 652 (1973).
2. A.B. Carter and A.I. Sanda, *Phys. Rev.* **D23**, 1567 (1981); I.I. Bigi and A.I. Sanda, *Nucl. Phys.* **B193**, 85 (1981).
3. H. Quinn and A.I. Sanda, *Eur. Phys. J.* **C15**, 626 (2000).
4. Belle Collab., K. Abe *et al.*(to be published in Nucl. Instrum. Methods), KEK Report No. 2000-4.
5. KEKB B Factory Design Report No.95-1, 1995 (unpublished).
6. Belle Collab., J. Suzuki, Proceedings of the 30th International Conference on High Energy Physics, Osaka, 2000 (to be published).
7. D.E. Groom *et al.* Particle Data Group, *Eur. Phys. J.* **C15**, 1 (2000); *Phys. Rev. Lett.* **79**, 4533 (1997) and CDF Collab., T. Affolder *et al.*, *Phys. Rev. Lett.* **85**, 4668 (2000).
8. Opal Collab., K. Ackerstaff *et al*, *Eur. Phys. J.* **C5**, 379 (1998); CDF Collab., T. Affolder *et al.* *Phys. Rev.* **D61**, 072005 (2000); ALEPH Collab., R. Barate *et al.* *Phys. Lett.* **B492**, 259 (2000).
9. For example, S. Mele, *Phys. Rev.* **D59**, 113011 (1999).

NONCUMMUTATIVE FIELD THEORIES AND SPONTANEOUS SYMMETRY BREAKING

BRUCE CAMPBELL and KIRK KAMINSKY
Department of Physics, University of Alberta
Edmonton, Alberta, T6G 2J1

We herein discuss the essential conclusions of our investigations into the quantum dynamics of spontaneous symmetry breaking in the framework of noncommutative field theories.

1 Motivation

Noncommutative (NC) spacetime geometry defined by the structure

$$[\hat{x}^\mu, \hat{x}^\nu] = i\theta^{\mu\nu} \quad , \quad \theta \in \mathbb{R}, \text{constant} \tag{1}$$

naturally arises in the zero-slope (i.e. field-theoretic) limit of string theory on a D-brane with a *constant* $B_{\mu\nu}$ background. As a consequence, noncommutative field theories with the structure inherited by (1) have been the subject of intense scrutiny in recent years. But independently of string theory, the class of infinitely nonlocal field theories are interesting in their own right. We will first briefly review how we proceed from (1) to building actual field theories, and then recount some of the early surprises emerging from this structure, before turning to our main topic of interest: spontaneous symmetry breaking in noncommutative field theories. This last discussion is based upon [1,2].

2 Formalism

Working directly with the structure (1) yields operator-valued functions, and while one can proceed this way, it turns out to be particularly fruitful to instead invoke the so-called Weyl-Moyal correspondence to replace operator-valued functions with ordinary functions over commutative spacetime background, at the expense of introducing a more complicated (i.e. non-pointwise, non-local) product between functions.

Thus consider the associative algebraic structure defined by (1), and following Weyl, define

$$O(f) = \frac{1}{(2\pi)^{n/2}} \int d^n k \; e^{ik_\mu \hat{x}^\mu} \; \tilde{f}(k) \tag{2}$$

where $\tilde{f}$ is the Fourier transform of f:

$$\tilde{f}(k) = \frac{1}{(2\pi)^{n/2}} \int d^n x \ e^{-ik_\mu x^\mu} \ f(x) \tag{3}$$

This uniquely associates an operator in the algebra with an ordinary function of classical variables in a symmetrical fashion. We now define the product of two such operators in the obvious way as

$$\begin{aligned} O(f) \cdot O(g) &= \frac{1}{(2\pi)^n} \int d^n k d^n p \ e^{ik_\mu \hat{x}^\mu} e^{ip_\nu \hat{x}^\nu} \ \tilde{f}(k)\tilde{g}(p) \\ &= \frac{1}{(2\pi)^n} \int d^n k d^n p \ e^{i(k_\mu + p_\mu)\hat{x}^\mu - \frac{i}{2} k_\mu \theta^{\mu\nu} p_\nu} \ \tilde{f}(k)\tilde{g}(p) \end{aligned} \tag{4}$$

where we have invoked the Baker-Campbell-Haussdorf lemma, and the fact that the commutator (1) is a c-number on the second line.

On the other hand, if we define the star product:

$$\begin{aligned} (f * g)(x) &= \frac{1}{(2\pi)^n} \int d^n k d^n p \ e^{i(k_\mu + p_\mu)x^\mu} e^{-\frac{i}{2} k_\mu \theta^{\mu\nu} p_\nu} \ \tilde{f}(k)\tilde{g}(p) \\ &= e^{+\frac{i}{2}\theta^{\mu\nu} \frac{\partial}{\partial y^\mu} \frac{\partial}{\partial z^\nu}} f(y) g(z)|_{y,z \to x} \end{aligned} \tag{5}$$

then we establish a homomorphism (algebra-preserving map) $O(f) \cdot O(g) = O(f * g)$, which allows us to replace NC $\mathcal{R}^n$ with ordinary $\mathcal{R}^n$, at the expense of replacing pointwise products with this infinitely nonlocal product.

The formal prescription for defining a quantum field theory (at this level) is then to simply replace ordinary products in the Lagrangian density with star products. Under the spacetime integral, we have the identity $\int A * B = \int B * A = \int AB$, so that quadratic mass and kinetic terms are oblivious to the noncommutativity at tree level. On the other hand, the interaction terms in momentum space are given by

$$\begin{aligned} \int d^4 x \phi_1(x) * \ldots * \phi_n(x) &= \int \left(\prod_{i=1}^{n} d^4 p_i \right) \delta \left(\sum_{i=1}^{n} p_i \right) e^{-\frac{i}{2} \sum_{i<j} p_i \times p_j} \times \\ &\quad \times \tilde{\phi}_1(p_1) \cdots \tilde{\phi}_n(p_n). \end{aligned} \tag{6}$$

Thus the Feynman rules are modified by introduction of a phase: $V(p_1 \ldots p_n) = e^{-\frac{i}{2} \sum_{i<j} p_i \times p_j}$, which means that the ordering of momenta into a vertex matters, since this factor is not invariant under arbitary momentum interchanges, though a residual cyclic invariance remains. Furthermore, Feynman graphs are generally split into a planar and a nonplanar piece.

3 NC ϕ^4 theory and UV/IR mixing

Consider specifically noncommutative ϕ^4 theory, whose symmetrized Feynman rule for the quartic self-interaction is

$$-2\lambda\left[\cos(\frac{p_1\times p_2}{2})\cos(\frac{p_3\times p_4}{2})+(2\leftrightarrow 3)+(2\leftrightarrow 4)\right]. \tag{7}$$

The one-loop correction to inverse propagator due to the noncommutative interaction splits as

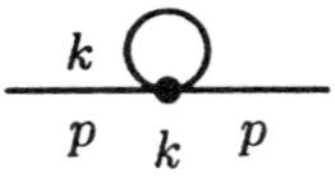

$$\equiv \Gamma^{(2)}_{planar} = -2\lambda\int\frac{d^4k}{(2\pi)^4}\frac{1}{k^2+m^2}$$

$$\equiv \Gamma^{(2)}_{nonplanar} = -\lambda\int\frac{d^4k}{(2\pi)^4}\frac{e^{ik\times p}}{k^2+m^2} \tag{8}$$

where $k\times p = k_\mu\theta^{\mu\nu}p_\nu$.

The first, planar piece has the usual divergence structure of ϕ^4 theory, except that it now carries only two thirds the weight of the equivalent commutative graph. To analyze the effect of the NC phase in the second, nonplanar diagram, parametrize the propagators exponentially by introducing Schwinger parameters and regulate by introducing a smooth UV cutoff $\exp[-1/(\Lambda^2\alpha)]$. Then the above diagrams are respectively given by

$$\Gamma^{(2)}_{planar} = -\frac{2\lambda}{16\pi^2}\Lambda^2+\ldots\,,\ \Gamma^{(2)}_{nonplanar} = -\frac{\lambda}{16\pi^2}\Lambda^2_{eff}+\ldots \tag{9}$$

where

$$\Lambda^2_{eff} = \frac{1}{\frac{1}{\Lambda^2}+p\circ p}\,,\ p\circ q = \frac{-p_\mu\theta^2_{\mu\nu}q_\nu}{4} \tag{10}$$

The central point to be gleaned here is that, perhaps somewhat counterintuitively, as $\Lambda\to\infty$, the nonplanar graph remains finite and is regulated by the noncommutativity of spacetime.

There are two regimes of interest: $p\circ p \ll 1/\Lambda^2$, and $p\circ p \gg 1/\Lambda^2$. In the former, $\Lambda_{eff}\approx\Lambda$ and we recover the effective action for the commutative theory, while in the latter, $\Lambda_{eff}\approx 1/(p\circ p)$, and $S^{(2)}_{1PI}$ diverges as $p\to 0$. Thus ultraviolet ($\Lambda\to\infty$) and infrared ($p\to 0$) limits do not commute[3].

Is this behaviour general? Consider a single complex scalar. Now we can have two distinct orderings of the quartic term:

$$A\phi^* * \phi * \phi^* * \phi + B\phi^* * \phi^* * \phi * \phi. \tag{11}$$

Structural renormalizability of the four-point amplitude at one loop requires either $A = B$ or $B = 0$[5]. For $B = 0$ there are no IR divergences of the type found above for ordinary ϕ^4, and furthermore only the first ordering is consistent with noncommutative $U(1)$ gauge invariance, under which $\phi \to U*\phi$, $\phi^* \to \phi^* * U^\dagger$.

4 SSB and the NC O(N) linear sigma model

The preceding discussion motivates at least two immediate reasons why we should be interested in spontaneous symmetry breaking in noncommutative theories. First, given the absence of the IR part of the spectrum seen above, what happens to the *massless* Goldstone modes associated with spontaneous symmetry breaking? Secondly, in commutative theories, Goldstone's theorem at the quantum level holds as a consequence of intricate graphical cancellations, and we have seen that already in noncommutative ϕ^4 theory, the graphs are re-weighted with respect to their commutative counterparts.

Thus consider first the spontaneously broken phase of the $O(N)$ linear sigma model, with scalars in the fundamental representation. We consider global symmetries and translationally invariant vacua, which explicitly do not see the star product. The tree level spectrum consists of $N-1$ pions, and one sigma.

Now consider the mass renormalization of the pions at one-loop. One-point tadpoles are oblivious to the noncommutativity at one-loop because of momentum conservation, and thus the one-point counterterm is the same as commutative case. In turn, the SSB structure implies the pion-mass counterterm is fixed. On the other hand, the 1PI graphs contributing to pion-mass renormalization are split into planar and nonplanar pieces in such a way that the sum of the 1PI graphs and the fixed counterterm is given by

$$\begin{aligned}\sum_{1-loop} &= \frac{\lambda\delta^{ij}}{16\pi^2}\left\{\Big[N(1-f)+f\Big]\Lambda^2\left(1-\frac{1}{1+\Lambda^2(p\circ p)}\right)\right.\\ &\left. -\Big(2-f\Big)m_\sigma^2\log(1+\Lambda^2(p\circ p)) + finite\cdot p^2\right\}. \end{aligned}\tag{12}$$

Here, the parameter f reflects the relative weighting of the two possible quartic terms consistent with the $O(N)$ global symmetry:

$$f\phi_i * \phi_i * \phi_j * \phi_j + (1-f)\phi_i * \phi_j * \phi_i * \phi_j. \tag{13}$$

Thus unless $f = 2$ and $N = 2$, not only do UV and IR limits not commute, the continuum limit does not exist, and Goldstone's theorem fails. The case $f = 2$, $N = 2$ corresponds to the Abelian $O(2)$ model, and written as a *complex* scalar corresponds to the ordering $\phi^* * \phi * \phi^* * \phi$. Furthermore, a detailed analysis of the four point amplitudes (in particular the pi-pi-sigma-sigma amplitude itself, and consistency between the 4-sigma and pi-pi-sigma-sigma amplitudes) also yields the consistency conditions $N = 2$ and $f = 2$.

5 Noncommutative linear sigma models

Since an attempt to build realistic models of particle physics on NC spacetimes would necessarily involve spontaneously broken gauge theories, which must have consistent global limits, it becomes imperative that we be able to evade the fatality of the previous section.

Consider instead the $U(N)$ linear sigma model, again with fundamental matter, in the SSB phase, choosing *a priori* the quartic ordering consistent with a possible gauging of the model $\phi^\dagger * \phi * \phi^\dagger * \phi$. Now the tree-level spectrum in the broken phase consists of $N-1$ complex pions, one real pion, and one sigma. At one-loop, the one-point functions again do not see noncommutativity, but now however, the complex pion mass renormalization 1PI graphs also do not see the noncommutativity (no phases are generated). Furthermore, the real pion mass renormalization 1PI graph phases cancel amongst themselves. Thus, we do not run into any conflict between continuum renormalizability and Goldstone symmetry realization for this model, at this order.

Since $U(N) \subset O(2N)$, one might wonder how the wildly different behaviour between the two models can occur. The answer at the algebraic level is that the $U(N)$ case generates new, purely noncommutative quartic interactions incompatible with $O(2N)$ symmetry as follows. First note that $(R-iI)(R+iI) = R^2 + I^2 + i[R,I]$. While such commutators in the quadratic term vanish when integrated over spacetime, the quartic invariant now picks up products of such commutators with other fields or commutators which, for $N > 1$, constitute new interactions between real components of two compolex ϕ's. Explicitly, we have (suppressing explicit star products):

$$\begin{aligned} \phi_i^*\phi_i\phi_j^*\phi_j &= \phi_{iR}^2\phi_{jR}^2 + \phi_{iR}^2\phi_{jI}^2 + \phi_{iI}^2\phi_{jR}^2 + \phi_{iI}^2\phi_{jI}^2 - [\phi_{iR},\phi_{iI}][\phi_{jR},\phi_{jI}] \\ &\quad + i(\phi_{iR}^2 + \phi_{iI}^2)[\phi_{jR},\phi_{jI}] + i(\phi_{jR}^2 + \phi_{jI}^2)[\phi_{iR},\phi_{iI}]. \end{aligned} \tag{14}$$

Thus we see the presence of new purely noncommutative terms with three or four distinct fields, which can occur for $i \neq j$, and $N > 1$, incompatible with $O(2N)$ symmetry. At the Feynman rule level, this argument essentially means that there are fewer symmetrizations occurring.

It is also natural to wonder about other scalar field representations, and in particular the adjoint representation. Other than allowing us to compare with the fundamental representation, adjoint scalars occur in $\mathcal{N} \geq 2$ SUSY gauge theories on noncommutative branes. Furthermore, in grand unified theories, the primary spontaneous symmetry breaking is accomplished with adjoint representation scalars.

Thus first consider the NC $U(2)$ model with adjoint matter Φ, taking the scalar potential

$$V(\Phi) = -\mu^2 \mathrm{Tr}(\Phi_*^2) + \lambda_1 \mathrm{Tr}(\Phi_*^4) + \lambda_2 [\mathrm{Tr}(\Phi_*^2)]^2 \tag{15}$$

where $\mathrm{Tr}(\Phi_*^4) \equiv \Phi^i_j * \Phi^j_k * \Phi^k_l * \Phi^l_i$, and $[\mathrm{Tr}(\Phi_*^2)]^2 \equiv \Phi^i_j * \Phi^j_i * \Phi^k_l * \Phi^l_k$. The tree-level spectrum consists of a complex pion, a sigma, and a $U(1)$ spectator. Defining $I(m^2) = \int \frac{d^4k}{(2\pi)^4} \frac{1}{k^2 - m^2}$ and $I_{\theta,p}(m^2) = \int \frac{d^4k}{(2\pi)^4} \frac{e^{ik \times p}}{k^2 - m^2}$, the one-loop mass corrections to pion from graphs which survive in the commutative limit sum to

$$\frac{\lambda_1}{2}[I(0) - I_{\theta,p}(0)] - \lambda_2 \left[I(m_\sigma^2) - I_{\theta,p}(m_\sigma^2)\right] - \frac{\lambda_1}{2}\left[I(m_s^2) - I_{\theta,p}(m_s^2)\right]. \tag{16}$$

Now however, there is a purely noncommutative graph involving the $U(1)$ component of Φ which precisely cancels the λ_1 dependent pieces above from NC commutator interaction. Furthermore, in a *noncommutative* gauge theory, products of trace invariants are *not* gauge invariant (even under $\int d^4x$). A simple, illustrative proof for $[\mathrm{Tr}(\Phi)]_*^2$ runs as follows:

$$\Phi^i_j * \Phi^j_i \rightarrow U^i_{i1} * \Phi^{i1}_{j1} * U^{\dagger j1}_j * U^j_{j2} * \Phi^{j2}_{i2} * U^{\dagger i2}_i. \tag{17}$$

Thus by taking $\lambda_2 = 0$, Goldstone's theorem holds for this model. Also, other internal index orderings (with respect to the star product) are forbidden, as well as the other neglected trace invariants. *If* we were to include $[\mathrm{Tr}(\Phi_*^2)]^2$ in the corresponding noncommutative gauge theory, we find that the physical Higgs receives a gauge-dependent, divergent mass renormalization, even on-shell, corroborating this claim. Thus, we yet again see that even the global models are consistent only if we take into account the restrictions imposed by noncommutative gauge invariance.

An immediate consequence of this argument for the adjoint representation is that the symmetry breaking pattern for adjoint $U(N)$'s are restricted, as compared with the possibilities allowed for commutative theories. As well, the infrared divergences reported by Seiberg and Van Raamsdonk in purely scalar theories with $N \times N$ matrix matter, are all precisely of the form $\mathrm{Tr}(O_1) * \mathrm{Tr}(O_2)^4$. Thus, in the corresponding gauge theory they would not occur for $N > 1$, since these potential terms are now forbidden by gauge invariance.

Finally, we considered the noncommutative $O(4)$ adjoint representation linear sigma model. The sum of the 1PI graphs, and one-point insertions is similar to the noncommutative $U(2)$ adjoint representation case, but now there is no purely noncommutative graph coming from a NC commutator interaction. Thus, as in the fundamental representaion $O(N)$ model, the continuum limit does not exist and Goldstone's theorem fails.

References

1. B.A. Campbell & K. Kaminsky, *Noncommutative field theory and spontaneous symmetry breaking*, *Nucl. Phys.* **B581**, 240 (2000). hep-th/0003137.
2. B.A. Campbell & K. Kaminsky, *Noncommutative linear sigma models*, *Nucl. Phys.* **B606**, 613 (2001). hep-th/0102022.
3. S. Minwalla, M. Van Raamsdonk & N. Seiberg, *Noncommutative Perturbative Dynamics*, *JHEP* **0002**, 020 (2000). hep-th/9912072.
4. M. Van Raamsdonk & N. Seiberg, *Comments on noncommutative perturbative dynamics*, *JHEP* **0003**, 035 (2000). hep-th/0002186.
5. I. Ya. Aref'eva, D.M. Belov & A.S. Koshelev, *A note on UV/IR for noncommutative complex scalar field*, hep-th/0001215.

THE ALEPH SEARCH FOR THE STANDARD MODEL HIGGS BOSON

JOHN KENNEDY on behalf of the ALEPH collaboration
Department of Physics and Astronomy, University of Glasgow
Kelvin building, University Avenue
Glasgow, G12 8QQ, UK

A search has been performed for the Standard Model Higgs boson in the data collected with the ALEPH detector in 2000. An excess of 3σ above the background expectation is found. The observed excess is consistent with the production of the Higgs boson with a mass close to 114 GeV/c^2.

1 Introduction

This note presents results on the search for the Standard Model Higgs boson by the ALEPH collaboration in the year 2000. This search primarily looks for the production of Higgs bosons via the Higgsstrahlung process, although some search channels also possess a limited sensitivity to Higgs production via W and Z vector boson fusion. In total, 216.2pb^{-1} of data was collected at centre of mass energies ranging from 200 GeV to 209 GeV, with the majority collected around 205.1 GeV (72pb^{-1}) and 206.7 GeV (107pb^{-1}).

2 Event Selection

The search analyses, described in detail elsewhere [1,2], are designed to detect specific "channels" or final states arising from the Higgsstrahlung production process:

- The four jet final state (h$q\bar{q}$)†.
- The missing energy final state (h$\nu\bar{\nu}$)†.
- The leptonic final state (h$\ell^+\ell^-$).
- The tau lepton final state (h$\tau\tau$ and h$\to \tau\tau$, Z $\to q\bar{q}$).

The Higgs boson search was conducted using both a Neural Network based stream (denoted "NN") and a cuts based stream ("cuts"). The two streams are formed from the above four final state analyses where final states marked with a † have alternate analyses based on NN's and cuts and the searches are identical in both streams for the leptonic and tau final states. The analysis selection criteria were fixed before the data taking period began thus ensuring that their application to the collected data was unbiased.

3 Results and Statistical Interpretation

The number of observed and expected events for each analysis channel and the total combined NN and cuts streams are given in Table 1.

Table 1: The number of expected signal and background events for each analysis channel with the expected significance and number of observed candidates.

Search channel	Expected background	Expected Signal	Events Observed	Expected Significance(σ)
4-jet (NN)	46.9	4.5	52	1.6
4-jet (Cut)	23.7	2.9	31	1.3
h$\nu\nu$ (NN)	37.5	1.4	38	0.8
h$\nu\nu$ (Cut)	19.7	1.3	20	0.7
h$\ell\ell$	30.6	0.7	29	0.8
$\tau\tau q\bar{q}$	13.6	0.4	15	0.4
Tot (NN)	128.7	7.0	134	2.1
Tot (Cut)	87.6	5.3	95	1.8

Figures 1a and 1b show the distributions of the reconstructed Higgs boson mass($m_{REC} = m_{12} + m_{34} - 91.2$) for the data and expected background in the NN and cuts based streams respectively. Both figures show good agreement between data (points) and the expected background (histogram) at low reconstructed Higgs masses while they both display a clear excess of data candidates at large reconstructed Higgs masses.

3.1 Confidence Level Results

The mass is not the only information which may be used to distinguish a Higgs boson signal from SM background. Further information (eg. b-tag, NN$_{output}$) is taken into account in the construction of the likelihood ratio $Q = L_{s+b}/L_b$, where L_b is the likelihood for the background-only hypothesis and L_{s+b} is the likelihood for the signal+background hypothesis for a given Higgs boson mass. The likelihood ratio measures the compatibility of the experiment with a given signal mass hypothesis and is defined as:

$$Q = \frac{L_{s+b}}{L_b} = \frac{exp^{-(s+b)}}{exp^{-b}} \prod_{i=1}^{n_{obs}} \frac{sf_s(X_i) + bf_b(X_i)}{bf_b(X_i)} \tag{1}$$

where s and b are the total number of signal and background events expected. The functions f_s and f_b are the probability densities that a signal

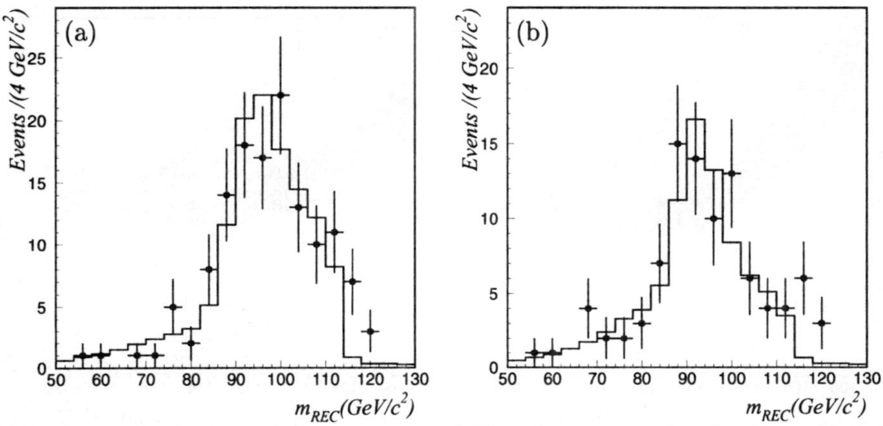

Figure 1: Distributions of the reconstructed Higgs boson mass for the data collected in 2000(points) and the expected background(histogram) for the NN(a) and cuts(b) streams.

or background event will be found in a given final state identified by a set of discriminating variables. By removing these functions, the likelihood ratio is returned to the ratio of the Poisson probabilities to observe n_{obs} events for the signal+background and background-only hypotheses.

The observed and expected distributions of the likelihood ratio, expressed as -2ln$Q$, are shown in Figures 2a and 2b for the NN and cuts streams respectively.

The compatibility of an experiment with a given hypothesis is derived by calculating the probability of obtaining a likelihood ratio smaller than that observed. This probability is referred to as the confidence level(CL). When observed results are tested against a background-only hypothesis we use the quantity CL_b. In the case of the expected SM background CL_b has a median value at 0.5 while the observation of a signal is expected to produce an excess relative to the background and the value of CL_b (1-CL_b) is expected to rise (drop).

Figures 3a and 3b show the observed and expected distributions for 1-CL_b as a function of the hypothesised Higgs boson mass. A large deviation from the expected background value of 0.5 can be seen in both the NN and cuts based streams. This deviation is consistent with an excess of events over the

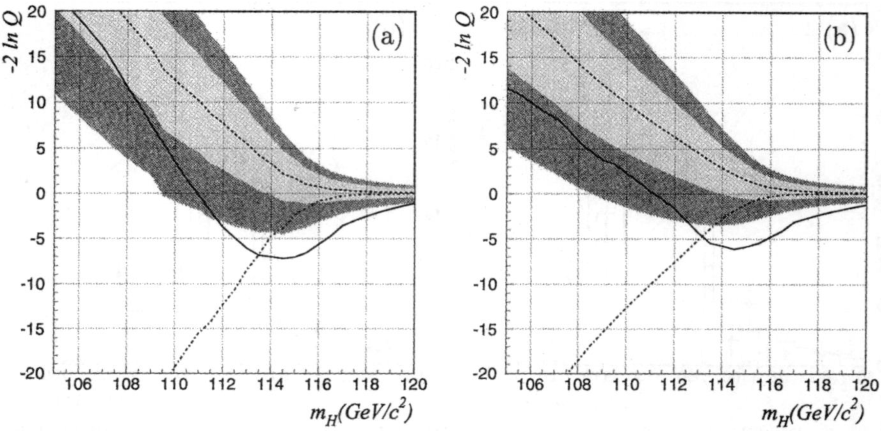

Figure 2: The log-likelihood estimator -2ln$Q$ for the NN(a) and cuts(b) streams as a function of the hypothesised Higgs boson mass. Observed(solid), background only expectation(dashed) and signal(dot-dash) are shown with light and dark bands around the background expectation representing the one and two sigma bands respectively.

background only hypothesis and is maximal for a Higgs boson mass of ≈ 116 GeV/c^2. The probability of deviations of this magnitude are 1.5 x 10^{-3} and 1.1 x 10^{-3} corresponding to significances of 3.0σ and 3.1σ [a] for the NN and cuts streams respectively.

3.2 Analysis of 'High Impact' Candidates

The 'Quality' of individual candidates may be assessed by two independent methods. These methods allow us to gain information about the individual candidates and also assess the origin and stability of the observed excess.

The impact of each candidate may be determined by calculating the contribution of the candidate to -2ln$Q$. In the calculation of -2ln$Q$, individual candidates contribute as a sum of event weights, $\ln(1+\frac{sf_s}{bf_b})$, which may then be analysed. Figures 4a and 4b show the event weights, as a function of hypothesised Higgs boson mass, for all candidates with weights larger than 0.4 at a mass of 114 GeV/c^2 for the NN and cuts analyses respectively. The details

[a]The ALEPH significances are calculated using single sided Gaussian distributions

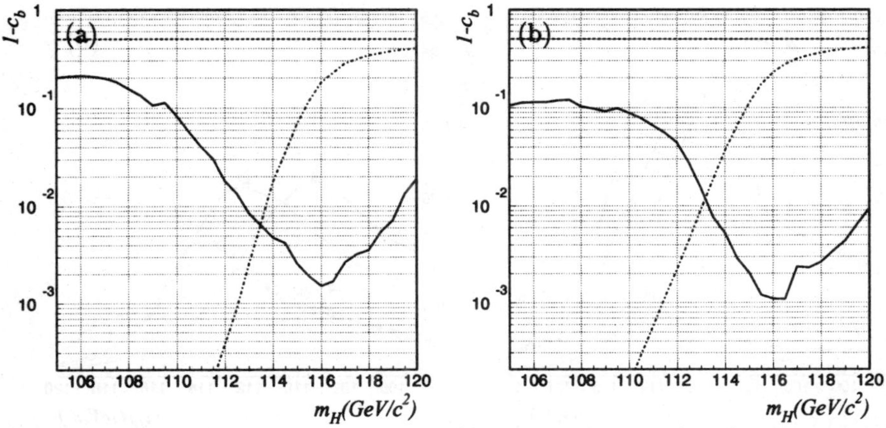

Figure 3: Distributions of the CL_b curves for the Observed(solid), expected background(dashed) and signal+background(dot-dash) for the NN(a) and cuts(b) streams.

of the highlighted 'high impact' candidates, a-e, all of which occur in the four jet final state, are found in Table 2.

Alternatively the 'Quality' of the candidates may be assessed by considering their 'purity'. Here we define purity to be the ratio of expected number of signal to background events with a reconstructed Higgs boson mass greater than 109 GeV/c^2, from here on denoted as $(s/b)_{109}$. Tightening the selection criteria of the analyses enables the purity to be increased and allows us to gain information about the candidate 'Quality' and also the stability of the observed excess. High purity values $(s/b)_{109}$=1.5 are achieved by tightening the NN cut on the four jet selection in the NN stream and tightening b-tag/Z-mass constraints on the four jets selection in the cuts stream. Figures 5a and 5b show the high purity distributions of the reconstructed Higgs boson mass for the NN and cuts streams respectively.

A large overlap between the high 'weight' and high 'purity' candidates adds confidence to the observed result and the stability of the excess.

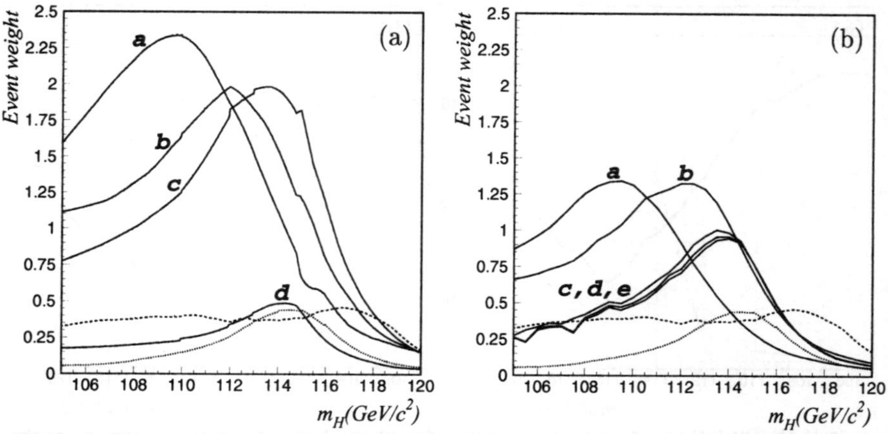

Figure 4: Event weights for the individual candidates as a function of the hypothesised Higgs boson mass for the NN(a) and cuts(b) based streams. Contributing candidates are from the four jet(solid), leptonic(dashed) and tau(dotted) channels.

4 Conclusion

The data collected with the ALEPH detector in 2000 have been analysed to search for the Standard Model Higgs boson. Both NN and Cuts based analysis streams show a 3σ excess beyond the Standard Model background expectation. The observed result is consistent with the production of a Higgs boson with a mass near 114 GeV/c^2.

Table 2: Details of the 'High Impact' candidates with weights above 0.4 at a Higgs boson mass of 114 GeV/c^2.

Event	Higgs Mass (GeV/c^2)	m_{12} (GeV/c^2)	m_{34} (GeV/c^2)	b-tag Jet 1	b-tag Jet 2	b-tag Jet 3	b-tag Jet 4	4-Jet NN
a	110.0	96.3	104.9	0.999	0.836	0.999	0.214	0.999
b	112.9	94.9	109.2	0.994	0.776	0.993	0.999	0.997
c	114.3	101.3	104.2	0.136	0.012	0.999	0.999	0.996
d	114.5	78.8	126.9	0.238	0..52	0.998	0.948	0.935
e	114.6	79.7	126.1	0.008	0.293	0.895	0.998	0.820

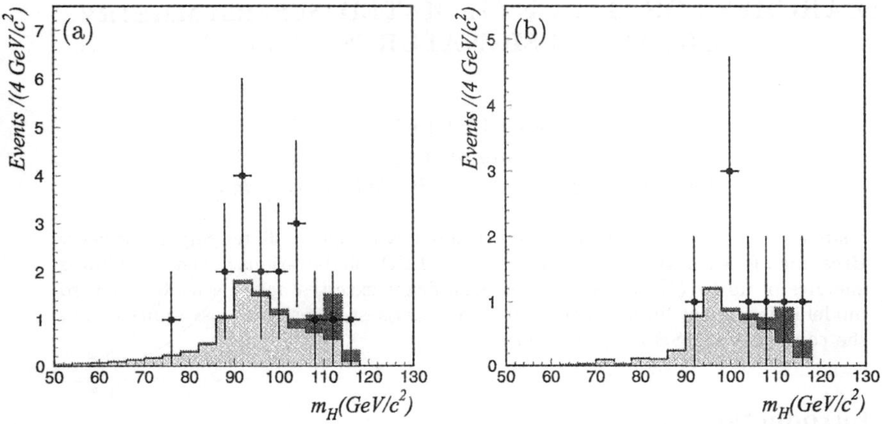

Figure 5: High 'Purity' $(s/b)_{109}$ reconstructed Higgs boson mass distributions for the NN(a) and cuts(b) streams. Data(points with errors), expected background(light histogram) and expected signal with $m_h = 114\text{GeV}/c^2$(dark histogram).

References

1. ALEPH Collaboration, Search for the neutral Higgs boson of the Standard Model and the MSSM in e^+e^- collisions at $\sqrt{s} = 189$ GeV, CERN-EP/2000-019, to be published in *Eur. Phys. J.* **C.**
2. ALEPH Collaboration, Observation of an excess in the search for the Standard Model Higgs boson at ALEPH' *Phys. Lett.* **B495**, 1 (2000).

SEARCHES FOR GAUGE MEDIATED SUPERSYMMETRY BREAKING SIGNATURES AT LEP2

KATJA KLEIN
University of Heidelberg
Philosophenweg 12, 69120 Heidelberg, Germany

Searches for neutralinos and sleptons predicted by Gauge Mediated Supersymmetry Breaking models have been performed by the LEP collaborations at center–of–mass energies of up to $\sqrt{s} = 209$ GeV. No significant excesses have been observed, so model independent limits on the production cross–sections and mass limits within the context of GMSB are presented.

1 Introduction

In supersymmetry (SUSY), an extension of the Standard Model (SM) designed mainly to solve the finetuning problem, every Standard Model particle has a supersymmetric 'partner' whose spin differs by half a unit.

If SUSY were an exact symmetry, the SUSY particles would have the same masses as their SM partners. From the fact that no SUSY particles have been observed yet it can be followed that the SUSY partners must be much heavier than the Standard Model particles and therefore SUSY must be a broken symmetry. Several breaking mechanisms have been considered. One possibility is that SUSY is broken via the usual gauge interactions in a 'hidden' sector, which couples to the visible sector of the SM and SUSY particles via a 'messenger sector' [1,2]. This model is called Gauge Mediated Supersymmetry Breaking (GMSB). In its minimal version there are five new parameters in addition to the SM parameters, usually chosen to be the effective SUSY breaking scale, Λ, which sets the mass scale of the SUSY particles, the messenger scale, M, the messenger index, N, the ratio of the vacuum expectation values of the two Higgs doublets, $\tan\beta$, and the sign of the Higgs sector mixing parameter, sign(μ). The scale of dynamical SUSY breaking, $\sqrt{F}$, depends on Λ and M and is typically of the order of 100 TeV.

In supersymmetric theories the phenomenology depends crucially on the nature of the lightest and next–to–lightest supersymmetric particles, the LSP and the NLSP. In GMSB the SUSY partner of the graviton, the gravitino $\tilde{G}$, is light due to the low SUSY breaking scale, and is the LSP. Depending on the choice of parameters of the model the NLSP can either be the lightest neutralino, $\tilde{\chi}^0_1$, which is a mixture of the SUSY partners of the γ, Z^0 and the neutral Higgs bosons, or the lightest SUSY partner of the leptons, a (right–handed) slepton, $\tilde{l}_R$.

One speciality of GMSB concerns the decay length $c\tau$ of the NLSP, which depends, apart from kinematical factors, only on $\sqrt{F}$:

$$c\tau = \frac{0.01}{\kappa_\gamma}\left(\frac{100\mathrm{GeV}}{m}\right)^5\left(\frac{\sqrt{F}}{100\mathrm{TeV}}\right)^4\left(\frac{E^2}{m^2}-1\right)^{1/2}\ \mathrm{cm} \qquad (1)$$

where κ_γ is a parameter of order unity depending on the photino component of the neutralino and m and E are the mass and energy of the NLSP. Taking into account the range allowed for $\sqrt{F}$ the NLSP decay length is basically unconstrained and therefore all possible decay lengths between zero and infinity have to be considered.

In the following it is assumed that R–parity is conserved, which implies that SUSY particles can be produced only in pairs and all decay chains have to terminate with the LSP, which is stable, plus SM particles.

The four LEP collaborations, ALEPH, DELPHI, L3 and OPAL, have searched for a variety of topologies expected from GMSB models. In this report a selection of these results [3,4,6,7,8], based mainly on the data recorded from 1998–1999 with center–of–mass energies between 189 GeV and 202 GeV, is presented. Some preliminary results from the data recorded in the year 2000 with center–of–mass energies of up to 209 GeV are also shown [5,9,10]. All presented limits are at 95% confidence level (C.L.).

2 The Neutralino NLSP Scenario

In GMSB, the main production channel for the $\tilde{\chi}_1^0$ in e^+e^- collisions is direct pair production in s–channel processes mediated by γ^*/Z^0 or in the t–channel via exchange of a selectron. Indirect production via slepton pair production followed by decays $\tilde{l}_R \to l\tilde{\chi}_1^0$, is also possible. In non–minimal versions of GMSB, chargino pair production followed by decays $\tilde{\chi}_1^+ \to W^+\tilde{\chi}_1^0$, and production of $\tilde{\chi}_1^0\tilde{\chi}_2^0$ with $\tilde{\chi}_2^0 \to Z^0\tilde{\chi}_1^0$, have to be considered as well.

If the $\tilde{\chi}_1^0$ is the NLSP, all final states will contain photons and missing energy, since the $\tilde{\chi}_1^0$ decays as $\tilde{\chi}_1^0 \to \gamma\tilde{G}$, and the gravitino, due to its weakly interacting nature, escapes detection.

2.1 Searches for Promptly Decaying Neutralinos

If the neutralinos are produced in pairs and decay promptly, the topology is two high momentum photons and significant missing energy and transverse momentum. The SM background in this channel is dominated by neutrino pair events with at least two radiated photons. All four LEP collaborations have

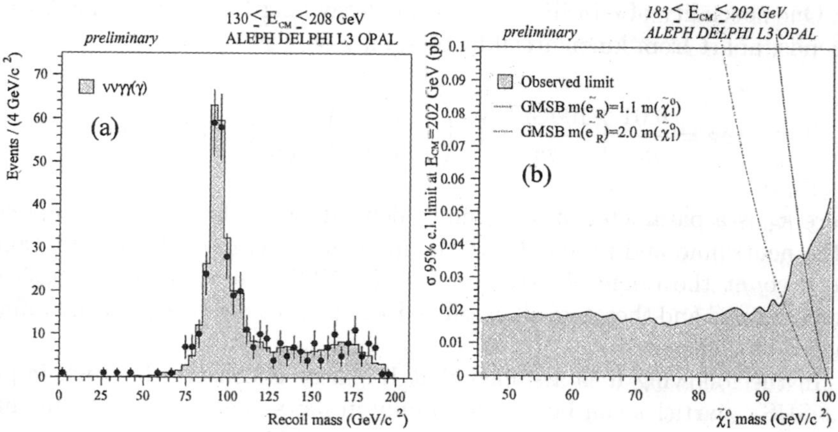

Figure 1: (a) Recoil mass distribution and (b) LEP combined cross–section limits at 95% C.L. for $\tilde{\chi}_1^0$ pair production. The theoretical expectations for two different $\tilde{e}_R$ masses are also shown.

searched for this signature and the results have been combined [9]. The distribution of the mass recoiling against the photonic system (Figure 1(a)) shows good agreement between data and the expectation from the neutrino background. As no excess over the SM expectation has been observed, cross–section limits at 95% C.L. have been derived (Figure 1(b)). The lower neutralino mass limit depends on the $\tilde{e}_R$ mass and is about 93 GeV for $m(\tilde{e}_R) = 2m(\tilde{\chi}_1^0)$, using the combined data from all four LEP experiments at $\sqrt{s} = 183 - 202$ GeV.

Promptly decaying neutralinos from indirect production, with a signature of photons, missing energy and leptons and/or jets from the decays of the $\tilde{l}_R$ and the $W^{\pm}$ and Z^0 bosons, have been searched for by the OPAL collaboration, but no excess over the estimated background has been found.

2.2 Searches for Long-lived Neutralinos

If the $\tilde{\chi}_1^0$ has a lifetime such that it decays in the detector, the signature consists of photons which do not point to the primary interaction point, and missing energy. To detect such non–pointing photons the direction of the photon must be reconstructed from the shower profile in the electromagnetic calorimeter. The ALEPH and DELPHI collaborations have searched for this signature but see no excess over the background expectation.

As stable neutralinos are invisible, only indirect searches are possible for decay lengths exceeding the detector dimensions. A relationship between the particle masses can be exploited to extract indirect limits on the $\tilde{\chi}_1^0$ mass from the results for MSSM sleptons and charginos.

3 The Slepton NLSP Scenario

In GMSB, sleptons can be produced in e^+e^- collisions either directly in pairs via s–channel processes (and via t–channel with $\tilde{\chi}_1^0$ exchange for the $\tilde{e}_R$) or indirectly via $\tilde{\chi}_1^0$ pair production followed by the decay $\tilde{\chi}_1^0 \to \tilde{l}_R l$. If the $\tilde{\tau}_R$ is significantly lighter than the other sleptons, which means that only the $\tilde{\tau}_R$ forms the NLSP, production might take place also via $\tilde{l}_R$ pair production followed by decays $\tilde{l}_R \to l\tilde{\tau}_R\tau$, where $\tilde{l}_R = \tilde{e}_R$ or $\tilde{\mu}_R$. This is called the stau NLSP scenario in contrast to the slepton co–NLSP scenario in which all sleptons are mass–degenerate.

If the sleptons form the NLSP, all final states will contain leptons and missing energy, since the $\tilde{l}_R$ decays as $\tilde{l}_R \to l\tilde{G}$.

3.1 Searches for Promptly Decaying Sleptons

If the sleptons are produced in pairs and decay promptly the topology is an acoplanar high–momentum lepton pair. This signature is completely identical to the topology expected in the 'Constrained Minimal Supersymmetric Standard Model' for a massless LSP (which is the $\tilde{\chi}_1^0$ in this model). All LEP collaborations have searched for this topology and no excess of candidates over the estimated background has been observed[10]. The LEP–combined mass limits using the data recorded at $\sqrt{s} = 183 - 208$ GeV are 100.5 GeV for the $\tilde{e}_R$, 95.4 GeV for the $\tilde{\mu}_R$, and 80.0 GeV for the $\tilde{\tau}_R$.

Promptly decaying sleptons from $\tilde{\chi}_1^0$ production have also been searched for by the ALEPH, DELPHI and OPAL collaborations, but good agreement between the data and the estimated SM background has been found. The same is true for promptly decaying staus from $\tilde{e}_R$ or $\tilde{\mu}_R$ pair production, searched for by the OPAL collaboration.

3.2 Searches for Long-lived Sleptons

Sleptons with a lifetime such that the decay happens in the tracking devices of the detector lead to topologies of tracks with large impact parameters or tracks with kinks, depending on the decay length. Long-lived sleptons have been searched for by the ALEPH, DELPHI and OPAL collaborations, but no excess over the expectation from SM events has been found. The limits on the slepton masses at 95% C.L. obtained by DELPHI using the data at

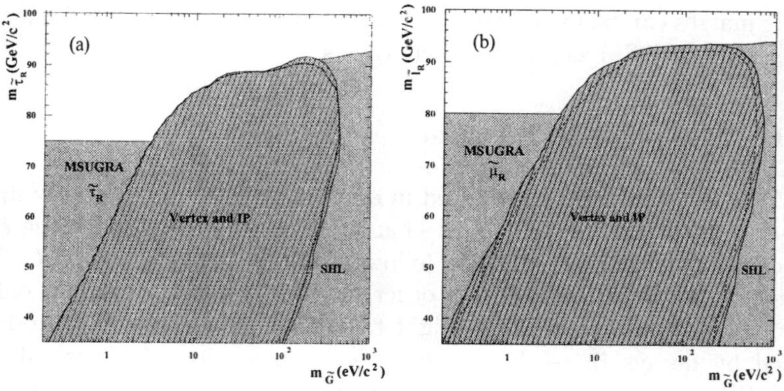

Figure 2: Excluded regions in the m_{NLSP} *vs.* $m_{\tilde{G}}$ plane ($m_{\tilde{G}}$ corresponds to the NLSP lifetime) obtained for (a) staus and (b) mass–degenerate sleptons by DELPHI using data at $\sqrt{s} = 130 - 202$ GeV by combining the results for long–lived sleptons (hatched) with results for promptly decaying und stable sleptons.

$\sqrt{s} = 130 - 202$ GeV are shown as hatched regions in Figure 2 for the stau NLSP scenario (a) and the slepton co–NLSP scenario (b).

3.3 Searches for Stable Sleptons

If the sleptons are stable, they, owing to their large masses, can be identified by their anomalous, high or low ionization energy loss (dE/dx) in the tracking chamber (Figure 3, left). Searches for this almost background–free signature have been performed by all LEP collaborations and good agreement between the data and the expectations from the SM background have been found. Therefore cross–section limits have been computed, e.g. by OPAL (Figure 3, right), and mass limits of 98 GeV for smuons and staus have been achieved using 220 pb^{-1} of data recorded in 2000 with the OPAL detector.

3.4 Lifetime Independent Slepton Mass Limits

The results for promptly decaying, long–lived and stable sleptons have been combined and lifetime independent lower limits of $m_{\tilde{l}_R} > 80$ GeV in the slepton co–NLSP scenario (DELPHI, $\sqrt{s} = 130 - 202$ GeV) and $m_{\tilde{\tau}} > 67$ GeV in the stau NLSP scenario (ALEPH, $\sqrt{s} = 189$ GeV) have been obtained.

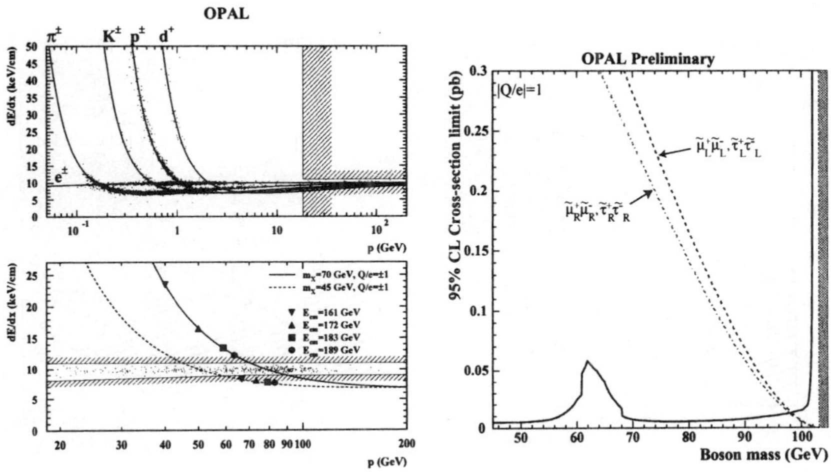

Figure 3: Left, top: dE/dx of SM particles measured in the OPAL jet chamber. The two hatched regions are the search regions. Left, bottom: Expanded view of the search regions with simulated dE/dx of heavy particles for different masses and $\sqrt{s}$. Right: Cross–section limits, and theoretical expectations for stable smuons and staus, from OPAL.

4 Interpretations

By combining the results for sleptons and neutralinos in the different search channels with corresponding results from LEP1 and indirect constraints on stable neutralinos general NLSP mass limits have been calculated (Figure 4). The lowest limit is obtained for long lifetimes if staus and neutralinos are almost mass–degenerate, as can be seen from Figure 4(b). A general NLSP mass limit independent of the nature of the NLSP of 45 GeV has been reported by the ALEPH collaboration using data recorded at $\sqrt{s} = 189$ GeV.

Scans over the GMSB parameter space have been performed by the LEP collaborations resulting in limits on the parameter Λ which sets the overall mass scale for the SUSY sector. For $N \leq 5$, a lower bound on Λ of 9 TeV has been found by ALEPH using the data at $\sqrt{s} = 189$ GeV, from which a lower limit on the gravitino mass of 0.02 eV has been derived.

5 Conclusions

Searches for neutralinos and sleptons in topologies predicted by GMSB have been performed by the four LEP collaborations at center–of–mass energies of

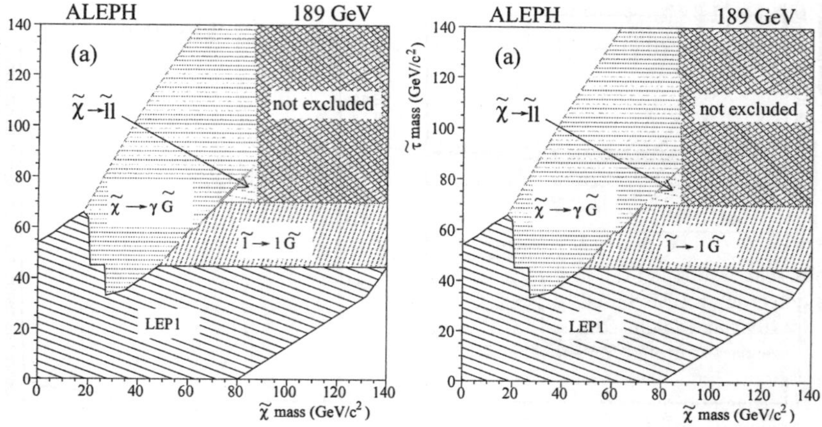

Figure 4: Excluded regions in the $m_{\tilde{\tau}}$ *vs.* $m_{\tilde{\chi}_1^0}$ plane for (a) short and (b) long lifetimes from ALEPH at $\sqrt{s} = 189$ GeV. The white regions are theoretically inaccessible.

up to 209 GeV. Since no evidence for the production of these particles has been observed, limits are placed on the production cross–sections and the particle masses in the framework of the GMSB model.

References

1. G.F. Giudice and R. Rattazzi, *Phys. Rept.* **322**, 419 (1999).
2. S. Dimopoulos, S. Thomas and J.D. Wells, *Nucl. Phys.* **B488**, 39 (1997).
3. OPAL Collaboration, G. Abbiendi *et al*, *Phys. Lett.* **B501**, 12 (2001).
4. OPAL Collaboration, *Searches for Light Gravitino Signatures in e^+e^- Collisions at $\sqrt{s}$=189 GeV*, OPAL Physics Note PN397, July 1999.
5. OPAL Collaboration, *New Particle Searches in e^+e^- Collisions at $\sqrt{s}$=200-209 GeV*, OPAL Physics Note PN470, February 2001.
6. DELPHI Collaboration, P. Abreu *et al*, *Eur. Phys. J.* **C16**, 211 2000.
7. DELPHI Collaboration, P. Abreu *et al*, *Phys. Lett.* **B503**, 34 (2001).
8. ALEPH Collaboration, R. Barate *et al*, *Eur. Phys. J.* **C16**, 71 (2000).
9. http://lepsusy.web.cern.ch/lepsusy/www/photons/acoplanar/acopl_public_moriond01.html
10. http://alephwww.cern.ch/~ganis/SUSYWG/SLEP/sleptons_2k01.html

SEARCH FOR THE STANDARD MODEL HIGGS BOSON AT $\sqrt{s} = 192 - 209\,\mathrm{GeV}$ WITH THE OPAL DETECTOR AT LEP°

THORSTEN KUHL

Physikalisches Institut der Universität Bonn, Nußallee 12, 53113 Bonn
E-mail: Thorsten.Kuhl@physik.uni-bonn.de

A search for the Standard Model Higgs boson has been performed with the OPAL detector at LEP based on the full data sample collected at $\sqrt{s} \approx$ 192–209 GeV in 1999 and 2000, corresponding to an integrated luminosity of 426 pb^{-1}. The data are examined for their consistency with the background-only hypothesis and various Higgs boson mass hypotheses. A lower bound of 109.7 GeV is obtained on the Higgs boson mass at the 95% confidence level. At higher masses, the data are consistent with both the background and the signal-plus-background hypotheses.

1 Introduction

In the Standard Model[1], the Higgs mechanism[2] gives mass to the electroweak gauge bosons, thus allowing the unification of the electromagnetic and weak interactions. Whether the Higgs boson exists is one of the most important open questions in particle physics.

Searches are performed for the "Higgs-strahlung" process $\mathrm{e^+e^-} \to \mathrm{H^0Z^0} \to \mathrm{H^0f\bar{f}}$, where $\mathrm{H^0}$ is the Standard Model Higgs boson, and $\mathrm{f\bar{f}}$ is a fermion-antifermion pair from the $\mathrm{Z^0}$ decay. For the $\mathrm{H^0}\nu\bar{\nu}$ ($\mathrm{H^0e^+e^-}$) final state, the contribution from the $\mathrm{W^+W^-}$ ($\mathrm{Z^0Z^0}$) fusion process is also taken into account. Only the decays of the Higgs boson into $\mathrm{b\bar{b}}$ and $\tau^+\tau^-$ are considered here. The search topologies (channels) taken into consideration are the 4-Jet channel ($HZ \to b\bar{b}q\bar{q}$), with a fraction of $\approx$ 52% of the production rate for a Higgs Boson of 115 GeV, the missing energy channel ($HZ \to b\bar{b}\nu\bar{\nu}$), 18%), the τ channel ($HZ \to b\bar{b}\tau\bar{\tau}$) and ($HZ \to \tau\bar{\tau}q\bar{q}$), 8%) and the lepton channels ($HZ \to b\bar{b}e^+e^-$) and ($HZ \to \tau\bar{\tau}\mu^+\mu^-$), 6%), which account for more than over 80% of the produced Higgs Bosons for a Higgs mass of 115 GeV.

The results shown here are published in Ref.[3]. OPAL has already reported results from the Standard Model Higgs boson search at $\mathrm{e^+e^-}$ centre-of-mass energies up to 189 GeV[4,5], where a lower mass limit of $m_H > 91.0$ GeV was obtained at the 95% confidence level.

°Work supported by the German Ministerrium für Bildung und Forschung (BMBF) under contracts no. 05 7BN18P 0.

2 Opal Detector, Data and Monte Carlo samples

Details of the OPAL detector can be found in Ref.[6].

The data used in the analyses correspond to integrated luminosities of approximately 216 pb^{-1} at 192–202 GeV, 80 pb^{-1} at 203–206 GeV, and 130 pb^{-1} at centre-of-mass energies higher than 206 GeV. The total luminosity used to search for the Higgs boson varies by $\pm 2\%$ from channel to channel, due to slightly different requirements on the operational status of different detector elements. During 2000 (1999), data were taken at $\sqrt{s}$ = 200-209 GeV (192-202 GeV) with a luminosity-weighted mean centre-of-mass energy of 206.1 (197.6) GeV. All data were processed with the most up-to-date detector calibrations available. For future publications more refined analyses and the final detector calibrations will be used.

A variety of Monte Carlo samples was generated at centre-of-mass energies between 192 and 210 GeV. Higgs boson production is modelled with the HZHA3 generator[7]. The process $(Z/\gamma)^* \rightarrow q\bar{q}(\gamma)$ is modelled with the KK2f [8] and the four-fermion processes (4f) are simulated using GRC4F [9]. The two-photon and other two-fermion processes have a negligible impact on the results. In each search channel, the estimates of the signal efficiency and the selected background rate depend strongly on the centre-of-mass energy and are thus interpolated on a fine grid. For each Monte Carlo sample, the full detector response is simulated in detail as described in Ref.[10].

3 Analysis Procedure

We use the same analysis techniques described in a previous publication[4], namely event reconstruction, b-flavour tagging, lepton (electron, muon and tau) identification and kinematic fits to reconstruct the Higgs boson mass.

The tracking and b-tagging performance in the Monte Carlo simulation are tuned using 8.2 pb^{-1} of calibration data collected at $\sqrt{s} \approx m_Z$ at intervals during 1999 and 2000 with the same detector configuration and operating conditions as the high-energy data.

The performance of the b-tagging for the data taken at $\sqrt{s} \geq 192$ GeV is checked with samples of $q\bar{q}(\gamma)$ events by selecting hadronic events with the mass of the $q\bar{q}$ system near m_Z. Figure 1(a) shows the b-tagging variable $\mathcal{B}$ for opposite b-tagged jets in the 2000 sample. The efficiency for tagging udsc flavours is also checked by computing $\mathcal{B}$ for the jets in a sample of $W^+W^- \rightarrow q\bar{q}\ell\bar{\nu}$ (ℓ=e or μ) obtained with the selection used to measure the W^+W^- cross-section[11] as shown in Figure 1(b). The expectation from the Standard Model Monte Carlo describes the data within the relative statistical

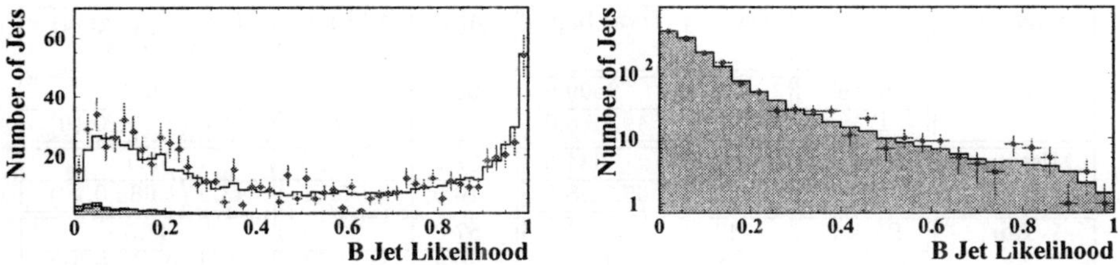

Figure 1: The b-tagging performance and modelling for data taken at $\sqrt{s}$ between 200–209 GeV in 2000. (a) The b-tagging output, $\mathcal{B}$, for jets opposite b-tagged jets in a sample of $q\bar{q}\gamma$ events, and (b) for jets in a sample of $W^+W^- \to q\bar{q}e^-\bar{\nu}_e$ and $W^+W^- \to q\bar{q}\mu^-\bar{\nu}_\mu$ events (and charge conjugates). The histogram in (b) shows the distribution from the four-fermion Monte Carlo samples.

uncertainty of 5–10%.

In each search channel, a pre-selection is applied to ensure that the events are well measured and are consistent with the desired signal topology. A likelihood selection combining 6 to 10 variables depending on the search channel is then used to further enrich the signal.

The numbers of events selected in each analysis after pre-selection and after the final likelihood selection are shown in Table 1 for the data taken at $\sqrt{s} \approx 192 - 209$ GeV. The distributions of the reconstructed masses of the selected events are shown in Figure 2. Note that all data taken in the wide $\sqrt{s}$ range from 192 to 209 GeV are summed in the figure, while the expected signal rates strongly depend on $\sqrt{s}$.

4 Confidence Level Calculation

After the event selection a testmass dependent discriminating variable is used to enrich a signal of a certain mass. In the 4-jet channel a testmass dependent discriminate is calculated using mass, kinematic and b-tag information. Other channels using the measured mass (tau channel) or a discriminate calculated from the selection likelihood output and the reconstructed mass (e, mu, neutrino channel). In order to compute the confidence levels, a test statistic is defined which expresses how signal-like the data are. The test statistic chosen is the likelihood ratio Q, the ratio of the probability of observing the data given the signal+background hypothesis to the probability of observing the data given the background-only hypothesis[12,13].

Channel	Cut	Data	Total bkg.	$q\bar{q}(\gamma)$	4-fermi.	Eff. [%] (signal) 115 GeV
Four-jet	Presel.	3820	3609.9	760.8	2859.4	86.9
	$\mathcal{L}^{HZ}$	60	49.8±6.0	12.8	37.0	41.8 (2.24±0.10)
Missing-E	Presel.	354	334.5	57.0	277.5	56.1
	$\mathcal{L}^{HZ}$	68	69.7±8.6	10.6	59.1	43.9 (1.68±0.10)
Tau	Presel.	343	334.5	57.0	277.5	48.3
	Final $\mathcal{L}$	8	11.1±1.2	0.4	10.7	22.9 (0.25±0.01)
Electron	Presel.	429	378.6	171.0	207.6	72.7
	$\mathcal{L}^{HZ}$	6	8.5±1.3	0.3	8.2	48.7 (0.17±0.004)
Muon	Presel.	79	66.2	36.8	29.4	70.3
	$\mathcal{L}^{HZ}$	10	7.0±1.0	0.2	6.8	59.9 (0.23±0.006)

Table 1: The number of events after preselection and after the final likelihood selection for the 192–209 GeV data and the expected background. The errors on the total background and the expected Higgs signal include all systematic errors. The last two columns show the detection efficiencies (and the numbers of expected signal events in parentheses) for a Higgs boson with m_H=115 GeV. For the four-jet channel, the efficiency is computed only for $H^0 \to b\bar{b}$ decays, and for the tau channel for the processes $H^0Z^0 \to \tau^+\tau^-(H^0 \to \ all)$ or $H^0Z^0 \to q\bar{q}\tau^+\tau^-$ assuming Standard Model branching fractions. For other channels, the efficiency is for all decays of the Standard Model Higgs boson.

The confidence levels are computed from the test statistic of the observed data and the expected distributions of the test statistic in a large number of simulated experiments under two hypotheses: the background-only hypothesis and the signal+background hypothesis.

5 Results

The observed $-2\ln Q$ is shown as a function of the test mass m_H in Figure 4 (a). Also shown are the 68% and 95% probability contours centred on the median expectation.

The signal rate limit n_{95} is shown as a function of m_H in Figure 4. Where the signal rate curve crosses the n_{95} curve is the 95% CL exclusion limit on m_H, and the expected limit is where the median expected n_{95} curve crosses the accepted signal rate curve. A lower mass bound of 109.7 GeV is obtained, and the expected limit is 112.5 GeV. In particular, the hypothesis m_H=107 GeV is excluded at the 98% CL (CL_s = 0.02) even in the presence of the excess candidates because the excess in the data is not large enough to be consistent with the expected signal rate from a Standard Model Higgs boson of that mass.

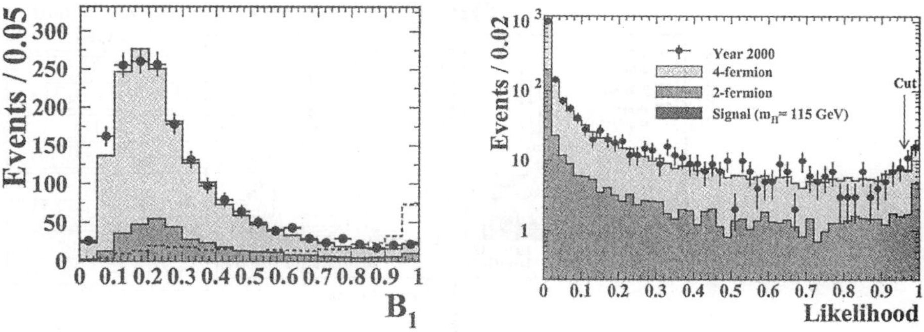

Figure 2: (a) The distributions of the B_1 likelihood input variable for the 1999 and 2000 data in the four-jet channel. The variable B_1 includes the B-tag of the jet with more energy from the two highest B-tags in the event. (b) The distributions of the likelihood output variables for the 1999 and 2000 data in the four-jet channel. OPAL data are shown with points, backgrounds by the shaded histograms, and the expectation from a signal with $m_H = 115$ GeV by the dashed histograms (scaled by a factor of 100 for visibility).

6 Conclusion

A search for the Standard Model Higgs boson has been performed with the OPAL detector at LEP based on the full data sample collected at $\sqrt{s} \approx$192–209 GeV in 1999 and 2000. For a Higgs boson with a mass of 115 GeV, the data are as consistent with the background hypothesis as with the hypothesis that a Higgs signal is present. A lower bound of 109.7 GeV on the mass of the Standard Model Higgs boson is obtained at the 95% confidence level while the median expectation for the background-only hypothesis is 112.5 GeV.

References

1. S. Glashow, *Nucl. Phys.* **22**, 579 (1961);
S. Weinberg, *Phys. Rev. Lett.* **19**, 1264 (1967);
A. Salam, ed. N. Svartholm, *Elementary Particle Theory*, Almquist and Wiksells, Stockholm (1968) 367.
2. P.W. Higgs, *Phys. Lett.* **12**, 132 (1964);
F. Englert and R. Brout, *Phys. Rev. Lett.* **13**, 321 (1964);
G.S. Guralnik, C.R. Hagen, and T.W.B. Kibble, *Phys. Rev. Lett.* **13**, 585 (1964).
3. OPAL Collaboration, G. Abbiendi *et al.*, it Phys. Lett. **B499**, 38 (2001).

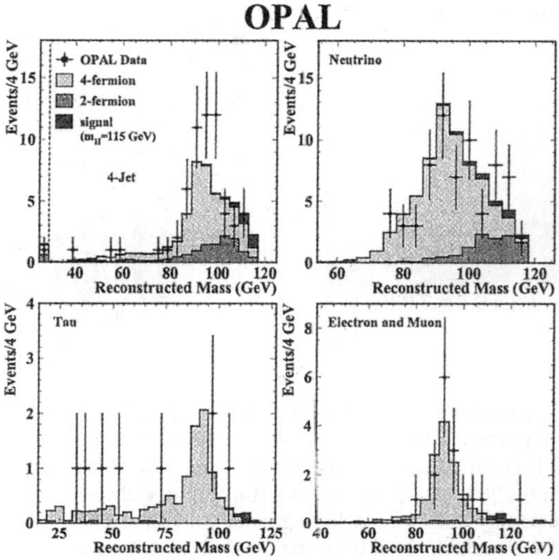

Figure 3: The reconstructed mass distribution for the selected events in the 1999 and 2000 data for (a) the four-jet channel, (b) the missing-energy channel, (c) the tau channels, and (d) the electron and muon channels combined. The first bin in (a) contains all events with χ^2 probability of the HZ 5C kinematic fit $< 10^{-5}$ for chosen jet-pairings. The dark (light) grey area shows the expected contribution from the $q\bar{q}(\gamma)$ (four-fermion) process. The Standard Model signal expectation for 115 GeV is shown by the very dark histograms on top of the Standard Model backgrounds.

4. OPAL Collaboration, G. Abbiendi *et al.*, *Eur. Phys. J.* **C12**, 567 (2000).
5. OPAL Collaboration, G. Abbiendi *et al.*, *Eur. Phys. J.* **C7**, 407 (1999); OPAL Collaboration, K. Ackerstaff *et al.*, *Eur. Phys. J.* **C1**, 425 (1998).
6. OPAL Collaboration, K. Ahmet *et al.*, *Nucl. Instr. and Meth.* **A305**, 275 (1991);
   S. Anderson *et al.*, *Nucl. Instr. and Meth.* **A403**, 326 (1998);
   B.E. Anderson *et al.*, *IEEE Trans. on Nucl. Science* **41**, 845 (1994);
   G. Aguillion *et al.*, *Nucl. Instr. and Meth.* **A417**, 266 (1998).
7. P. Janot *et al.*, in *Physics at LEP2,* edited by G. Altarelli, T. Sjöstrand and F. Zwirner, CERN 96-01 Vol. 2, 309.
   For HZHA3 and HZHA2, see *http://alephwww.cern.ch/~janot/Generators.html.*
8. S. Jadach, B.F. Ward and Z. Was, *Phys. Lett.* **B449**, 97 (1999).
9. J. Fujimoto *et al.*, *Comp. Phys. Comm.* **100**, 128 (1997);

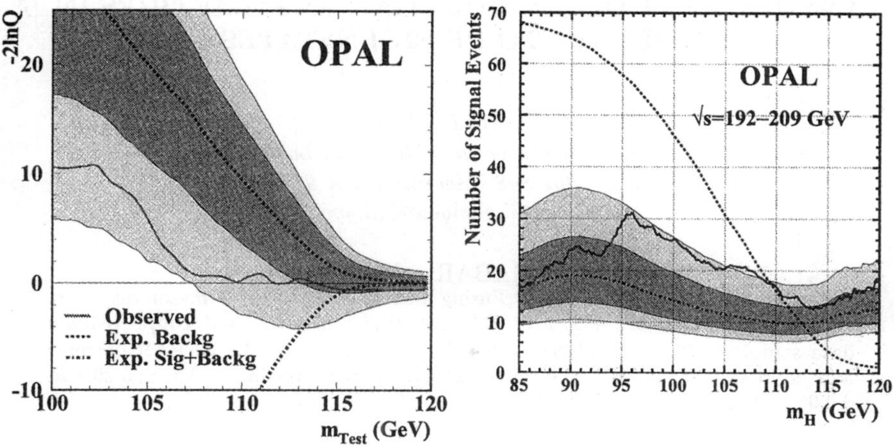

Figure 4: (a) The log-likelihood ratio $-2\ln Q$ comparing the relative consistency of the data with the signal+background hypothesis and the background-only hypothesis, as a function of the test mass m_{TEST}. The observation for the data is shown by the a solid line. The dotted line indicates the median background expectation and the dark (light) shaded band shows the 68% (95%) probability intervals centred on the median. The median expectation in the presence of a signal is shown by a dot-dashed line where the hypothesized signal mass is the test mass. (b) Upper limits on the signal counts at the 95% confidence level (n_{95}), as observed (solid line) and the expected median (dot-dashed line) for background-only experiments, as a function of the Higgs boson test mass. The expected rate of the accepted signal counts for a Standard Model Higgs boson with a mass equal to the test mass is shown by the dotted line. The shaded bands are the 68% and 95% probability intervals centred on the median background expectation.

Physics at LEP2, J. Fujimoto *et al.*, CERN 96-01, Vol. 2, 30.

10. J. Allison *et al.*, *Nucl. Instr. and Meth.* **A317**, 47 (1992).
11. OPAL Collaboration, G. Abbiendi *et al.*, *Phys. Lett.* **B493**, 249 (2000).
12. A. Stuart and J.K. Ord, *Kendall's Advanced Theory of Statistics*, Vol. 2, Ch. 23 5th Ed., Oxford University Press, New York, (1991).
13. ALEPH, DELPHI, L3, OPAL Collaborations, and the LEP working group for Higgs boson searches, *Search for Higgs bosons: Preliminary combined results using LEP data collected at energies up to 202 GeV*, CERN-EP/2000-055 (2000)

MEASUREMENT OF CP-VIOLATING ASYMMETRIES IN B^0 DECAYS TO CP EIGENSTATES

DAVID J. LANGE
L-50, Lawrence Livermore National Laboratory,
P.O. Box 808, Livermore CA 94551
E-mail: lange6@llnl.gov

For the BABAR collaboration

We present a measurement of sin2β using a sample of neutral B meson decays to the CP eigenstates $B^0 \to J/\psi K^0_s, B^0 \to \psi(2S)K^0_s$ and $B^0 \to J/\psi K^0_L$. Using a data sample of 23×10^6 $\Upsilon(4S) \to B\bar{B}$ decays collected by the BABAR detector located at the PEP-II asymmetric B factory, we measure sin$\beta = 0.34 \pm 0.20$ stat ± 0.05 (syst).

1 Introduction

Measurements[1,2] of time dependent decay rates in $\Upsilon(4S) \to B^0\bar{B}^0$ decays have been used to determine sin2β and Δm_d using the BABAR[3] detector. CP-violating asymmetries arise from the interference of multiple decay amplitudes and have a variety of possible experimental signatures. In the case described here, the decay rates for B^0 and $\bar{B}^0$ (at $t = 0$) to a common final state f have a different time dependence

$$\frac{d\Gamma(B^0 \to f)}{dt} \neq \frac{d\Gamma(B^0 \to f)}{dt}. \tag{1}$$

In $\Upsilon(4S) \to B^0\bar{B}^0$ decays, the magnitude of the interfering amplitudes are comparable, which lead to possibly large asymmetries in the Standard Model. While the branching fractions for common final states are small ($\approx 10^{-3}$), sizable samples of these states have been reconstructed in approximately 23×10^6 $B\bar{B}$ decays collected in between October 1999 and 2000 by the BABAR experiment.

CP-violation is accommodated in the Standard Model through a complex phase within the CKM matrix[4]

$$V_{CKM} \equiv \begin{pmatrix} V_{ud} & V_{us} & V_{ub} \\ V_{cd} & V_{cs} & V_{cb} \\ V_{td} & V_{ts} & V_{tb} \end{pmatrix}, \tag{2}$$

that describes the coupling for charged weak transitions $q \to W^{*+}q'$ ($\propto V^*_{qq'}$).

The orthogonality of the first and third columns

$$V_{ud}V_{ub}^{*} + V_{cd}V_{cb}^{*} + V_{td}V_{tb}^{*} = 0 \qquad (3)$$

gives the "Unitarity Triangle" shown in Figure 1, and defines the angles α, β, and γ. Together with other measurements [5] (such as ϵ_k , $|V_{ub}|, |V_{cb}|$, Δm_d, and Δm_s), measurements of CP-violating asymmetries over-constrain the Unitarity Triangle, and thus are a test of the Standard Model. Unlike other constraints, the measurement of $\sin 2\beta$ using the Charmonium modes discussed here is essentially free of theoretical uncertainties.

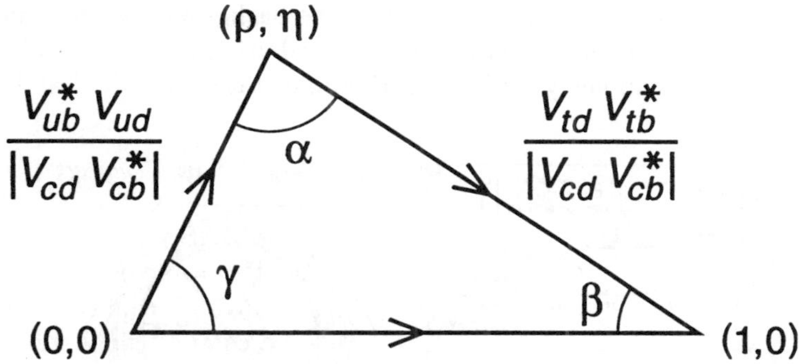

Figure 1. The normalized Unitarity Triangle determined from the orthogonality of the first and third columns of the CKM matrix.

$B^0\bar{B}^0$ pairs are produced in a coherent $L = 1$ state at the $\Upsilon(4S)$. Thus, the decay distribution for $B \rightarrow f$, where f is a CP eigenstate, depends on Δt, the difference between the decay time of the B that decays to f (B_{CP}) and the other B in the event (B_{tag}).

We present the measurement of $\sin 2\beta$ using $b \rightarrow c\bar{c}s$ decays to the final states $B^0 \rightarrow J/\psi K_s^0, B^0 \rightarrow \psi(2S)K_s^0$, and $B^0 \rightarrow J/\psi K_L^0$, where

$$f_{\pm}(\Delta t) = \frac{\Gamma e^{-\Gamma|\Delta t|}}{4}[1 \mp \eta_f sin2\beta sin(\Delta m_d \Delta t)] \qquad (4)$$

describes the decay distribution when B_{tag} is a B^0 (+) or B^0 (−) at its decay. $\Gamma = 1/\tau_{B^0}$ and η_f is the CP eigenvalue of the final state. For $B^0 \rightarrow J/\psi K_s^0$ and $B^0 \rightarrow \psi(2S)K_s^0, \eta_f = -1$ and for $B^0 \rightarrow J/\psi K_L^0$, $\eta_f = +1$.

2 Experimental method

Experimental effects modify Eq. 4. These include corrections for background events, imperfect determination of the $B_{\rm tag}$ flavor and Δt.

Figure 2 shows the samples of CP events that have been reconstructed. We find $273 B \to J/\psi(\psi(2S))K_s^0$ and 256 $B \to J/\psi K_L^0$ candidate (signal plus background) events where Δt and the flavor of the $B_{\rm tag}$ can be determined. In both cases, in particular for the K_L^0 event sample, backgrounds must be properly accounted for in the determination of the Δt distribution of the CP events to avoid bias in the measurement of sin2β. While the K_s^0 sample is $96 \pm 1\%$ pure, only $39 \pm 6\%$ of the K_L^0 sample is signal. For the latter case, Monte Carlo samples are used to determine the background sources and their Δt distributions. The largest background contribution is $B \to J/\psi K^*(K_L^0\pi)$, which has a CP asymmetry itself when the final state is $J/\psi K^{*0}$ and $K^{*0} \to K_L^0\pi^0$.

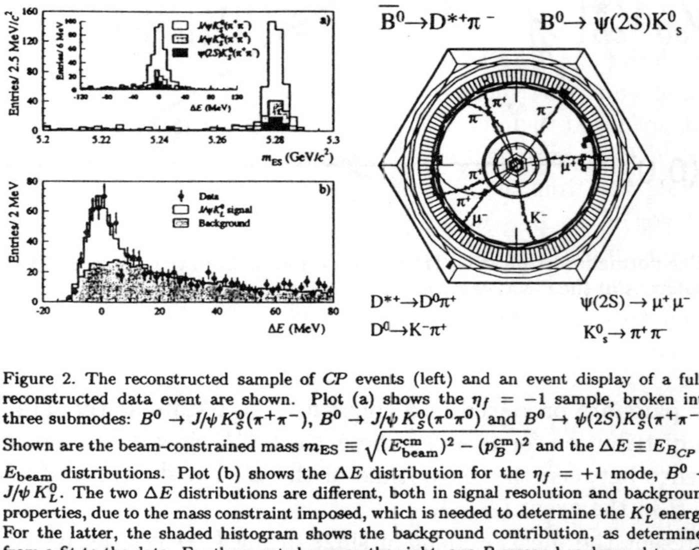

Figure 2. The reconstructed sample of CP events (left) and an event display of a fully reconstructed data event are shown. Plot (a) shows the $\eta_f = -1$ sample, broken into three submodes: $B^0 \to J/\psi K_S^0(\pi^+\pi^-)$, $B^0 \to J/\psi K_S^0(\pi^0\pi^0)$ and $B^0 \to \psi(2S)K_S^0(\pi^+\pi^-)$. Shown are the beam-constrained mass $m_{\rm ES} \equiv \sqrt{(E_{\rm beam}^{\rm cm})^2 - (p_B^{\rm cm})^2}$ and the $\Delta E \equiv E_{B_{CP}} - E_{\rm beam}$ distributions. Plot (b) shows the ΔE distribution for the $\eta_f = +1$ mode, $B^0 \to J/\psi K_L^0$. The two ΔE distributions are different, both in signal resolution and background properties, due to the mass constraint imposed, which is needed to determine the K_L^0 energy. For the latter, the shaded histogram shows the background contribution, as determined from a fit to the data. For the event shown on the right, one B meson has decayed to a CP eigenstate ($B \to \psi(2S)K_S^0$) and the other to a flavor eigenstate. Both the slow pion from the D^{*+} and the kaon from the D^0 can be used to tag its flavor of the latter as a $\overline{B}^0$.

In addition, we reconstruct a larger sample of B^0 decay modes $B_{\rm flav}$ in the flavor eigenstates $B^0 \to D^{(*)-}\pi^+$, $B^0 \to D^{(*)-}\rho^+$, $B^0 \to D^{(*)-}a_1^+$, and $B^0 \to J/\psi K^{*0}(K^+\pi^-)$. We find approximately 4600 signal events with 86±1% purity. These events have two uses in this analysis. First, these events measure Δm_d using a technique similar to that used to measure sin2β. Second, they are used to determine the performance of the algorithms that measure Δt and the flavor of $B_{\rm tag}$, as described below.

Δt is determined from the distance between the decay points of B_{CP} and $B_{\rm tag}$. As B mesons are produced nearly at rest in the $\Upsilon(4S)$ rest frame, the $\Upsilon(4S)$ must be produced with a boost (conventionally along the z axis) so that $\Delta z \equiv z_{CP} - z_{\rm tag} \approx \beta\gamma c\Delta t$ can be measured. z_{CP} is the vertex position of B_{CP}. $z_{\rm tag}$ is determined from an iterative algorithm using the remaining tracks in the event after the B_{CP} has been reconstructed. Large contributors to the vertex χ^2 are dropped to reduce the bias from charm decay. The determination of $z_{\rm tag}$ dominates the resolution of Δz. The fraction of events with a B_{CP} candidate where Δz can be determined is approximately 86%. We parametrize the experimental resolution function as the sum of three Gaussians. Two of these are a function of the determined event by event error and the third is an outlier contribution of fixed width (8 ps). In the unbinned maximum likelihood fit described below, the core Gaussian is determined to contain 88% of the events with a scale factor of 1.1 ± 0.1 with respect to the event by event error.

The flavor of the $B_{\rm tag}$ must be determined for each event. The charge of leptons, kaons, and slow pions from B mesons are correlated with the flavor of the B that produced them. For example, Figure 2 shows a fully reconstructed event, with a B_{CP} candidate decay and a $\bar{B}^0 \to D^{*+}\pi^-$ decay. In this case, both the slow pion from the D^{*+} decay and the K^- from the D decay indicate that the flavor of the $B_{\rm tag}$ was a $\bar{B}^0$.

The determination of the $B_{\rm tag}$ flavor can be incorrect due to imperfect particle identification and because the correlation between B flavor and the charge of the lepton, kaon, or slow pion is not perfect. The fraction of mistagged events w dilutes the observed asymmetry so that Eq. 4 becomes

$$f_{\pm}(\Delta t) = \frac{\Gamma e^{-\Gamma|\Delta t|}}{4}\big[1 \mp \eta_f \mathcal{D} sin2\beta sin(\Delta m_d \Delta t)\big], \tag{5}$$

where $\mathcal{D} \equiv 1 - 2w$.

Similarly for the $B_{\rm flav}$ sample, the decay time distribution is

$$f_{\pm}(\Delta t) = \frac{\Gamma e^{-\Gamma|\Delta t|}}{4}\big[1 \pm \mathcal{D}\cos(\Delta m_d \Delta t)\big], \tag{6}$$

for unmixed (+) and mixed (−) events. The amplitude of the observed asymmetry in the $B_{\rm flav}$ sample gives $\mathcal{D}$, which cannot be disentangled from $\sin 2\beta$ using the B_{CP} sample alone.

We use a hybrid algorithm that consists of four exclusive categories, to determine the flavor of $B_{\rm tag}$. Two cut based categories, one for lepton tags

($\varepsilon = (10.9\pm0.4)\%$, $w = (11.6\pm2.0)\%$), and one for kaon tags ($\varepsilon = (36.5\pm0.7)\%$, $w = (17.1 \pm 1.3)\%$), contribute most of the effective tagging efficiency. A neural network algorithm is used to tag the remaining events. This algorithm identifies leptons and kaons that are missed by the cut based selection and finds slow pion candidates. Based on the output of the neural network, events are separated into more certain ($\varepsilon = (7.7 \pm 0.4)\%$, $w = (21.2 \pm 2.9)\%$) and less certain ($\varepsilon = (13.7 \pm 0.5)\%$, $w = (31.7 \pm 2.6)\%$) tags, or are not tagged at all. The total effective tagging efficiency $Q \equiv \Sigma\epsilon_i\mathcal{D}_i^2$ is $(26.7 \pm 1.6)\%$.

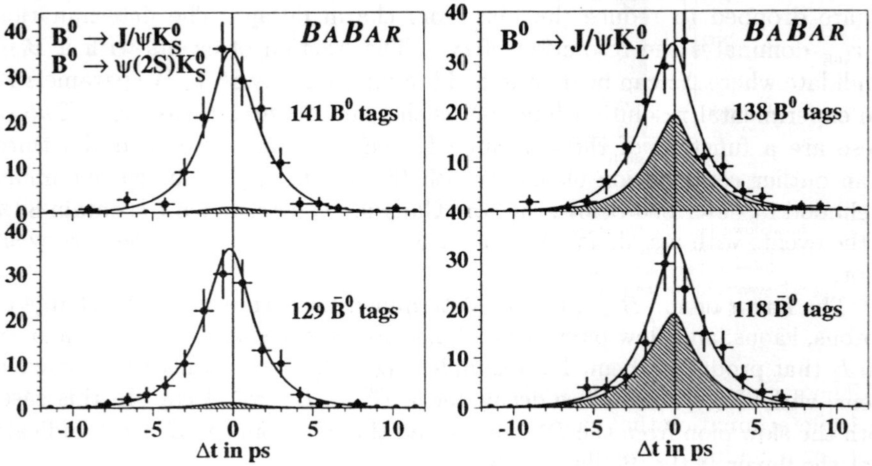

Figure 3. The projection of the likelihood fits on to Δt for the $\eta_f = -1$ (left) and $\eta_f = +1$ (right) *CP* channels. The points are the data and the solid curve is the fit projection. The contribution from background events is shown by the hatched region.

3 Results

We perform unbinned maximum likelihood fits to determine sin2β and Δm_d, using Eqs. 5 and 6, corrected for finite resolution and backgrounds, for CP events and the $B_{\rm flav}$ sample, respectively. For each of these fits, the parameters of the resolution function, dilutions, and background parameters are determined simultaneously, for a total of 35 (34) free parameters for the sin$2\beta(\Delta m_d)$ fit. We determine

$$\begin{aligned} sin2\beta &= 0.34 \pm 0.20 \pm 0.05, \\ \Delta m_d &= 0.519 \pm 0.020 \pm 0.016 \text{ ps}^{-1} \text{ (preliminary)}, \end{aligned} \tag{7}$$

where the first error is statistical and the second systematic. Figure 3 shows the Δt distribution for the CP samples with the fit result overlaid. Subsamples of the data, divided by reconstructed decay mode or by flavor tag category, are consistent as shown in Figure 4 for the CP data sample.

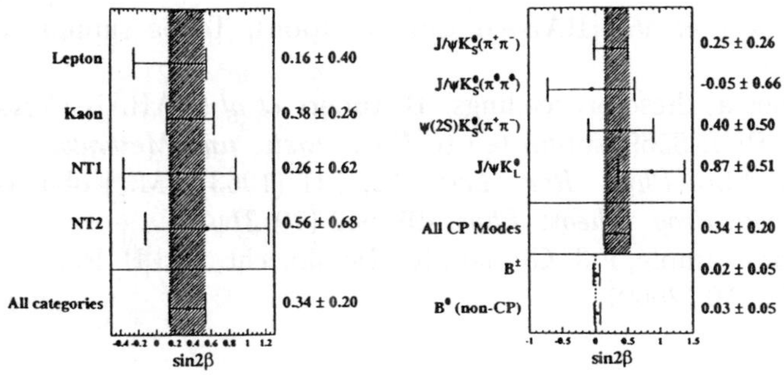

Figure 4. sin2β measurements on subsamples of the data set. The left column shows a breakdown into the four tagging categories, and the right into the submodes of the CP sample. We find that the data are self consistent. In addition, the right column gives the results for the asymmetry measured on the B_{flav} sample and a sample of fully reconstructed B^+ events, where the asymmetry is expected to be 0. For both samples, we find a result consistent with 0.

The systematic error on sin2β is dominated (± 0.04) by possible effects in the Δt determination that cannot be completely parametrized by our resolution function, including those due to SVT alignment. In addition, uncertainties in the level, composition, and CP asymmetry of the background in the B_{CP} event sample contributes an error of ± 0.02. Other uncertainties, including those due to Δm_d and τ_{B^0}, are 0.01 or smaller.

4 Conclusion

Using data collected during the first run of the PEP-II asymmetric B Factory at SLAC, we have measured sin2$\beta = 0.34 \pm 0.20 \pm 0.05$. This result is currently the world's best measurement and is consistent both with Standard Model expectations and with a null asymmetry.

Acknowledgments

This work was performed under the auspices of the U.S. Department of Energy by Lawrence Livermore National Laboratory under Contract W-7405-Eng-48.

References

1. B. Aubert *et al.* (BABAR Collaboration), *Phys. Rev. Lett.* **86**, 2515 (2001).
2. B. Aubert *et al.* (BABAR Collaboration), To be submitted to *Phys. Rev.* **D**.
3. J. Panetta, these proceedings. B Aubert *et al* (BABAR Collaboration), SLAC-PUB-8569, submitted to *Nucl. Instr. and Methods.*
4. N. Cabibbo, *Phys. Rev. Lett.* **10**, 531 (1963). M. Kobayashi and T. Maskawa, *Prog. Theor. Phys.* **49**, 652 (1973).
5. See, for example, F.J. Gilman, K. Kleinknecht, and B. Renk, *Euc. Phys. J* **C15**, 110 (2000).

ELECTROMAGNETIC INTERACTIONS IN STRONG MAGNETIC FIELDS

D.A. LEAHY
Dept. of Physics and Astronomy, University of Calgary, Calgary, Alberta, Canada T2N 1N4
E-mail: leahy@iras.ucalgary.ca

Electrodynamic interactions in strong magnetic fields have only been studied rather recently. In strong magnetic fields, new processes such as one-photon pair creation and one-photon pair annihilation can occur. The motivation for these studies was the existence of neutron stars with strong magnetic fields, of order 10^{12} Gauss. More recently, neutron stars with extreme magnetic fields, $\sim 10^{14} - 10^{15}$ Gauss, have been discovered. This has necessitated revisiting previous calculations, as they were carried out with assumptions implying $B << B_{cr} = 4.414 \times 10^{13}$ Gauss. Here, a brief review is given of quantum electrodynamics in a strong magnetic field, and a few of the new calculations are summarized.

1 Introduction

Radio pulsars were discovered in 1968, and X-ray pulsars discovered in 1971. Both types of pulsars had pulse periods of order 1 second, and high values of rate of change of period. There properties were consistent only with a rotating compact object of order 10 km in radius, a solar mass in mass, and a magnetic field of order 10^{12} Gauss (required to provide the torque for slowing down or speeding up). That is, the pulsars were identified with rotating magnetic neutron stars emitting a beam of radiation along an axis different from the rotation axis.

The question arises as to how we go about understanding the emission process for either x-ray or radio pulsars. As was noted in Daugherty and Harding[1], quantum electrodynamic processes such as magnetic pair production and synchrotron radiation in strong magnetic fields have come to play an important role in radio pulsar models. Also, one-photon pair production is likely to be the dominant source of e^+-e^- pairs in fields exceeding 10^{12} Gauss (Harding[2]). For X-ray pulsars the emission region is much more optically thick, but the radiation transfer is anisotropic so a detailed understanding of the microscopic processes is needed (e.g. see Meszaros[3]).

With the observational discovery of magnetars (Vasisht and Gotthelf[6], Kouveliotou *et al.*[7]), which are pulsars with magnetic fields in the range $\sim 10^{14} - 10^{15}$ Gauss, calculations which are valid for very high fields are of great interest. The critical value of the magnetic field is defined as $B_{cr} = \frac{m^2c^3}{e\hbar} =$

4.414×10^{13} Gauss. Most previous calculations are valid only for $B << B_{cr}$ as they use photon normal modes (to include the effects of vacuum polarization and plasma) which apply only in the weak field limit ($B << B_{cr}$). They also use the Johnson-Lippmann [4] electron wavefunctions which makes their calculation valid only in weak fields, as shown by Graziani [5].

This paper presents a very brief introduction to electron states in strong magnetic fields, then presents the results of some new calculations which are valid for very strong fields ($B \geq B_{cr}$) as well as for ($B < B_{cr}$) The calculations summarized in section 3 below utilize the electron wavefunctions of Sokolov and Ternov [8], and are valid for strong fields. Electron spin is also treated correctly, unlike most previous work, when the parallel momentum of the electron is non-zero.

2 Electron energy states in a strong magnetic field

A brief summary only is given here. Much more complete discussions can be found in the PhD thesis of Frangodimitraki-Georgiadou [9] (with emphasis on the physics) or in Meszaros [3] (with more emphasis on the astrophysics). The first reference above includes a good discussion of the various electron spinors that have been used in the past and which ones are approximate, and valid only under certain conditions.

In general the electron energy is given by:

$$E_N = [p_z^2 + m^2.(1 + 2N\gamma)]^{1/2}, \tag{1}$$

with $\gamma = \frac{B}{B_{cr}}$. So, the energy of the electron is characterized by the principal quantum number $N = l + \frac{1}{2}(s+1) = 0, 1, 2, \ldots$, where ($l = 0, 1, 2, \ldots$). In each Landau state, the electron may have spin-up ($s = +1$) or spin-down ($s = -1$) along the field direction, except in the ground state ($N = 0$), where only the spin-down state is allowed (Harding and Preece [10]).

E_N can also be written:

$$E_N = m[1 + 2N\gamma + (p_z/m)^2]^{1/2}, \tag{2}$$

The energy as a function of p_z consists of hyperbolas which come closer together as p_z grows, and are known as Landau levels. The electron energy levels, E_N, consist of a discrete and continuous spectrum, since the perpendicular momentum $p_\perp = m\sqrt{2N\gamma}, N = 0, 1, 2 \ldots$, is discrete, whereas p_z is continuous. Figure 1 presents the dependence of E (divided by the electron mass energy) on p_z/mc for values of $N = 0, 4, 8, 12$ and for two values of magnetic field, $B = 0.05B_{cr}(\gamma = 0.05)$ and $B = B_{cr}(\gamma = 1)$. The weak field limit ($\gamma << 1$) reduces to the $N = 0$ line.

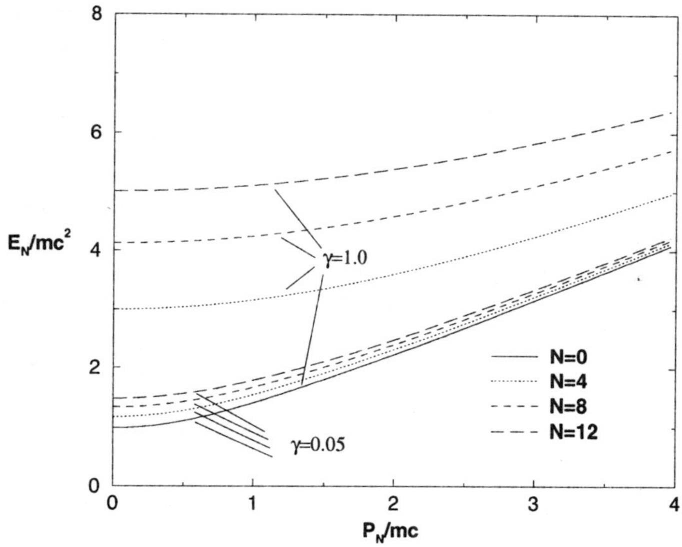

Figure 1: The energy of an electron as a function of parallel momentum, for different $\gamma = B/B_{cr}$ and Landau level N.

The non-relativistic approximation of E_N is:

$$E_N \approx m(1 + N\gamma + \frac{1}{2}(p_z/m)^2) \tag{3}$$

If we define $E^{NR} = E_N - m$, then

$$E^{NR} \approx p_z^2/2m + NeB/m \tag{4}$$

It is seen that the energy splitting between two successive Landau levels, with same p_z, is ${E^{NR}}_{N+1} - {E^{NR}}_N = eB/m = \omega_B$. The critical magnetic field is seen as the value of B for which this splitting is equal to the electron rest mass energy, m.

3 Electron-Photon Interactions in an External Magnetic Field

3.1 *First Order Pair Creation*

Pair creation can be understood as a photon causing the jump of an e^- from a state with negative energy to a state with positive energy, so for this process to occur the photon must have energy no less than $2mc^2$. Daugherty and

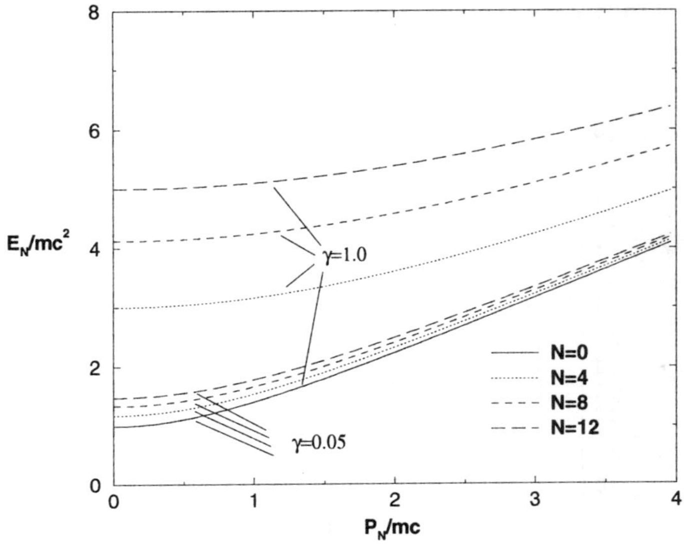

Figure 2: The Feynman diagram for the pair creation process.

Harding[1] have a list of references relevant to this process. Their calculations are valid only for $B << B_{cr}$. Frangodimitraki-Georgiadou[9] reanalyzed pair production using wave functions valid for large B. Semionova and Leahy[11] generalized this work to include electron and photon polarizations.

The basic process is summarized by the Feynman diagram (Fig. 2), where one can read time as increasing upward here. The photon, if it is not parallel to the magnetic field, can create a pair since the strong external field can absorb or give transverse momentum (i.e. only z-momentum is conserved). Given a particular photon energy only a given set of final electron and positron states are possible. This set is largest for a photon transverse to the field ($\theta = 90°$). The allowed final states are shown for a photon energy of 5 MeV in Fig. 3. The threshold energy for $\theta = 90°$ is twice the electron mass, and for photons at smaller angles to the field the threshold increases as $1/sin(\theta)$.

The pair creation rate is calculated using the standard S-matrix formulation but with the electron wave functions appropriate for a strong magnetic field. Details of the calculations are given in Semionova and Leahy[11]. The rate can be calculated for the different possible photon polarizations for the initial state and can be calculated for the different channels of the final state for electron and positron spin. For the two cases of photon polarization perpendicular

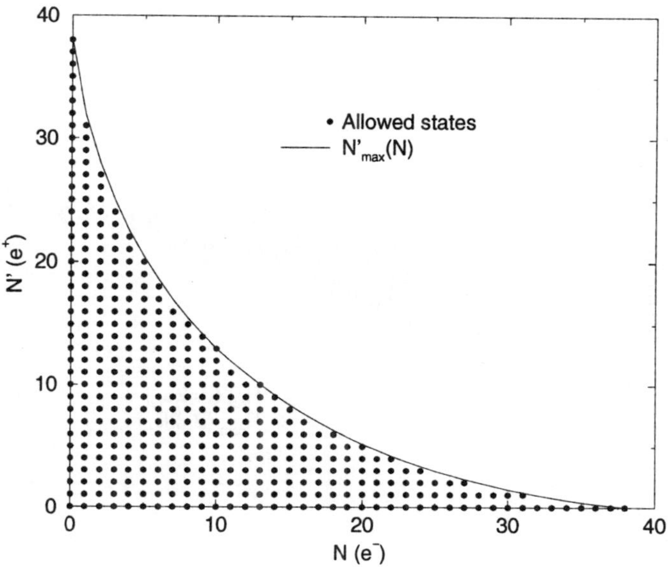

Figure 3: Kinematically allowed Landau states for pair creation for the case $B = B_{cr}$, ω =5 MeV and $\theta = 90°$.

to the magnetic field (labelled 'perp') and the orthogonal polarization (labelled 'par'), the dependence of pair creation rate on photon energy is given in Fig. 4. The field strength is $2B_{cr}$ here, the photon is propagating perpendicular to the field, and the final state electron and positron are spin-down and spin-up respectively. The rate exhibits many resonances- each one occurs due to new final allowed states as photon energy increases (e.g. in Fig. 3, as one moves the kinematic boundary line to the upper right, states which were previously energy-forbidden become allowed suddenly).

3.2 Other processes

Several other electromagnetic processes are important for radio and x-ray pulsars such as one- or two-photon pair annihilation, one and two photon emission, and Compton scattering.

Two photon annihilation can occur at low magnetic field strengths, but for B approaching B_{cr} the one photon process becomes dominant. Early calculations were valid only for $B << B_{cr}$ and also did not treat electron spin. Semionova and Leahy [12] present work correcting these problems. The pair an-

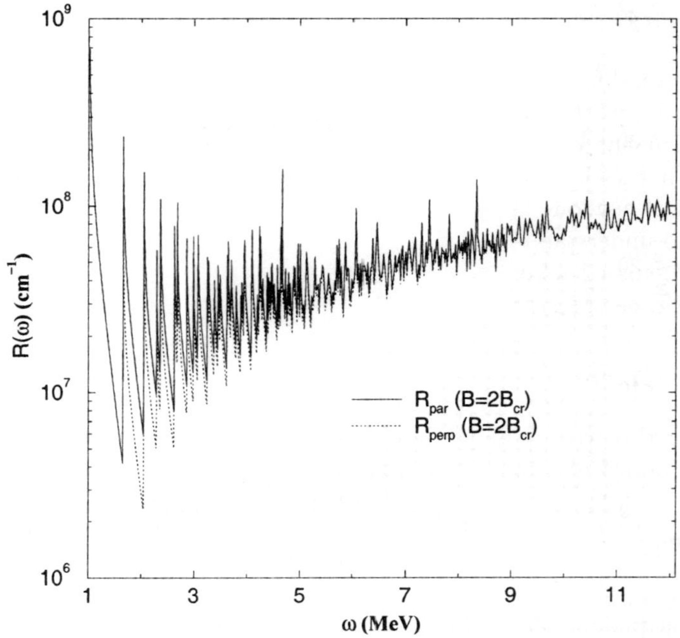

Figure 4: The attenuation coefficient for pair production, R, as a function of photon energy, ω. The two curves are for the two photon polarizations and for electron spin-up, positron spin-down, with $\theta = \pi/2$ and $B = 8.8x10^{13}$ $(2B_{cr})$ Gauss.

nihilation rate is strongly spin dependent with the electron spin-down, positron spin-up rate the largest and the electron spin-up, positron spin-down rate the smallest of the four possible cases. The one-photon annihilation rate rapidly increases with B for $B < B_{cr}$, reaches a maximum near B_{cr} and declines slowly for larger B.

Semionova and Leahy [13] discuss two-photon emission and also give references for the one-photon emission case. Second order processes like two photon emission and Compton scattering require use of the electron propagator for strong magnetic fields. The resulting emission rate depends on the emission angles of the two final state photons, and exhibits resonances which are fairly complex.

4 Summary

The understanding of electron-photon interactions in strong magnetic fields has increased greatly with the earlier calculations valid for $B << B_{cr}$, and now with current work valid for $B \geq B_{cr}$, which has been motivated by the discovery that such strong field objects actually probably exist in nature. The whole field of how one applies the fundamental physical interactions, however complex, to understand the observed emission of radio and x-ray pulsars has not been discussed here. That involves the physics of radiation transfer and plasmas (e.g. see Meszaros[3]).

Acknowledgments

Support for this work was provided by the Natural Sciences and Engineering Research Council of Canada.

References

1. J. Daugherty and A. Harding, *ApJ*, **273**, 761 (1983).
2. A. Harding, in *Proceedings of the International Conference on "Coherent Radiation Processes in Strong Fields"* (the Catholic University, 1990).
3. P. Meszaros, *High Energy Radiation from Magnetized Neutron Stars* (University of Chicago Press 1992).
4. M. Johnson and B. Lippmann, *PRD* **76**, 828 (1949).
5. C. Graziani, *ApJ* **412**, 351 (1993).
6. G. Vasisht and E. Gotthelf, *ApJ* **486**, L129 (1997).
7. C. Kouveliotou, *et al.*, *ApJ* **510**, L115 (1999).
8. A. Sokolov and I. Ternov, *Synchrotron Radiation* (Berlin: Akademie 1968).
9. M. Frangodimitraki-Georgiadou, Ph.D. Thesis, University of Tübingen (1991).
10. A. Harding and R. Preece, *ApJ* **319**, 939 (1987).
11. L. Semionova and D. Leahy submitted to *A&A* 2001.
12. L. Semionova and D. Leahy, *A&A Supp.* 144, 307 (2000).
13. L. Semionova and D. Leahy, *PRD* **60**, 073011 (1999).

CLEO MEASUREMENTS OF THE CKM ELEMENTS $|V_{ub}|$ and $|V_{cb}|$

T.O. MEYER
Cornell University, Ithaca, NY 14853, USA
E-mail: tm35@cornell.edu

We describe a preliminary measurement at CLEO of the CKM matrix element $|V_{cb}|$ using the exclusive decay $B \to D^*\ell\nu$. We find a branching fraction of $\mathcal{B}(\bar{B}^0 \to D^{*+}\ell\nu) = (5.66\pm0.29\pm0.33)\%$ and $|V_{cb}| = 0.0464\pm0.0020\pm0.0021\pm0.0021$, where the first error is statistical, the second is systematic, and the third is theoretical. We also describe a new analysis of the exclusive decays $B \to \pi/\rho/\omega/\eta\ \ell\nu$ from which branching fractions, $|V_{ub}|$, and information on the q^2 distribution are expected.

1 Introduction

The Cabibbo-Kobayashi-Maskawa matrix describes the mixing between quark mass eigenstates in the charged-current weak interactions of the Standard Model.[1] The elements of this unitary matrix provide the sole mechanism for CP violation within the Standard Model, and so understanding the entries in the CKM matrix sheds light on the matter-antimatter asymmetry observed in the universe today. In addition, by checking that independent measurements of the various elements are consistent with the overall requirements of unitarity, we hope to probe for physics beyond the Standard Model.

The familiar "unitarity triangle" is a graphical way of depicting such consistency checks. The unitarity constraints on the CKM matrix are effectively orthogonality conditions on the rows and columns of the matrix. These can rather naturally be represented as triangles in the complex plane. Of the six triangles that result, the one formed from the d and b columns is interesting theoretically since all three sides are of comparable size, and interesting experimentally since B decays are very accessible. If this particular triangle is appropriately normalized, the CKM element $|V_{cb}|$ specifies the length of the base and $|V_{ub}|$ determines the location of the apex.

Study of this particular triangle benefits our general understanding of the Standard Model in two ways. First, since the CKM matrix is fully specified by only four independent real parameters, understanding one unitarity triangle can lead to powerful constraints on the rest of the matrix. Second, of these four parameters, one leads to the complex phase that is responsible for CP violation. Indeed, the area of the unitarity triangle directly measures the amount of CP violation allowed for in the Standard Model.

Semileptonic B decays lend themselves to experimental determinations of

$|V_{cb}|$ and $|V_{ub}|$. In these decays, the b quark within the B meson decays weakly to a c or u quark and the virtual W from that vertex produces a charged-lepton and neutrino pair. The amplitude for the flavor-change $b \to q$ is governed by the CKM element V_{qb}, making the decay rate $B \to X_q \ell \nu$ proportional to $|V_{qb}|^2$.

2 Extracting $|V_{cb}|$ from $B \to D^* \ell \nu$

The decay $B \to D^* \ell \nu$ is a typical semileptonic decay deriving from a $b \to c$ transition. The rate for this decay, however, depends not only on $|V_{cb}|$ and well-known weak-decay physics, but also on strong interaction effects, which are parameterized by form factors. These effects are difficult to quantify, but Heavy Quark Effective Theory[2] (HQET) offers a framework for calculating them at the kinematic endpoint where the final-state D^* is at rest with respect to the initial B meson, often called the point of zero recoil. The kinematic variable w is typically used to describe the departure from this limit: $w \equiv v_B \cdot v_{D^*}$ is the relativistic boost γ of the D^* in the rest frame of the B. The decay rate may then be written

$$\frac{\mathrm{d}\Gamma}{\mathrm{d}w} = \frac{G_F}{48\pi^3} |V_{cb}|^2 \mathcal{F}^2(w) \mathcal{G}(w) \tag{1}$$

where $\mathcal{G}(w)$ is a known kinematic factor and $\mathcal{F}(w)$ is the combination of form factors appropriate for the $B \to D^*$ transition. Fundamentally, $\mathcal{F}(w)$ is incalculable from first principles since it involves non-perturbative strong physics, but model-independent results from HQET can provide $\mathcal{F}(1)$. Since the phase space factor $\mathcal{G}(w)$ vanishes at $w = 1$, the analysis proceeds by measuring the rate $\mathrm{d}\Gamma/\mathrm{d}w$ over the full range range of $w \in (1, 1.51]$ and then extrapolating to $w = 1$ to obtain the product $|V_{cb}|\mathcal{F}(1)$. Combined with the theoretical results, this provides $|V_{cb}|$.[3]

2.1 Signal Selection

The analysis uses 3.33 million $B\bar{B}$ events (3.1 fb^{-1}) produced at the $\Upsilon(4S)$ resonance at the Cornell Electron Storage Ring and collected with the CLEO II detector.[4] The analyses of the charged-B mode $B^- \to D^{*0} \ell \nu$ and the neutral-B mode $\bar{B}^0 \to D^{*+} \ell \nu$ are performed separately, due to differences in signal efficiency, systematics, and details of event reconstruction. (Charge conjugate modes are implied throughout.)

We reconstruct the D^* through the decay chain $D^* \to D^0 \pi$, with $D^0 \to K^- \pi^+$. The D^0 candidate must have a mass within 20 MeV of the nominal D^0 mass. A D^* candidate is formed with the addition of a slow pion, and we require $\Delta M = M(K\pi\pi) - M(K\pi)$ to lie within 2(3) MeV of the $D^* - D^0$ mass difference for the charged (neutral) D^* mode.

Lepton candidates are either electrons identified using the CsI calorimeter ($0.8 < p_e \leq 2.4$ GeV) or muons that penetrate the muon system more than five interaction lengths ($1.4 < p_\mu \leq 2.4$ GeV).

2.2 Signal Yield

Candidates are divided into 10 equally-sized bins of w. In each bin we extract the yield of $D^*\ell\nu$ decays using a fit to the distribution of $\cos\theta_{B-D^*\ell}$, where

$$\cos\theta_{B-D^*\ell} = \frac{2E_B E_{D^*\ell} - m_B^2 - m_{D^*\ell}^2}{2|\vec{p}_B||\vec{p}_{D^*\ell}|}. \tag{2}$$

This angle is just the reconstructed angle between the $D^*\ell$ combination and the B, computed with the assumption that the only missing mass is that of the neutrino. This distribution distinguishes $B \to D^*\ell\nu$ decays from backgrounds such as $B \to D^{**}\ell\nu$, which extends into the unphysical region $\cos\theta_{B-D^*\ell} < -1$ because of its larger missing mass. Exact reconstruction of w for an event requires knowledge of the flight direction of the signal B, which is unknown without observing the neutrino. However, knowing $\cos\theta_{B-D^*\ell}$ limits the B flight direction relative to that of the $D^*\ell$ combination. We compute the two extreme values for w for each event and use the average.

Extracting the decay rate as a function of w requires separating the signal $D^*\ell\nu$ from other $D^*X\ell\nu$ decays and various backgrounds in each w bin. Here we use $D^*X\ell\nu$ generically to include both resonant $B \to D^{**}\ell\nu$ and non-resonant $B \to D^*\pi\ell\nu$ decays. The backgrounds are divided into five classes: continuum background from $e^+e^- \to q\bar{q}$, combinatoric background resulting from mis-reconstructed D^*'s, uncorrelated background where the D^* and ℓ come from different B's, correlated background where the D^* and ℓ are not from the signal mode, and fake or mis-identified leptons. Together, these backgrounds contribute less than 15% of the events in the signal region $-1 < \cos\theta_{B-D^*\ell} < 1$, integrated over w.

We employ a binned maximum likelihood fit to the $\cos\theta_{B-D^*\ell}$ distribution to extract the signal yields. In the fit, the normalizations of the various backgrounds are fixed, and we allow the normalizations of both the signal $D^*\ell\nu$ and competing $D^*X\ell\nu$ to float. The distributions for these last two come from signal Monte Carlo samples, while the distribution for the other backgrounds are determined from data and/or Monte Carlo. A sample fit is shown in Figure 1 for the D^{*+} analysis. The quality of the fits is good, as is agreement between the data and the fit distributions outside the fitting region.

2.3 Fitting for $|V_{cb}|$

Recent theoretical work has used dispersion relations to constrain the shape of the form factor $\mathcal{F}(w)$.[5,6] The parameterization has a single shape parameter ρ^2, the slope at $w = 1$. It also depends on the previously-measured form factor ratios R_1 and R_2; we use the values measured by CLEO,[7] consistent with theoretical expectations.[8] Using this parameterization, we employ a simple χ^2 fit to extract the best-fit values for the parameters ρ^2 and $\mathcal{F}(1)|V_{cb}|$, while accounting for efficiency and smearing in w.

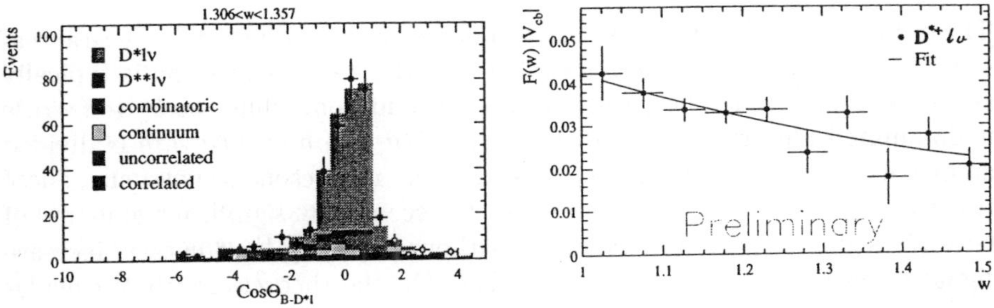

Figure 1: **Left**: The distribution in $\cos\theta_{B-D^*\ell}$ observed in data in the seventh w-bin (circles). The results of the fit to separate the signal $D^{*+}\ell\nu$ decays from various backgrounds are superimposed (histograms). The fit range is -8 $\leq \cos\theta_{B-D^*\ell} \leq$ 1.5 and is indicated with the solid circles. **Right**: The results of the fit to the w distribution measured in $\bar{B}^0 \to D^{*+}\ell\nu$ decays. The figure displays the product $\mathcal{F}(w)|V_{cb}|$: the data points are the yields in each w-bin, corrected for efficiency, smearing, and all other terms in the differential decay rate apart from $\mathcal{F}(w)|V_{cb}|$. The curve shows the result of the maximum-likelihood fit.

The result of the fit in the $D^{*+}\ell\nu$ analysis is shown in Figure 1. We find

$$|V_{cb}|\mathcal{F}(1) = 0.0424 \pm 0.0018 \pm 0.0019 \quad (3)$$

$$\rho^2 = 1.67 \pm 0.11 \pm 0.22 \quad (4)$$

$$\chi^2/\text{dof} = 3.1/8\,, \quad (5)$$

where the first uncertainty is statistical, and the second is systematic. These results are consistent with LEP measurements[10] of $\mathcal{F}(1)|V_{cb}|$ and ρ^2, but are somewhat higher. These parameters give a branching fraction

$$\mathcal{B}(B \to D^{*+}\ell\nu) = (5.66 \pm 0.29 \pm 0.33)\%. \quad (6)$$

When combined with the HQET result for $\mathcal{F}(1) = 0.913 \pm 0.042$,[9] the fit results imply

$$|V_{cb}| = 0.0464 \pm 0.0020 \pm 0.0021 \pm 0.0021, \tag{7}$$

where the first error is statistical, second systematic, and third theoretical, arising from the uncertainty in $\mathcal{F}(1)$.

A combined fit to extract $|V_{cb}|$ from both the D^{*0} and D^{*+} analyses is currently underway. Both analyses benefit from small backgrounds and good resolution in w, and promise the best single measurement of $|V_{cb}|$ from $B \to D^* \ell \nu$ decays.

3 Technique for Extracting $|V_{ub}|$

The extraction of $|V_{ub}|$ from exclusive B decays is more difficult for several reasons. First, $|V_{ub}|/|V_{cb}| \approx 0.08$, meaning that the $b \to u$ transition is typically swamped by the much larger $b \to c$ rate. In addition, while the $b \to c\ell\nu$ rate is dominated almost entirely by $B \to D/D^* \ell \nu$, the $b \to u\ell\nu$ rate is divided among many channels, including a significant non-resonant portion. These two factors combine to make it difficult to reconstruct significant amounts of $b \to u\ell\nu$ signal in particular channels without imposing hard cuts to increase signal-to-background as much as possible. On the theoretical front, calculations are hard to make in a model-independent fashion because the lightness of the u quark spoils the heavy quark symmetry. Restricting predictions to small regions of phase space where the decays are experimentally accessible often incurs a crippling increase in theoretical uncertainty.

CLEO has made several measurements of both $|V_{ub}|$ and the branching fractions for $B \to \pi \ell \nu$ and $B \to \rho \ell \nu$.[11] We describe here a new analysis that uses the full CLEO dataset, 9.1 fb^{-1} of data from the $\Upsilon(4S)$, equivalent to 9.6 million $B\bar{B}$ decays. In addition to the increase in raw statistics, this new analysis also benefits from a lower lepton momentum cut, an increase in the number of modes reconstructed, and improved signal efficiency. Consequently, in addition to measurements of the $\pi\ell\nu$ and $\rho\ell\nu$ branching fractions and a competitive measurement of $|V_{ub}|$, its goals include the extraction of decay distributions in the kinematic variable $q^2 \equiv (p_\ell + p_\nu)^2$, which describes the momentum transfer to the $\ell\nu$ system. This analysis is not yet complete.

3.1 Event Reconstruction

We fully reconstruct candidates in the modes $B \to \pi/\rho/\omega/\eta\ \ell\nu$. The analysis of these semileptonic decays is complicated by the undetected neutrino, but we take advantage of the hermeticity of the CLEO detector and reconstruct the neutrino via the missing energy ($E_{\rm miss} \equiv 2E_{\rm beam} - \sum E_i$) and missing momentum ($\vec{p}_{\rm miss} \equiv -\sum \vec{p}_i$) in each event.

In the production process $e^+e^- \to \Upsilon(4S) \to B\bar{B}$, the total energy of the beams is imparted to the $B\bar{B}$ system; at CESR, that system is essentially at rest. Hence the neutrino combined with the signal lepton (ℓ) and hadron (X) should satisfy the usual constraints on energy, $\Delta E \equiv (E_\nu + E_\ell + E_X) - E_{\rm beam} = 0$, and momentum, $M_{X\ell\nu} \equiv (E^2_{\rm beam} - |\vec{p}_\nu + \vec{p}_\ell + \vec{p}_X|^2)^{1/2} = M_B$. The data in each mode are binned coarsely over the region $(5.1075 \leq M_{X\ell\nu} < 5.2875\,{\rm GeV}, |\Delta E| < 0.75\,{\rm GeV})$

Backgrounds arise from the $e^+e^- \to q\bar{q}/\tau^+\tau^-$ continuum, fake leptons, $b \to c\ell\nu$, and other $b \to u\ell\nu$ modes. Continuum backgrounds are reduced by comparing the thrust axis of the candidate $X\ell$ pair to that of the other particles. In signal decays, the random B daughter orientations determine the axes, while for jetty continuum, the axes are nearly parallel. We determine the level of remaining continuum background from the off-resonance data; similarly, we account for fake leptons by applying fake rates measured in nonleptonic data. To control both $b \to c\ell\nu$ and $b \to c \to s\ell\nu$ backgrounds, we apply a cut on the lepton momentum $|p_\ell| > 1.0(1.5)\,{\rm GeV}$ for pseudoscalar (vector) modes; this is significantly looser than in past analyses. Candidates in the vector modes are also required to have $\cos\theta_\ell > 0$, where θ_ℓ is the angle between the lepton flight direction in the W rest frame and the W direction in the lab frame. Due to the $V-A$ nature of the weak current, vector modes have a $\sin^2\theta_\ell$ distribution in this variable while backgrounds tend to be flat.

The backgrounds, particularly $b \to c\ell\nu$, can smear into the signal $(M_{X\ell\nu}, \Delta E)$ region when the measured $\vec{p}_{\rm miss}$ misrepresents $\vec{p}_\nu$. They can be highly suppressed by rejecting events with multiple leptons, with $M^2_{\rm miss} \equiv E^2_{\rm miss} - |\vec{p}_{\rm miss}|^2$ inconsistent with zero, or other indications of missing particles. In this analysis, the requirement that the total charge be zero is relaxed to include events with a net charge of ± 1; this improves signal efficiency without significantly worsening backgrounds. Monte Carlo studies reveal that the remaining $b \to c\ell\nu$ background is dominated by events containing either a K_L or a second neutrino, *e.g.* from $b \to c \to s\ell\nu$ with the lepton unidentified.

3.2 Fitting Technique

The continuum- and fake-subtracted data in the seven modes are fit simultaneously in $(M_{X\ell\nu}, \Delta E)$-space to account for "cross feed" between the modes due to mis-reconstruction. The isospin and quark symmetry relations $\Gamma(B^0 \to \pi^-\ell^+\nu) = 2\,\Gamma(B^+ \to \pi^0\ell^+\nu)$ and $\Gamma(B^0 \to \rho^-\ell^+\nu) = 2\,\Gamma(B^+ \to \rho^0\ell^+\nu) \approx 2\,\Gamma(B^+ \to \omega\ell^+\nu)$ are used to constrain the rates for B^+ relative B^0. Hence

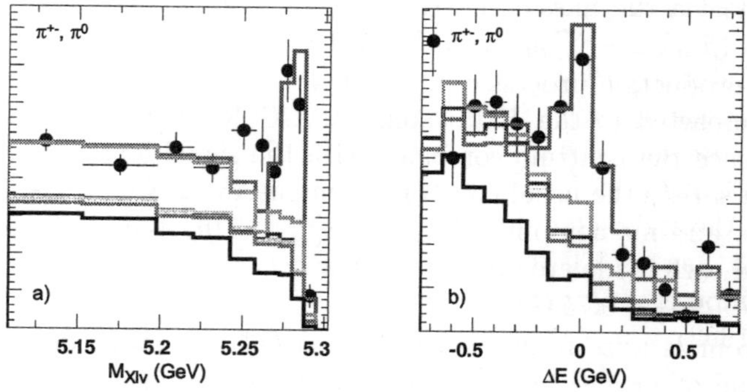

Figure 2: The distribution of (*a*) $M_{X\ell\nu}$ and (*b*) ΔE for the $\pi\ell\nu$ modes in the signal bin of the larger $(\Delta E, M_{X\ell\nu})$ fit region. Data appear as points with error bars, and the histograms are the results of the fit. The uppermost histogram is the signal yield, followed by continuum backgrounds, fake lepton backgrounds, "other" $b \to u\ell\nu$ modes, cross feed, and finally continuum backgrounds. For illustration purposes, the contributions from continuum and fake lepton backgrounds are shown here along with the data before subtraction. The ISGW2 model was used to simulate the other backgrounds. These results are preliminary.

we only fit for four independent yields, N_π, N_ρ, N_ω, and N_η. A preliminary fit for the $\pi\ell\nu$ modes is shown in Figure 2.

Monte Carlo simulation provides the distributions in each mode for signal, the $b \to c$ background, the cross feed among the modes, and the feeddown from higher X_u mass $B \to X_u\ell\nu$ decays. The $b \to c$ normalization in the fit varies independently for each mode. The signal and cross feed distributions are simulated using several different theoretical models, including results from QCD sum rules, lattice gauge calculations, and phenomenological quark models.

A later iteration of the fitting procedure will fit the data in bins of q^2, with the signal and background yields in each bin varying separately. This additional information should provide useful discrimination between various phenomenological models as well as feedback on lattice QCD results for $b \to u\ell\nu$ decays. It will also provide for extracting an improved measurement of $|V_{ub}|$.

Acknowledgments

We gratefully acknowledge the effort of the CESR staff in providing us with excellent luminosity and running conditions. This work was supported by the National Science Foundation, the U.S. Department of Energy, and the Natural Sciences and Engineering Research Council of Canada.

References

1. N. Cabibbo, *Phys. Rev. Lett.* **10**, 531 (1963); M. Kobayashi and T. Maskawa, *Prog. Theor. Phys.* **49**, 652 (1973).
2. N. Isgur and M.B. Wise, *Phys. Lett.* **B232**, 113 (1989); *Phys. Lett.* **B237**, 527 (1990).
3. J.P. Alexander *et al.* [CLEO Collaboration], ICHEP 00-70, CLEO CONF 00-03, hep-ex/0007052.
4. Y. Kubota *et al.* [CLEO Collaboration], *Nucl. Instrum. Methods* **A320**, 66 (1992).
5. C.G. Boyd, B. Grinstein, and R.F. Lebed, *Phys. Rev.* **D56**, 6895 (1997), hep-ph/9705252.
6. I. Caprini, L. Lellouch, and M. Neubert, *Nucl. Phys.* **B530**, 153 (1998), hep-ph/9712417.
7. J. Duboscq *et al.* [CLEO Collaboration], *Phys. Rev. Lett.* **76**, 3898 (1996).
8. M. Neubert, *Physics Reports* **245**, 259 (1994).
9. BaBar Physics Book, P.F. Harrison and H.R. Quinn, editors, SLAC-R-504 (1998).
10. E. Barberio, ICHEP 2000 proceedings.
11. J.P. Alexander *et al.* [CLEO Collaboration], *Phys. Rev. Lett.* **77**, 5000 (1996); B.H. Behrens *et al.* [CLEO Collaboration], *Phys. Rev.* **D61**, 052001 (2000), hep-ex/9905056.

RECENT RESULTS FROM K2K

T. NAKAYA

Department of Physics, Faculty of Science, Kyoto University, Kyoto 606-8502, Japan

for K2K collaboration

K2K is the first accelerator-based long baseline neutrino oscillation experiment to investigate the neutrino oscillation discovered by the atmospheric neutrino observation in Super-Kamiokande. The K2K neutrino beam with the energy of $\sim$1.3 GeV is produced by KEK 12 GeV PS, and is directed toward the worlds largest water Čerenkov detector, Super-Kamiokande, at 250 km away. For the first year from June 1999 to June 2000, 2.3×10^{19} protons were delivered on the target. During this period, 28 accelerator-origin neutrino events are observed with negligible background events. The expectation is estimated to be $37.8^{+3.5}_{-3.8}$ based on the measurement of neutrino events at near neutrino detectors at KEK and also with the far-to-near extrapolation calculated by the beam Monte Carlo simulation. The beam Monte Carlo simulation is experimentally validated by the measurement of pion distributions downstream of horns. The probability that 28 events are observed with $37.8^{+3.5}_{-3.8}$ expected events is approximately 10%. The deficit might be due to neutrino oscillation, though statistics are not good enough yet to establish it.

1 Introduction

The discovery of neutrino oscillation[1] is one of the most exciting news in physics in the last decade. The neutrino oscillation is the only discovered phenomena beyond the standard model in particle physics so far. Existence of neutrino oscillation is the evidence of finite neutrino mass and lepton flavor violation[2]. In the field of particle physics, the firm and undoubted confirmation of neutrino oscillation is one of the most important issues. For the purpose, K2K was proposed in 1995[3], and has been running from 1999. The K2K is the first accelerator-based long baseline neutrino oscillation experiment from KEK to Kamioka with 250 km baseline. At KEK, almost pure muon neutrinos are produced with the average energy of 1.3 GeV. At Kamioka, there is the worlds largest water Čerenkov detector, Super-Kamiokande (SK). The SK looks at the muon neutrino originating from KEK identified by the GPS time stamp. The signal of the neutrinos oscillation is the deficit of muon neutrinos and the distortion of the energy spectrum at SK. In the paper, we report on the observed deficit of muon neutrinos.

2 K2K neutrino beam

K2K neutrino beam is produced by KEK 12 GeV PS with single turn extraction every 2.2 sec. The beam spill is 1.1 μsec long in which there are nine bunches. One spill contains $5 \sim 6 \times 10^{12}$ protons at the target. The target is 3 cmϕ $\times$ 66cm long Aluminum rod. The secondary pions from the target are focused by dual horn system[4] operated by 250 kA pulsed current, and decay in the 200 m long decay pipe. The target is embedded in the first horn to maximize focusing the low energy pions. The horn system amplifies the neutrino flux by a factor of 20 at SK. The expected neutrino beam spectrum at SK for on-axis and off-axis (1km away) are shown in Figure 1.

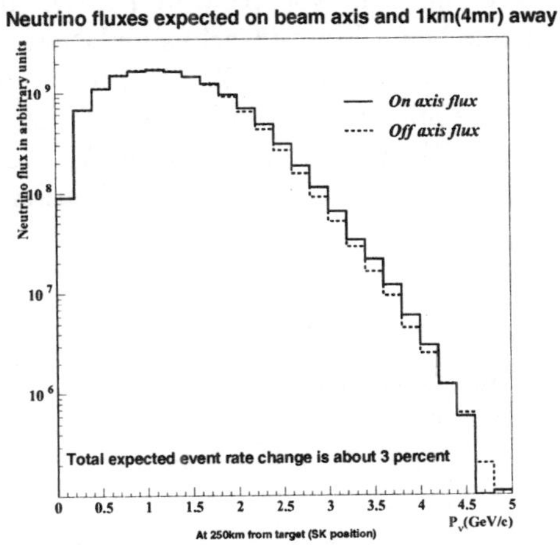

Figure 1: The expected neutrino spectrum at SK (on-axis) and at 1 km away horizontally from SK (off-axis).

Since the beam profile at SK is broad up to a few km, the pointing accuracy of the neutrino beam is required to be less than 1 mrad (= 0.25km/250km). At KEK, the GPS survey guarantees the beam line geometry down to $<$ 0.01 mrad. The civil construction was done with the precision of $<$ 0.1 mrad. These numbers well satisfy the experimental requirement. The actual beam direction and the profile are measured spill by spill by using the muons from pion decays. Downstream of the beam dump, there are muon profile detectors, called MUMON, which are sensitive to high energy muons with energy greater than 5 GeV. MUMON consists of 2 m $\times$ 2 m segmented ionization chambers and an array of silicon pad detectors. The detector monitors the mean of the muon profile with the accuracy better than 0.01 mrad.

The stability of the beam is monitored by various detectors including neutrino detectors explained in Section 3. Muon Range Detector explained in the Ref. [5] monitors the profile and the energy spectrum of muons generated by neutrino interaction for the long run period.

The neutrino beam spectrum is experimentally confirmed by measuring the momentum and the angular distribution of secondary pions from the target. The ring image Čerenkov detector, PIMON, is used for this measurement. PIMON uses Freon gas in order to be insensitive to the primary 12 GeV protons. By varying the gas pressure, PIMON changes the Čerenkov threshold to pions. The Čerenkov image from many pions is measured, which has angular information of the pions. Once the momentum and the angular distributions are known, the neutrino beam spectrum and the profile at any distance can be estimated. The calculated neutrino spectrum by PIMON information and the spectrum expected by our beam Monte Carlo simulation (MC) are compared as shown in Figure 2, and agree very well. Another important function of PIMON is to estimate the spectrum ratio between far and near sites, called "far-to-near ratio". Since PIMON data and beam MC agree quite well, the beam MC is used to calculate far-to-near ratio in Section 4. Figure 2 shows the far-to-near ratio.

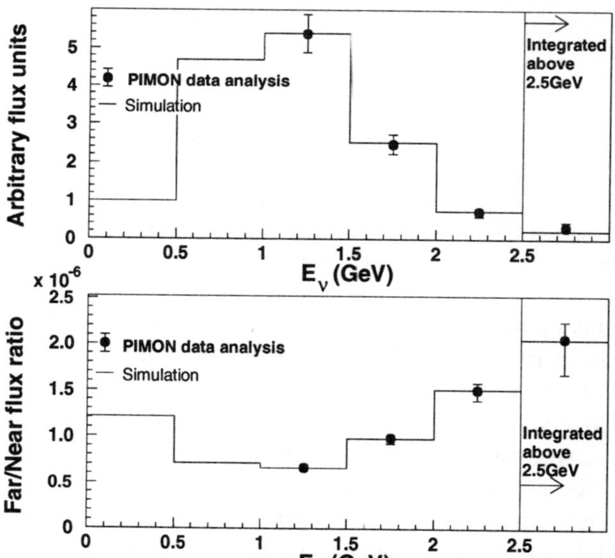

Figure 2: (Top) The neutrino spectrum at SK estimated by PIMON information and calculated by K2K beam MC. (Bottom) "Far-to-Near" ratio.

3 K2K neutrino detectors

The K2K neutrino detectors consist of near neutrino detectors at KEK and the far detector, Super-Kamiokande. The near detectors are located at 300 m downstream from the target. The detectors consist of 1 kt Water Čerenkov detector (1KT), Scintillation-Fiber Tracker (SciFi), Lead Glass(LG), and Muon range detector (MRD) as shown in Figure 3. The main purpose of the near

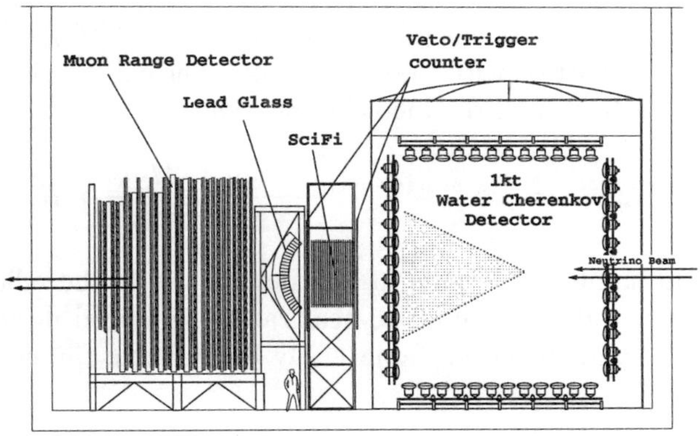

Figure 3: A schematic picture of K2K near neutrino detectors.

detectors is to measure the neutrino beam flux for this analysis (note that we will eventually need information from all detectors to measure neutrino spectrum, neutrino interactions, ν_e flux, etc..). In this paper, we only explain 1KT. Other detectors are well explained in Ref's. [3,5,6]

1KT is a ring image water Čerenkov detector similar to SK except for the size. 1KT and SK use the same 20-inch PMTs with the same photo-coverage of 40%, and use the same analysis programs and MC simulation code, which results in a very small systematic difference between two detectors. Twenty-five tons of the fiducial volume is defined in 1KT center.

The far detector, SK, is a ring image water Čerenkov detector located 1000 m underground (2700 m water equivalent) at 250 km away from KEK. SK consists of the inner detector (ID) and the outer detector (OD). ID is a cylindrical water tank of 36 m in diameter and 34 m in height. There are 11146 20-inch PMT's in ID and 2000 8-inch PMT's in OD. The function of ID is to measure the type and the velocity of the particles produced in neutrino interactions. ID can identify a particle as an electron type or a muon type, and can count the number of particles with the momentum. The fiducial volume of 22.5 ktons is defined as 2 m away from the wall for neutrino events. The function of OD is to identify entering and out-going events, which are rejected offline.

4 Analysis and Result

A signal of neutrino oscillation is the deficit of the number of observed neutrino events at SK in comparison with the expectation. The expectation is estimated with 1KT and PIMON with the following relation:

$$N_{SK}^{expect} = N_{1KT}^{measurement} \cdot \frac{\varepsilon_{SK}}{\varepsilon_{1KT}} \cdot \frac{M_{SK}}{M_{1KT}} \times R, \qquad (1)$$

where N is the number of events at the detector, ε is the efficiency, M is the fiducial mass, and R is the far-to-near ratio introduced in Section 2. All the numbers are summarized in Table 1. By normalizing the neutrino events

Table 1: Summary of information to estimate the number of events at SK. # Events of 1KT is not the number with correcting the multiple-event in one spill, but the raw number.

Detector	# Events	Efficiency	Fiducial Mass (tons)
1KT	$\sim 32,000$	0.71	25
SK	28	0.81	22,5000

at 1KT, N_{SK}^{expect} is free from the absolute neutrino flux which is much more difficult to be measured accurately. The detailed description of the selection criteria is found in the Ref.[3]. In this paper, only a few important selection criteria are explained. At 1KT, we selected in-time events on beam spill with more than 1,000 photo-electrons, which roughly corresponds to 100 MeV threshold. The events with only one in-time activities above the threshold are selected. The events are also required with the vertex inside of the 22.5 kton fiducial volume. There are approximately $32,000$ remaining events. Then, the number is corrected by considering multiple-interaction in a spill. The background events by cosmic rays or neutrinos interacting with rock are also subtracted. The efficiency, 72%, is calculated by MC simulation. An inefficiency is mainly due to the number of photo-electrons requirement, by which some Neutral Current events are discarded. By including the systematic uncertainties of $^{+9}_{-10}$ %, N_{SK}^{expect} is estimated to be $37.8^{+3.5}_{-3.8}$ events. The main sources of systematic uncertainties are $^{+6}_{-7}$ % from the far-to-near ratio calculation and ±4% from the fiducial volume error of 1KT.

The number of observed SK events, $N_{SK}^{measurement}$, are selected as follows. The events are fully contained in ID, which means that the starting point and stopping point of a particle should be inside of ID. There are no activities in OD. The vertex of the events is 200 cm away from the wall. The number of photo-electrons is greater than 200 within 300 nsec time window. The time difference of the events between SK and KEK after subtracting time-of-flight, ΔT, is between −0.2 and 1.3 μsec, where the time is synchronized by GPS. The measured uncertainty of GPS is < 200 $nsec$. With these selection criteria, we find 28 accelerator origin events with negligible background level of $O(10^{-3})$. The 28 events are summarized in Table 2 with the expected number of events.

Table 2: Summary of SK observed events

Category	Observed	Expected
Total	28	$37.8^{+3.5}_{-3.8}$
Single Ring μ-like	14	20.9
Single Ring e-like	1	2.0
Multi Rings	13	14.9

The probability that 28 events are observed with $37.8^{+3.5}_{-3.8}$ expected events is approximately 10%. The deficit might be due to neutrino oscillation, though statistics is not good enough yet to establish it. The spectrum distortion is another signal of neutrino oscillation, while we have not finished the spectrum analysis yet. The neutrino energy can be reconstructed assuming quasi-elastic (QE) interaction, which leads to the following relation:

$$E_\nu = \frac{m_N E_\mu - m_\mu^2/2}{m_N - E_\mu + p_\mu cos\theta_\mu}, \quad (2)$$

where m_N and m_μ is the mass of the neutron and the muon, and E_μ, p_μ and θ_μ is the energy, the momentum, and the angle of the muon relative to the neutrino beam, respectively. The observed neutrino energy spectrum for single ring μ-like events is shown in Figure 4 with the MC expectation of null oscillation. In order to confirm the neutrino oscillation by the spectrum analysis with Figure 4, we need more statistics.

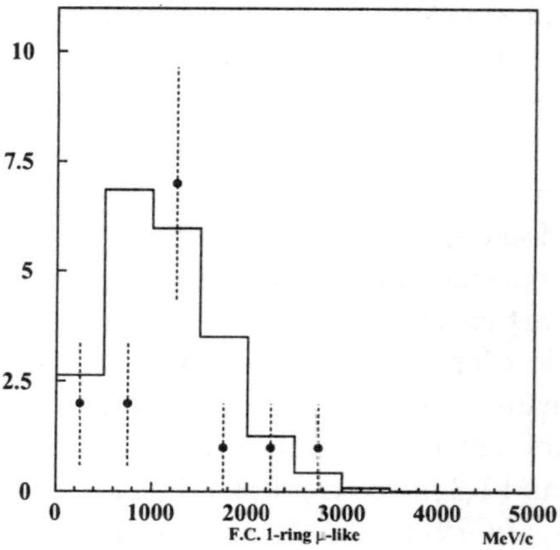

Figure 4: Reconstructed neutrino momentum for single ring μ-like events.

5 Future Prospect and Upgrade

The goal of K2K is to accumulate 10^{20} protons on target (POT) which is by a factor of four more than this sample. With the sample of 10^{20} POT, K2K will establish neutrino oscillation with 3σ level for $\Delta m^2 > 2.5 \times 10^{-3}$ eV^2, or rule out the oscillation at 90% C.L. for $\Delta m^2 > 2 \times 10^{-3}$ eV^2.

At the next step, the spectrum analysis is a definite way with more statistics. The spectrum analysis requires both the precise neutrino energy spectrum and a detailed knowledge of neutrino interactions. Though SciFi is expected to contribute to these subjects, we also plan to install a new detector as a part of K2K upgrade. The new detector is a full-active scintillator tracker. The schematic picture of the detector is shown in Figure 5. The detector is $3 \times 3 \times 1.8$ m^3 and the mass is 15 tons. The detector contains a liquid scintillator, and is segmented to $2 \times 2 \times 300$ cm^3 area by plastic cells, each of which is readout by WLS fibers attached to multi-anode PMTs. The detector can measure the deposit of energy at every segment with the fast timing information. With the energy and the range information, the detector can perform particle identification for a proton, a pion and a muon. The positive ID of a proton and a muon is the key to identify the QE interaction with high purity. With QE sample, we can measure the neutrino spectrum very precisely. The detector has high efficiency down to low energy neutrinos of 500 MeV. The measurement of the neutrino flux at the energy below 1 GeV is important to investigate Δm^2 suggested by SK. As an example, the neutrino with energy of 600 MeV is most sensitive to Δm^2 at 3×10^{-3} eV^2. The non-QE contribution

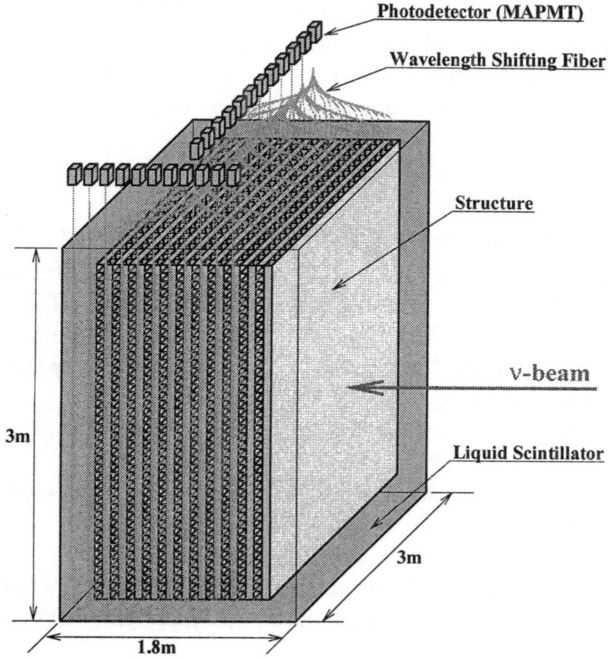

Figure 5: A schematic picture of the full active scintillator tracker for K2K upgrade.

in the energy spectrum at SK has to be well understood to be correctly subtracted, which requires a detailed study of neutrino interaction. The detector is sensitive to all particles from neutrino interactions to study non-QE interaction with advantages of full activity and with fine segment. The detector can track the particle 5 cm long which corresponds to 400 MeV momentum for a proton or 100 MeV for a pion.

The detector is planned to be installed in the summer of 2003, and the active R&D is currently undergoing. The detector is expected to improve information of neutrino energy spectrum in QE interactions and knowledge of neutrino interaction at a few GeV region.

6 Conclusion

The K2K started as the first long baseline neutrino oscillation experiment to investigate the atmospheric neutrino oscillation. We accumulate 2.3×10^{19} protons on target from June 1999 and June 2000, which is a quarter of the proposal. We observed 28 accelerator-origin neutrino events with negligible background events. The expectation is estimated to be $37.8^{+3.5}_{-3.8}$. The probability that 28 events are observed with $37.8^{+3.5}_{-3.8}$ expected events is approximately 10%. We might start looking at the deficit as the evidence of neutrino oscillation though statistics is not good enough as yet.

References

1. Super-Kamiokande collaboration, *Phys. Rev. Lett.* **81**, 1562 (1998).
   K. Kaneyuki, talk presented at 2001 Aspen winter conference on particle physics *"PARTICLE PHYSICS AT THE MILLENNIUM"*, Jan. 2001.
2. H.V. Klapdor eds., *"Neutrinos"*, Springer-Verlag, 1998.
   M. Fukugita and A. Suzuki eds., *"Physics and Astrophysics of Neutrinos"*, Springer-Verlag, 1994.
   S.M. Bilenky, C. Giunti and W. Grimus, *"Phenomenology of Neutrino Oscillations"*, *Prog. Part. Nucl. Phys.* **43**, 1 (1999).
3. K. Nishikawa *et. al.*, KEK-PS E362 proposal, March, 1995.
   Nucl. Phys. **B59** (Proc. Suppl.), 289 (1997).
   K2K collaboration, *Phys. Lett.* **B511**, 178 (2001).
4. For example, Yamanoi Y. et al., KEK Preprint 97-225, November 1997.
5. T. Ishii, T. Inagaki, *et. al.*, hep-ex/0107041, Julu 2001, submitted to *Nucl. Instrum. Meth.*
6. A. Suzuki, H. Park, *et. al.*, *Nucl. Instrum. Meth.* **A453** 165 (2000).

MEASUREMENTS OF B^0 AND $B^{\pm}$ LIFETIMES USING FULLY RECONSTRUCTED B DECAYS AT BABAR

JAMES H. PANETTA

2575 Sandhill Rd. MS41

Menlo Park, CA 94025

E-mail: panetta@slac.stanford.edu

(For the Babar Collaboration)

The Babar detector has been recording data at and around the $\Upsilon(4S)$ resonance since May, 1999. In this paper, we briefly describe the Babar detector and the PEP-II accelerator. With the first year's data, lifetimes of the charged and neutral B mesons are studied in a sample of fully reconstructed B mesons.

1 Introduction

Preliminary measurements of B^0 and $B^{\pm}$ lifetimes have been performed in the Babar detector. This analysis exploits the copious production of B meson pairs in $\Upsilon(4S)$ decays, produced in asymmetric e^+e^- collisions at the PEP-II B-Factory at SLAC.

2 The PEP-II B-Factory

PEP-II is an e^+e^- colliding beam storage ring facility designed to produce a luminosity of 3×10^{33} cm^{-2}s^{-1} at center of mass energies near the $\Upsilon(4S)$ resonance (10.58 GeV). The PEP-II B-Factory was constructed by a collaboration between SLAC, LLNL and LBNL[1] on the SLAC site. The machine is asymmetric with energies of 3.1 GeV for the positron beam and 9.0 GeV for the electron beam, with design currents of 1.5 A and 0.8 A respectively. As the beam energies are different, a two ring configuration is required.

The asymmetric collisions provided by PEP-II produce a center of mass energy of 10.58 GeV with a boost of $\beta\gamma = 0.56$ in the lab frame. This boost provides an average separation of $\beta\gamma c\tau = 250$ μm between the two B vertices, allowing measurements of time dependence in the B decays.

PEP-II became operational in July 1998, with first collisions seen shortly thereafter. In October 2000, peak luminosities reached 3.1×10^{33} cm^{-2}s^{-1}. A detailed status report on PEP-II performance as of July 2000 can be found in reference [2].

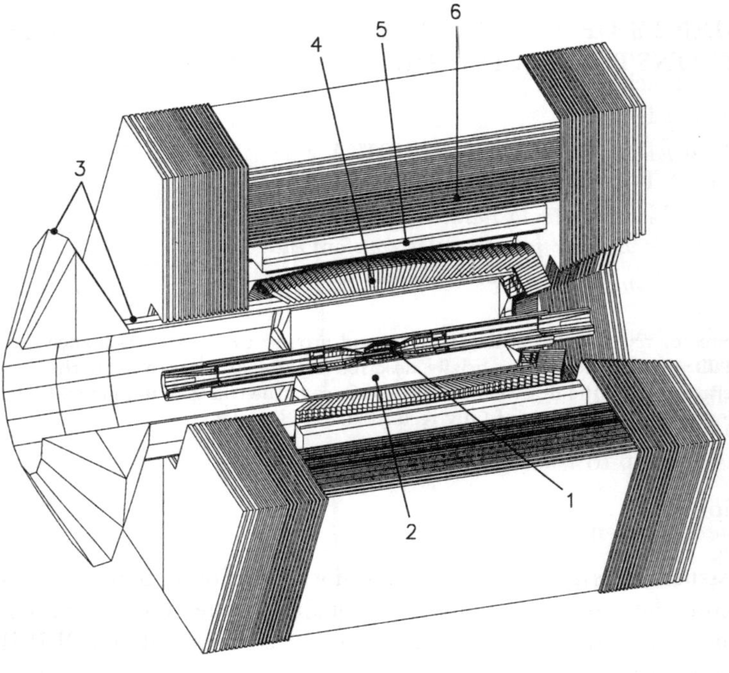

Figure 1: The Babar Detector. 1. Silicon Vertex Tracker (SVT), 2. Drift Chamber (DCH), 3. Detector of Internally Reflected Cherenkov Light (DIRC), 4. Electromagnetic Calorimeter (EMC), 5. Magnet, 6. Instrumented Flux Return (IFR).

3 The Babar Detector

The magnetic spectrometer Babar (Figure 1) was constructed at SLAC by a collaboration between nine countries. Construction was approved in November 1995, and first data with colliding beams was taken in May 1999.

The 1.5 T superconducting solenoid contains a set of nested detectors: Innermost is a five layer silicon-strip vertex tracker (SVT); the central drift chamber (DCH) used for tracking and energy loss measurements; a quartz bar cherenkov radiation detector (DIRC) used for particle identification; and a CsI crystal electromagnetic calorimeter (EMC), used for photon detection and separating electrons from charged pions. Two layers of cylindrical resistive plate chambers (RPCs) are located between the EMC and the magnet cryostat. The instrumented flux return (IFR) outside the magnet contains 18 layers of steel and 19 layers of RPCs (18 in the endcaps), used for muon and K_L identification. While a detailed description of the Babar detector can be found in reference[3], a brief overview of the subsystems is presented here.

3.1 Silicon Vertex Tracker

The Silicon Vertex Tracker (SVT) provides the vertex resolution needed for CP violation studies at PEP-II. It is also capable of independent charged particle tracking down to a $p_\perp$ of 60 MeV/c. Single hit resolution is $\sim$ 15 μm in the inner layers. The SVT is a five layer silicon strip design, with an active area of 0.9 m^2.

3.2 Central Drift Chamber

The Central Drift Chamber (DCH) is the primary tracking device used for particles with transverse momenta of $p_\perp > 120$ MeV/c. It consists of 40 layers of hexagonal cells, grouped into 10 superlayers, with single hit resolution in data of $\sim$ 125 μm. The DCH is also capable of dE/dx measurements, providing K/π separation of 2σ up to 700 MeV/c.

3.3 Particle Identification System

The primary particle identification system in the Babar detector is the Detector of Internally Reflected Cherenkov Light (DIRC). This system consists of 144 fused silica bars, arranged in 12 boxes, placed just outside the DCH. Charged particles traveling through the bars can emit a Cherenkov cone, which is transmitted via total internal reflection to an array of photomultiplier tubes. The bars cover 87% of the polar angle in the center of mass frame. The DIRC provides 4σ K/π separation up to 3 GeV/c, and has a 2.5 mrad θ_C resolution.

3.4 Electromagnetic Calorimeter

The Electromagnetic Calorimeter (EMC) contains 6580 CsI(Tl) crystals, arranged quasi projectively with 48 rows of 120 crystals in the barrel, and 8 rows of crystals in the forward endcap. Crystal thicknesses range from 16 to 17.5 radiation lengths. The EMC provides a 5% resolution in m_π and a 1.9% resolution in E_{Bhabha}. The EMC is also used for the separation of hadrons from electrons, and in conjunction with the IFR for K_L and muon identification.

3.5 Instrumented Flux Return

The Instrumented Flux Return (IFR) is used for the identification of muons and neutral hadrons. It consists of nearly 900 bakelite based Resistive Plate Chambers (RPCs) interleaved with the iron flux return for the 1.5 Tsuperconducting solenoid. There are 19 layers in the barrel sections, and 18 layers in the endcaps, with an additional 2 layers of cylindrical RPCs arranged between the

EMC and the solenoid. Muon efficiency for a loose selection is approximately 90% for $1 < p_\mu < 3$ GeV/c. Pion misidentification fraction is approximately 5% over this range.

3.6 Level 1 Trigger

The Level 1 Trigger (L1) system consists of the DCH track trigger (DCT), the EMC energy trigger (EMT), the IFR trigger (IFT), and the global trigger (GLT). The GLT operates on the basic trigger primitives given by the DCT, the EMT and the IFR, combines them, and makes decisions based on the topology of the primitives. Combined efficiency for $B\bar{B}$ events is $> 99.9\%$.

3.7 Level 3 Trigger

The Level 3 Trigger (L3) is the first component of the system to see complete events. It processes data from the DCT, the DCH and the EMC using two independent algorithms to provide track and cluster objects. The L3 DCH algorithm provides fast tracking down to a transverse momentum of ~ 250 MeV/c. The L3 EMC algorithm provides a fast two dimensional clustering over the entire calorimeter using a lookup table. L3 clusters are accepted for $E > 100$ MeV.

The track and cluster objects returned by the above algorithms are combined in basic topologies to determine the logging decision. An exception is made for Bhabhas, which at design luminosity of $3 \times 10^{33}/\text{cm}^2/s$ occur at a rate of 150 Hz, and thus are vetoed to reduce the rate.

4 *B* Lifetime

The data used in this analysis were collected in the period from January 2000 to June 2000. Integrated luminosity for the sample is 7.4 fb^{-1} on the $\Upsilon(4S)$ resonance, and 0.9 fb^{-1} collected 40 MeV below the $B\bar{B}$ threshold. This corresponds to approximately 8.6×10^6 $B\bar{B}$ pairs produced at threshold.

At PEP-II the B meson pairs are moving in the lab frame along the beam axis ($+z$ direction) with a velocity of $\beta\gamma = 0.56$. One B (B_{REC}) is fully reconstructed in an all hadronic mode ($B^0 \to D^{(*)-}\pi^+$, $D^{(*)-}\rho^+$, $D^{(*)-}a_1^+$, $J/\psi K^{*0}$ and $B^- \to D^{(*)0}\pi^-$, $J/\psi K^-$, $\psi(2S)K^-$ or charge conjugate). A total of about 2600 each of neutral and charged B mesons is reconstructed in these modes, with a purity of $\sim 90\%$ (Figure 2). The signal region for each decay mode in the selected sample is defined by the three standard deviation bands in the two dimensional distribution of the kinematic variables ΔE and m_{ES}[a]. The resolution on m_{ES} is about 3 MeV/c^2, and that on ΔE varies from 12 to 40 MeV, depending on the decay mode.

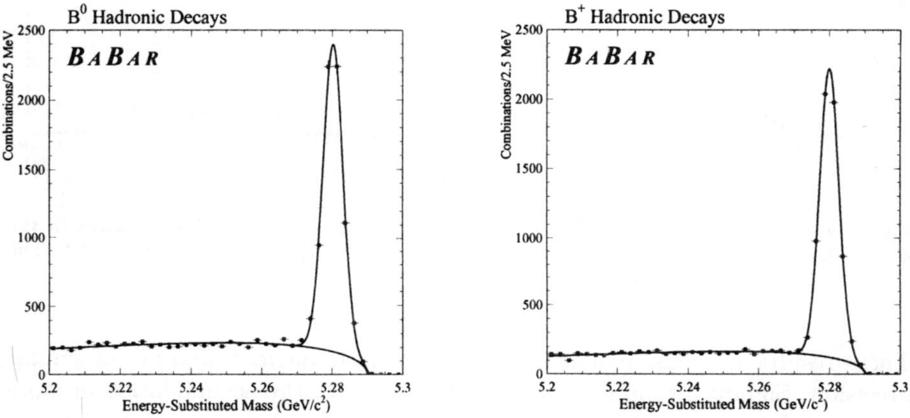

Figure 2: m_{ES} distributions for all the hadronic modes in the B^0 (Left) and the B^- (Right) samples, fit to the sum of a gaussian and the ARGUS background parameterization [4]. Total yields in each sample are 2577 ± 59 for B^0 and 2636 ± 56 for B^-.

The separation between the two B vertices along the boost direction, $\Delta z \equiv z_{\rm REC} - z_{\rm TAG}$, is measured and used to estimate the decay time, $\Delta t \approx \Delta z / \beta\gamma c$. The vertex $B_{\rm TAG}$ is determined by grouping tracks not associated with the $B_{\rm REC}$ vertex together and applying a procedure detailed in reference [5]. The typical separation between the two vertices is $\Delta z \approx 260\ \mu$m, while the experimental resolution is $\sim 110\ \mu$m.

The B^0 and B^+ lifetimes are extracted from a simultaneous unbinned maximum likelihood fit to the Δt distributions of the signal candidates, assuming a single common resolution function. An empirical description of the Δt background shapes is assumed, using the m_{ES} sidebands independently for the charged and neutral B mesons. Figure 3 shows the Δt distributions with the fit and the background functions superimposed.

The preliminary results for the B meson lifetimes are

$$
\begin{aligned}
\tau_{B^0} &= 1.506 \pm 0.052 \pm 0.029 \text{ ps}, \\
\tau_{B^+} &= 1.602 \pm 0.049 \pm 0.035 \text{ ps},
\end{aligned}
$$

and that for their ratio is

$$\tau_{B^+} / \tau_{B^0} = 1.065 \pm 0.044 \pm 0.021.$$

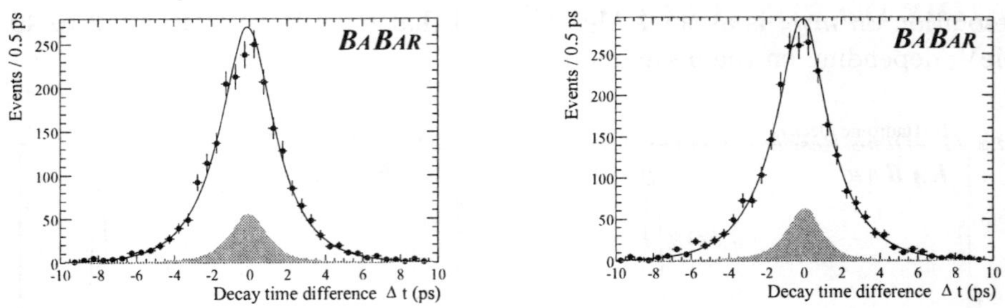

Figure 3: Δt distributions for the $B^0/\overline{B}^0$ (Left) and B^+/B^- (Right) candidates in the signal region ($m_{\rm ES} > 5.27$ GeV$/c^2$, with the results of the lifetime fits superimposed. The background is shown by the shaded region.

These results are consistent with previous measurements [6], and are of similar precision. Significant improvement is expected with the accumulation of more data, and further examination of the systematic uncertainties.

References

1. An Asymmetric *B* Factory Based on PEP, The Conceptual Design Report for PEP-II, LBNL Pub 5303, SLAC-372 (1991).
2. Status Report on PEP-II Performance, J. Seeman *et al.*, to appear in the Proceedings of the VIIth European Part. Acc. Conf. - EPAC 2000, Vienna, Austria (June 2000).
3. The Babar Detector, B. Aubert *et al.*, SLAC-PUB-8569, to be published in *Nuc. Inst. Meth.*
4. ARGUS Collaboration, H. Albrecht *et al.*, *Z. Phys.* **C48**, 543, (1990)
5. A Measurement of Charged and Neutral *B* Meson Lifetimes Using Fully Reconstructed Decays, Babar Collaboration, B. Aubert *et al.*, SLAC-PUB-8529, (July 2000). Preprint hep-ex/0008060.
6. Review of Particle Properties, D.E. Groom, *et al.*, *Eur. Phys. Jour.* **C15**, 1 (2000).

THE DØ RUN II DETECTOR AND PHYSICS PROSPECTS

NEETI PARASHAR
111 Dana Research Center, Northeastern University, 360 Huntington Avenue, Boston, MA 02115, USA
E-mail: neeti@fnal.gov
FOR THE DØ COLLABORATION

The DØ Detector at Fermilab is currently undergoing an extensive upgrade to participate in the Run II data taking which shall begin on March 1, 2001. The design of the detector meets the requirements of the high luminosity environment provided by the accelerator. This paper describes the upgraded detector subsystems and gives a brief outline of the physics prospects associated with the upgrade.

1 Introduction

The DØ detector was designed to study collisions between protons and antiprotons in the Tevatron collider at Fermilab. DØ took its first data run in the period 1992-1996, called Run I. The successes of Run 1, including the discovery of the top quark, and the physics potential of high-luminosity running at the Tevatron have additionally motivated the present upgrade of the detector. We would now like to pursue a detailed top quark physics study, search for the Higgs boson and for supersymmetry, investigate b-physics, and look for physics beyond the Standard Model (SM).

In order to enhance the physics reach of these processes in Run II, the Tevatron also had to undergo two major upgrades. First, the Tevatron for Run II is designed to achieve a luminosity of $5\text{x}10^{32}$ cm^{-2}s^{-1}, which is a factor of 10 more than in Run I. The second upgrade involves a decrease in the bunch crossing time, which for Run I was 3.5 μs, while Run II will begin with 396 ns and eventually reach 132 ns as the number of bunches is increased. There is also an increase in the center-of-mass (CM) energy from 1.8 TeV to 2.0 TeV. To take full advantage of the new physics opportunities and to contend with the high radiation environment and shorter bunch crossing times, an extensive upgrade of the DØ detector was undertaken, which is now in its final stages.

Figure 1 shows an elevation view of the upgraded detector. The upgrade consists of an addition of a solenoid, a tracker with silicon and fiber detectors, muon scintillation trigger counters, new forward muon system, preshower detectors, readout electronics and new trigger and data acquisition systems.

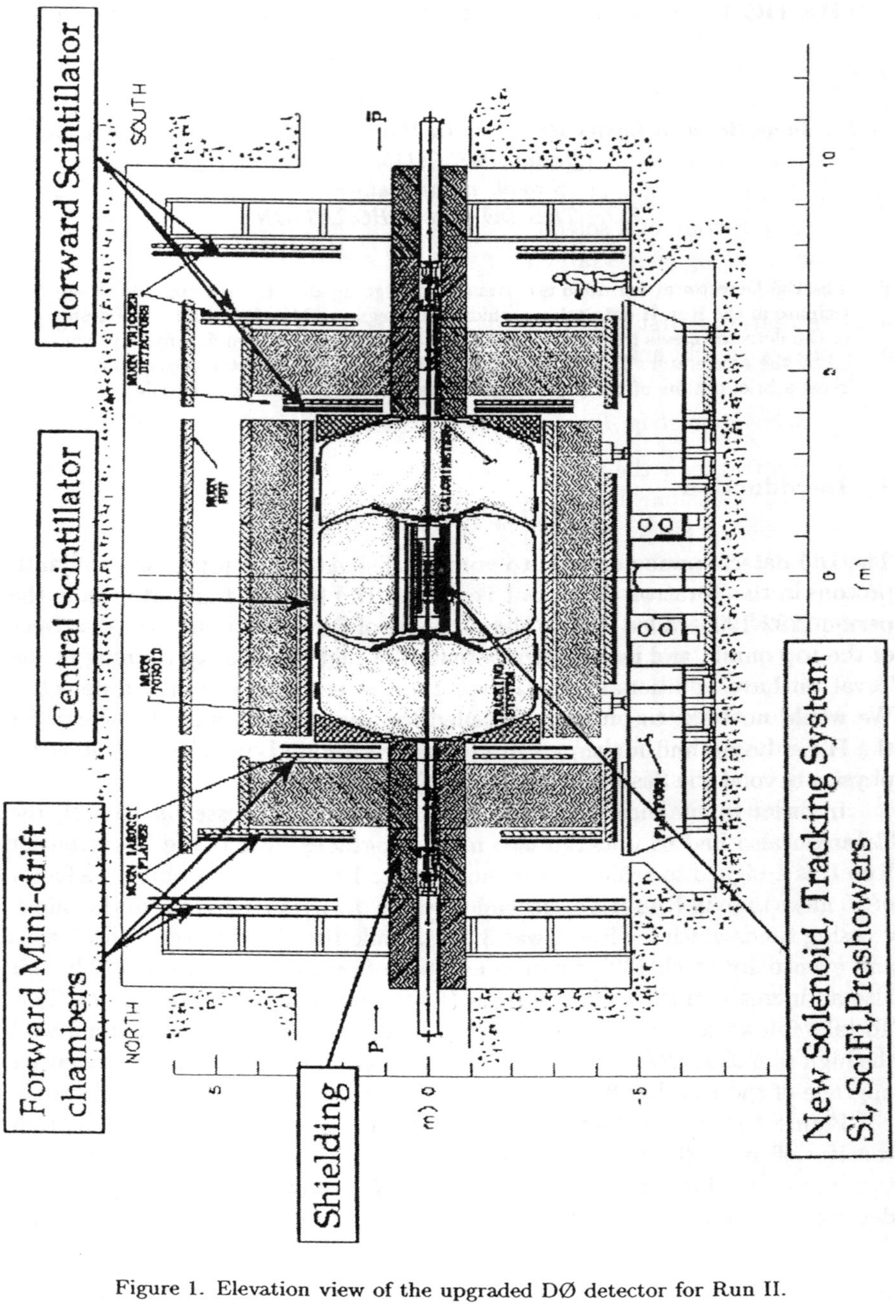

Figure 1. Elevation view of the upgraded DØ detector for Run II.

2 Tracking

The upgraded tracking system (Figure 2) consists of an inner silicon vertex detector, surrounded by eight superlayers of scintillating fiber tracker. These detectors are located inside a 2 Tesla superconducting solenoid, which is surrounded by a scintillator based preshower detector. The upgraded tracking system has been designed to meet several goals: momentum measurement by the introduction of a solenoidal field; good electron identification and e/π rejection; tracking over a large range in pseudo-rapidity ($\eta \approx \pm 3$); secondary vertex measurement for identification of b-jets from Higgs and top decays and for b-physics; first level tracking trigger; fast detector response to enable operation with a bunch crossing time of 132 ns; and radiation hardness.

2.1 Silicon Microstrip Tracker (SMT)

The silicon tracker is the first set of detectors encountered by particles emerging from the collision. The detector design consists of interspersed disks and barrels, based on single and double-sided silicon microstrip detectors, with a total of 793,000 channels and is radiation hard up to about 2 Mrad. The SVX IIe chip is used for readout [1]. The combination of small-angle and large-angle stereo provides good pattern recognition and allows good separation of primary vertices in multiple interaction events. The expected hit position resolution in $r\phi$ is 10 μm.

A silicon track trigger preprocessor is being built which will allow the use of SMT information in the Level 2 trigger. This will add the capability for triggering on tracks displaced from the primary vertex, as well as sharpening the p_T threshold of the Level 2 track trigger and of the electron and jet triggers at Level 3.

2.2 Central Fiber Tracker (CFT)

The detector just outside the SMT is the 8-layered CFT, which is based on scintillating fiber ribbon doublets with visible light photon counter (VLPC) readout [2]. Each layer contains 2 fiber doublets in a zu or zv configuration (z = axial fibers and $u, v = \pm 3°$ stereo fibers). Each double, consists of two layers of 830 μm diameter fibers with 870 μm spacing, offset by half the fiber spacing. This configuration provides very good efficiency and pattern recognition and results in a position resolution of $\approx$ 100 μm in $r\phi$. There are a total of 77,000 channels.

The CFT serves two main functions. First, with the SMT it enables track reconstruction and momentum measurement for $\eta < \pm 1.7$, second it provides

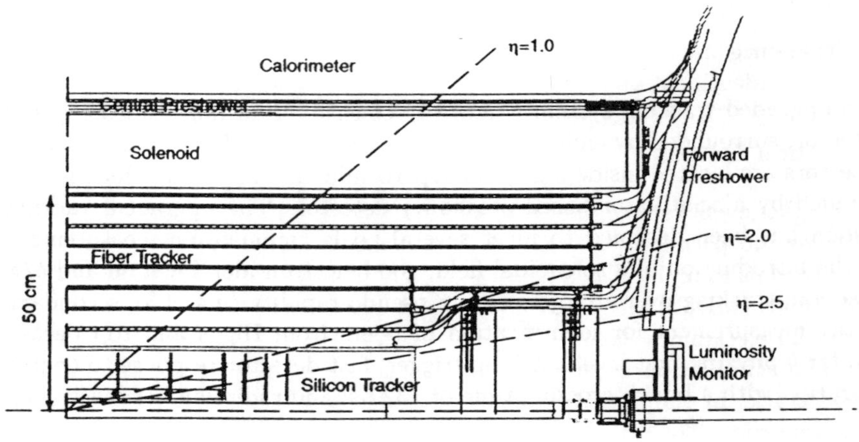

Figure 2. $r - z$ view of the DØ tracking system.

fast Level 1 triggering on charged track momentum.

The fibers are up to 2.5 m long and the light is piped out by clear fibers of length 7-11 m to the VLPCs situated in cryostats outside the tracking volume, which are maintained at 9°K. The VLPCs are solid state devices with a pixel size of 1 mm, matched to the fiber diameter. The fast risetime, high gain and excellent quantum efficiency of these devices make them ideally suited to this application.

2.3 Superconducting Solenoid

The momenta of the charged particles will be determined from their curvature in the 2T magnetic field provided by a 2.8 m long, 1.42 m in diameter and 1.1 radiation length thick solenoid magnet. From the value of $\sin\theta \times \int B_z dl$ and the space point precision provided by the silicon and fiber tracking system (located inside the solenoid) a momentum resolution of $\delta p_T/p_T = 0.002 p_T$, ($p_T$ in GeV) is expected.

2.4 Preshower detectors

The central and forward preshower detectors (CPS and FPS) are based on a similar technology, using triangular scintillator strips (axial and 20° stereo) with wavelength shifter readout. These detectors provide fast energy and position measurements for the electron trigger and facilitate offline electromagnetic identification.

3 Muon Detectors

The higher event rates in Run II have led us to add new muon trigger detectors covering full pseudorapidity range and the harsh radiation environment has prompted us to replace the forward proportional drift tubes(PDTs) with minidrift tubes (MDTs). In the central region, the Run I PDTs are used and the front-end electronics have been replaced to ensure dead-timeless operation.

The scintillation counters provide the time information and match the muon tracks in the fiber tracker, and consists of three layers to reduce hit combinatorics. The drift tubes provide enhanced muon momentum resolution and pattern recognition. The design of the muon system reduces backgrounds and trigger rates with additional shielding and we have the ability to trigger on inclusive single muons with $p_T > 7$ GeV and dimuons with $p_T > 2$ GeV.

4 Trigger and Data Acquisition

The DØ trigger and DAQ systems have been completely restructured to handle the shorter bunch spacing and new detector subsystems in Run II. The Level 1 and 2 triggers utilize information from the calorimeter, preshower detectors, central fiber tracker, and muon detectors. The Level 1 trigger reduces the event rate from 7.5 MHz to 10 KHz and has a latency of 4 μsec. The trigger information is refined at Level 2 using calorimeter clustering and detailed matching of objects from different subdetectors. The Level 2 trigger has an output rate of 1 KHz and a latency of 100 μsec. Level 3, consisting of an array of PC processors, partially reconstructs event data within 50 msec to reduce the rate to 50 Hz. Events are then written to tape.

5 Run II Physics Prospects - Some Highlights

We have seen that the design of the Run II DØ detector makes important additions to the DØ physics capabilities, namely: energy/momentum matching for electron identification, improved muon momentum resolution, charged sign and momentum determination, calorimeter calibration and displaced vertex identification (b-tags).

One of the crucial physics goals of Run II is to search for the Higgs boson. In Run II the reach for Higgs at the Tevatron suggests a possibility of discovering light Higgs, given sufficient integrated luminosity from both experiments

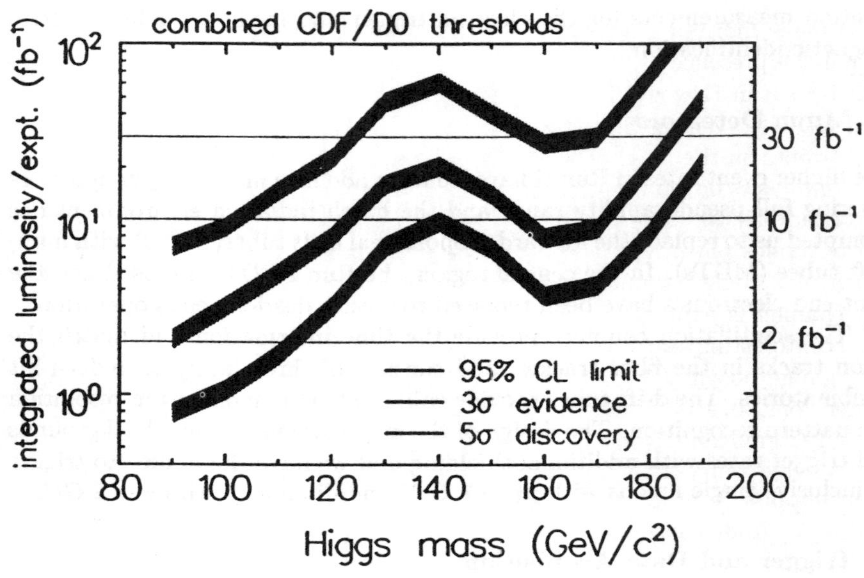

Figure 3. Integrated luminosity required as a function of Higgs mass for a 95% C.L. exclusion, 3σ evidence and 5σ discovery.

(DØ and Collider Detector at Fermilab, CDF) at the collider [3]. The dominant decays can be classified into $M_{Higgs} < 135\,GeV$ and $M_{Higgs} > 135\,GeV$. The light mass range is of great interest, since minimal supersymmetric versions of SM always predict a light Higgs boson lying in this region. If there is no SM Higgs, 95% confidence level exclusion limits can be set up to $\approx$120 GeV mass with 2 fb^{-1}. However if there is a Higgs, discovery at the 5σ level can be obtained with 20 fb^{-1} up to similar mass. This sensitivity is shown in Figure 3.

A detailed study of the top quark is another important physics interest. Since the CM energy is expected to increase from 1.8 TeV to $\approx$2.0 TeV, the $t\bar{t}$ production cross section will increase by about 38% (the production is dominated by $q\bar{q} \to t\bar{t}$). For single top production the increase will be about 22% for the s-channel ($q\bar{q} \to t\bar{b}$), and about 44% for the W-gluon fusion process ($qg \to qt\bar{b}$). The expected uncertainty in the top quark mass measurement is $\delta m_t \simeq 2$ GeV, and we should also be able to observe single top production for the first time.

SM has been very successful so far and in the electroweak sector, the goal is to make more precise tests of the SM and search for deviations that may signal the presence of new physics. A prime measurement is of the W mass, which in Run II would be done with a precision of 30 MeV from both DØ and CDF [4]. Within SM, the mass of the top quark and the W boson set constraints on the mass of the Higgs boson.

In physics beyond the SM, there is potential for observing supersymmetry. For example, the supersymmetric gaugino pair production in the trilepton decay modes $p\bar{p} \rightarrow \tilde{\chi}_1^{\pm}\tilde{\chi}_1^{\mp}, \tilde{\chi}_1^{\pm}\tilde{\chi}_2^{0} \rightarrow 3\ell + X$, can be discovered with a mass reach of 220 GeV. Supersymmetric squark/gluino production in the jets + $\not{E}_T$ channel is expected to probe masses up to about 400 GeV for 2 fb^{-1}. Other interesting searches include a wide range of topics, namely, leptoquarks, new gauge bosons (W', Z'), topcolor, compositeness, technicolor and extra dimensions. In Run II there is an excellent chance to discover new physics or exclude significant regions of parameter space.

A wide range of b-physics and QCD physics will also be pursued, but their details are beyond the scope of this article. Details of the DØ detector and its upgrade can be found elsewhere [5].

Acknowledgments

I would like to thank the Lake Louise Winter Institute for arranging a stimulating set of talks and for their warm hospitality. I would also like to thank Darien Wood for careful reading of this paper.

References

1. T. Zimmerman *et al*, *IEEE Trans. Nucl. Sci.* **42**, 803 (1995).
2. M.D. Petroff and M.G. Stapelbroek, *IEEE Trans. Nucl. Sci.* **36**, 158 (1989);M.D. Petroff and M. Atac, *IEEE Trans. Nucl. Sci.* **36**, 163 (1989).
3. M. Carena *et al*, Report of the Tevatron Higgs Working Group, hep-ph/0010338.
4. R. Brock *et al*, Report of the Working Group on Precision Measurements, hep-ex/0011009.
5. Report on "The DØ upgrade: The Detector and its Physics, Fermilab Pub-96/357-E; J. Ellison, "The DØ Detector Upgrade and Physics Program, hep-ex/0101048.

SEARCHES FOR R-PARITY VIOLATION PROCESSES IN DELPHI

V. POIREAU
CEA Saclay DSM/DAPNIA/SPP
91191 Gif sur Yvette cedex, France
E-mail: vincent.poireau@cern.ch

We present a review of searches of supersymmetric signal with R-parity violation in DELPHI. An introduction to R-parity is given, followed by a presentation of the different signatures at LEP. The experimental searches are described for each channel and the results are summarized in a final table.

1 Introduction on R-parity violation

The most general superpotential of the minimal supersymmetric standard model (MSSM) can be expressed as a sum of two terms:

$$W_{MSSM} = W_{R_p} + W_{\not R_p},$$

where W_{R_p} is the classical term of the MSSM [1] and where $W_{\not R_p}$ reads[a]:

$$W_{\not R_p} = \lambda_{ijk} L_i L_j E_k^c + \lambda'_{ijk} L_i Q_j D_k^c + \lambda''_{ijk} U_i^c D_j^c D_k^c.$$

The numbers i, j and k are the generation indices; L and E represent respectively the left-handed doublet and the right-handed singlet of the lepton superfield; Q, U and D denote respectively the left-handed doublet, the up-type right-handed singlet and the down-type right-handed singlet of the quark superfields; λ_{ijk}, λ'_{ijk} and λ''_{ijk} are Yukawa couplings shown on Figure 1.

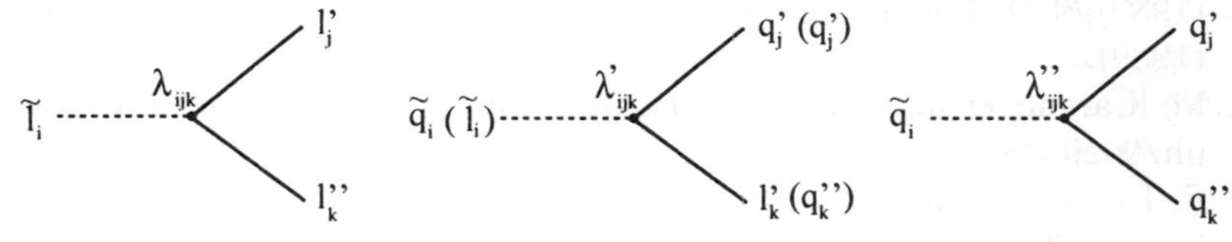

Figure 1: Couplings involved in the $W_{\not R_p}$ superpotential.

From the superpotential, we observe that λ and λ' couplings violate the lepton number (L) conservation and that λ'' couplings violate the baryon number (B) conservation (since a superparticle carries the same quantum numbers

[a]We made here the assumption that the term $\mu'_i H_2 L_i$ of the superpotential could be rotated away.

as its standard partner, except for the spin). A direct consequence of these violations is a possible decay of the proton (via for example the λ' and λ'' couplings). In 1975 [2], a new discrete symmetry, the R-parity associated with a multiplicative quantum number, R_p, was introduced in order to suppress the $W_{\not R_p}$ term and consequently to avoid the proton decay. The R-parity was defined as $R_p = (-1)^{3B+L+2S}$, where S is the spin of the particle. With this definition, a standard particle obeys $R_p = 1$ and a supersymmetric one obeys $R_p = -1$. If we ask for the conservation of R-parity, the coupling constants of Figure 1 are suppressed, and only the W_{R_p} term of the superpotential remains. This model may be called the *MSSM with R-parity conservation.*

The introduction of the R-parity was motivated for phenomenological reasons: conservation of L and B, and proton stability. However, the proton decay can be avoided if we assume that only one coupling dominates at a time. Furthermore, L and B violation are allowed if their level are below experimental limits. Finally, no strong theoretical reason can be given for the suppression of couplings which appear naturally in the most general superpotential. Thus, we consider here models where we keep the $W_{\not R_p}$ term in the superpotential. This term does not conserve the R-parity, and therefore this model is called *MSSM with R-parity violation.*

At the phenomenological level, the main consequence is that the lightest supersymmetric particle (LSP) can decay into standard particles and therefore may be visible within the detector volume. Furthermore, one single superparticle can be produced in the collision of two standard particles.

The working hypotheses in this paper are the following. First, we use the framework of the constrained MSSM, which depends on 5 parameters [1]: M_2 the SU(2) gaugino mass at the electroweak scale, m_0 the sfermion mass at the grand unified scale, A_0 the trilinear term at the grand unified scale, $\tan\beta$ the ratio of the vacuum expectation values of the two Higgs doublets and μ the mixing mass term of the Higgs doublets. The LSP is assumed to have a negligible lifetime so that the production and decay vertices coincide. Finally, the assumption is made that only one R-parity violating coupling is dominant at a time.

2 R-parity violation signatures at LEP

2.1 Production

The pair production of superparticles is exactly similar to the R-parity conserved case. The R-parity violating couplings are only involved in the subsequent decays of the produced superparticles (see below). Different kinds of pair productions were studied in DELPHI: slepton, squark, neutralino and chargino pair production. In addition, on the contrary of the R-parity conserved case, one superparticle may be produced from a standard particle collision. At LEP, a sneutrino can be singly produced via λ_{121} or λ_{131} couplings.

2.2 Decay

Two kinds of decay of the produced superparticles are possible: the direct and the indirect decays. If the superparticle is the LSP, it will decay into standard particles, possibly via an off-shell superparticle: this is the so-called direct decay. If the superparticle is not the LSP, it will first decay into the LSP and a standard particle (for example a W boson), then the LSP will decay directly into standard particles. The difference of mass between the produced superparticle and the LSP is called ΔM. The final state depends strongly on the dominant coupling constants. For λ couplings, the final state is mainly composed of leptons. For λ' couplings, the final state consists of a mixed state of jets and leptons whereas for λ'' couplings it contains a large number of jets.

3 Searches in DELPHI

The DELPHI experiment [3], located on the LEP ring at CERN, recorded an integrated luminosity of 226 pb^{-1} in 1999 and of 224 pb^{-1} in 2000, with energy ranging from 192 GeV to 208 GeV. Although lots of topologies were treated by DELPHI, only λ and λ'' R-parity violating[b] couplings were covered[4]. The results of the analyses are presented in the following sections. Excluded domains in the MSSM parameter space and limits presented below are given at the 95 % confidence level.

3.1 Charginos and neutralinos

If we assume that the dominant coupling is λ, then the final state consists of leptons and missing energy coming from neutrinos, and additional jets or leptons if the superparticle decay is indirect. The nature of leptons depends on the type of the λ coupling constant. Two analyses have been performed: one with a λ_{122} coupling which involves electrons and muons (most efficient case) and one with a λ_{133}, which represents the worse case due to the taus in the final state. The analyses were based on missing quantities, lepton identification and kinematic properties, and jet characteristics.

[b]Spontaneous R-parity violating signals were also searched for in DELPHI [5], but are not described here.

If we assume that the dominant coupling is λ'', the final state contains 6 to 10 jets. These multijet events are present in the standard model via four-fermion processes with gluon radiations. A jet algorithm was used to characterize the signal and to provide powerful variables for the discrimination between signal and background events.

For both analyses, no excess was found in the data with respect to the standard model expectations. These results translate into limits on the parameters μ, M_2, $\tan\beta$ and m_0 (notice that the cross-section does not depend on the parameter A_0). Figure 2 shows examples of exclusion domains obtained from these analyses for $\tan\beta = 1.5$ and $m_0 = 90$ GeV$/c^2$. From a scanning on the parameters, we obtained limits on the superparticle masses. The chargino mass limit was 103 GeV$/c^2$ for dominant λ couplings and 102.5 GeV$/c^2$ for λ'' couplings. Concerning the neutralino mass, the limits were 40 GeV$/c^2$ for λ couplings and 38 GeV$/c^2$ for λ'' couplings.

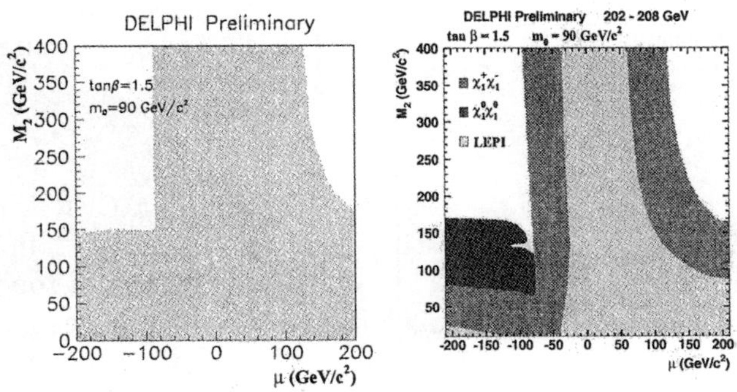

Figure 2: Exclusion plots in the (μ, M_2) plane for $\tan\beta = 1.5$ and $m_0 = 90$ GeV$/c^2$ (left-hand side: λ couplings; right-hand side: λ'' couplings).

3.2 Sleptons

In the case of a dominant λ coupling, the sneutrino pair production was carefully studied. The direct decay of the two sneutrino gives a 4 lepton final state, whereas the indirect decay gives 4 leptons and missing energy. Furthermore, the indirect decay of charged slepton pair was studied. The final state of this signal is composed of 6 leptons and missing energy. For all these topologies, different analyses were performed depending on the slepton flavor and the indices of coupling. No excess of candidate events was found in the 1999 and 2000 data.

In the hypothesis of a λ'' dominance, only the indirect decay is possible via this coupling. The selectron and the smuon were searched for in the data. These pair productions give 2 leptons (electrons or muons) and 6 jets. These analyses combined the use of lepton identification and jet algorithm, resulting in a good rejection power. No excess of data over standard model backgrounds was observed.

These results can be interpreted in the framework of the constrained MSSM. The analyses from 1999 and 2000 data were combined to exclude an area in the plane of the neutralino mass versus the slepton mass. Figure 3 presents examples of exclusion domains, from which limits on slepton masses can also be derived. For the λ couplings, a limit of 87 GeV/c^2 was obtained on the slepton masses (charged sleptons and sneutrinos). Concerning λ'' couplings, limits were 92 GeV/c^2 for the $\tilde{e}_R$ and 85 GeV/c^2 for the $\tilde{\mu}_R$.

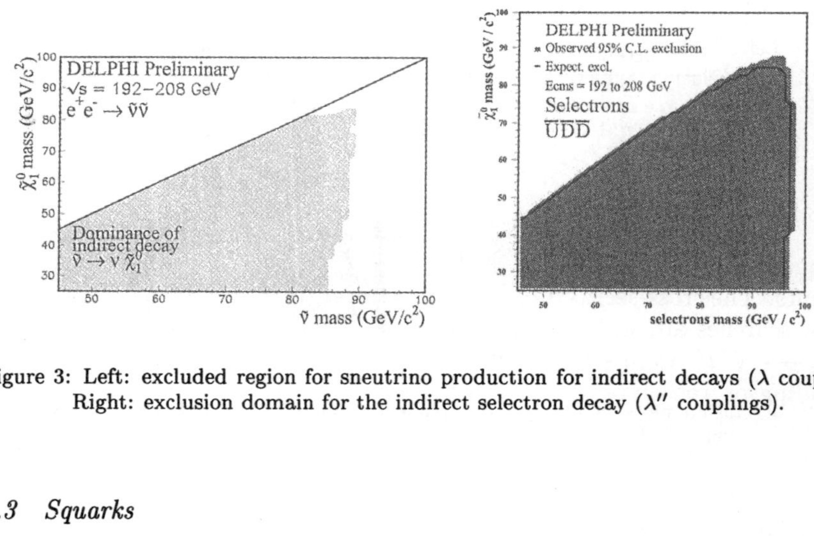

Figure 3: Left: excluded region for sneutrino production for indirect decays (λ couplings). Right: exclusion domain for the indirect selectron decay (λ'' couplings).

3.3 Squarks

The mixing between right- and left-handed stops can be important, so that the mixed state $\tilde{t}_1$ is expected to be the lightest squark. Each of the produced stop decays into a charm quark and a neutralino, giving a signature with jets, charged leptons and neutrinos in the case of λ dominance, and a signature with 8 jets in the case of dominant λ'' couplings. For both couplings, agreement between data and standard model expectations was satisfactory. Thus, exclusion regions were derived from these results, with the conservative hypothesis of maximal decoupling to the Z boson (figure 4). Limit on $\tilde{t}_1$ mass was found to be 87 GeV/c^2 for λ and 75 GeV/c^2 for λ'' couplings (in the case of maximal

Z-decoupling).

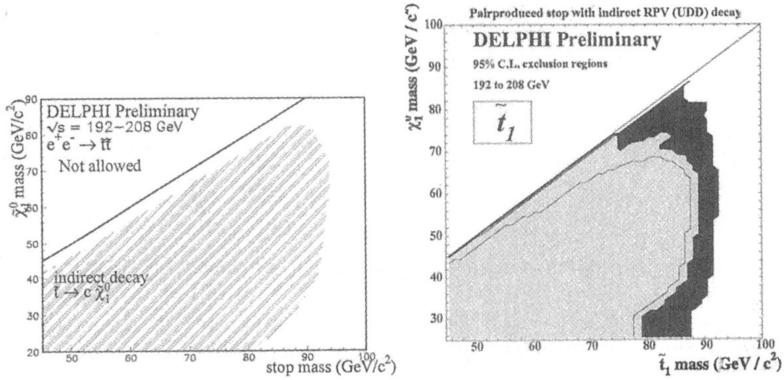

Figure 4: Left: exclusion domain for the $\widetilde{t}_1$ production for λ couplings (maximal decoupling). Right: exclusion domain for the $\widetilde{t}_1$ production for λ'' couplings (light grey: maximal decoupling; dark grey: no mixing).

3.4 Sneutrino single production

A resonant single production is allowed at LEP via the λ_{121} or λ_{131} coupling constants. The cross-section of the process depends directly on these couplings, so that searches are sensitive to their values. The decay channels $\widetilde{\nu} \longrightarrow \widetilde{\chi}^0\nu$ and $\widetilde{\nu} \longrightarrow \widetilde{\chi}^{\pm}l^{\mp}$ were searched for in the data (only for the λ_{121} coupling). Depending on the decay of the neutralino and chargino, the final state is either purely leptonic or semi-leptonic (with additional jets). After a complete analysis, the number of observed events was in agreement with the number of expected events. This result allows us to derive limits on the λ_{121} coupling: Figure 5 shows this exclusion for $\tan\beta = 1.5$ (with the hypothesis that $\Gamma_{\widetilde{\nu}} \geq 0.1$ GeV$/c^2$).

4 Conclusions

R-parity violating signatures have been studied in the DELPHI detector: a large number of topologies was covered. No excess was observed with respect to the standard model predictions, which was interpreted to constrain the MSSM parameter space. The mass limits derived from these searches are summarized in table 1 for different R-parity violating couplings.

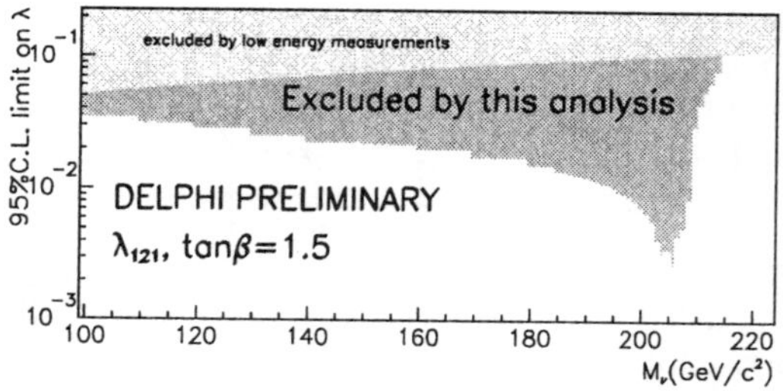

Figure 5: Upper limit on λ_{121} as a function of the sneutrino mass.

Table 1: Superparticle mass limits in GeV/c^2 from the DELPHI searches.

Superparticle	λ	λ''
$\widetilde{\chi}_1^0$	**40**	**38**
$\widetilde{\chi}_1^{\pm}$	**103**	**102.5**
$\widetilde{e}_R$	**87**[1,2]	**92**[1,3]
$\widetilde{\mu}_R$	**87**[1,2]	**85**[1,3]
$\widetilde{\tau}_R$	**87**[1,2]	-

Superparticle	λ	λ''
$\widetilde{\nu}_e$	**87**[1]	-
$\widetilde{\nu}_\mu$	**87**[1]	-
$\widetilde{\nu}_\tau$	**87**[1]	-
$\widetilde{t}_1$	**87**[1,3]	**75**[1,3]

[1] Obtained for $\mu = -200$ GeV/c^2 and tan$\beta = 1.5$.
[2] This result is valid for ΔM > 3 GeV/c^2.
[3] This result is valid for ΔM > 5 GeV/c^2.

References

1. M. Drees, *An introduction to supersymmetry*, hep-ph/9611409.
2. P. Fayet, *Nucl. Phys.* **90**, 104 (1975).
3. DELPHI collaboration, *Nucl. Instum. Methods* **A378**, 57 (1996).
4. R. Barbier, C. Bérat, P. Jonsson and V. Poireau, DELPHI 2001-083 CONF 511;
   F. Ledroit-Guillon, DELPHI 2001-019 CONF 460.
5. D. Moraes, L. de Paula and M. Gandelman, DELPHI 2001-084 CONF

DOUBLEY RESONANT Z PAIR PRODUCTION WITH DELPHI

J. REHN[a]
Institut fuer experimentelle Kernphysik, D-76131 Karlsruhe, Germany
E-mail: Jens.Rehn@cern.ch

Since 1997 until the end of its operation in 2000, the LEP accelerator was working at a center of mass energy above the kinematic threshold for doubly resonant Z pair production. This made it possible to measure this cross-section with high precision and statistical significance. The results obtained in different final states analyzed by DELPHI are briefly presented here. Combining all different final states and energies, an averaged cross-section with respect to the Standard Model (SM) expectation is derived and gives $1.04 \pm 0.12(stat.) * SM$ which agrees very well within statistical uncertainty.

1 Introduction

A measurement of doubly resonant Z pair production in e^+e^- annihilation by t–channel electron exchange firstly became possible in 1997, when LEP operated at a center of mass energy of 183 GeV. Main backgrounds arise from WW and QCD processes. Since the signature of a WW decay is very similar to that of a ZZ, it is a challenging task to separate this background. In addition, WW and QCD cross-sections are roughly 25 and 215 times larger, respectively.

The ZZ production channel enabled us to perform additional cross-checks on the Standard Model by comparing the observed production rate with the expectation. Since LEP has increased its operating energy during the runs in 1999 and 2000, the cross-section evolution at different center of mass energies can also be compared with the corresponding Standard Model predictions. Another very interesting reason to study this channel is its great importance for the Higgs searches. Since ZZ forms an irreducible background for the Higgsstrahlungs process HZ, especially if the Higgs boson mass is close to that of the Z boson. Hence powerful selection and discrimination techniques are required in order to clearly identify doubly resonant Z pair processes. Measuring the ZZ cross-sections also contributes to the searches for new physics beyond the standard model, such as anomalous neutral current triple gauge couplings (TGC). Here a Z pair couples directly to another Z or photon and hence forms a TGC which is forbidden in the Standard Model.

A doubly resonant ZZ can decay in several channels with typical signatures and branching fractions. Therefore a separate analysis for every decay channel is performed, trying to exploit the kinematical and topological properties of

[a]on behalf of the DELPHI collaboration

Table 1: Overview of the decay channels analyzed by DELPHI. Concerning the $qq\nu\nu$ channel two analyses are available using different discrimination techniques. The $\tau\tau$ channels are covered by the visible fermion analyses.

	channel	BR ratio	technique
vis. fermions	qqqq	48.8 %	probabilistic
	qqll	9.4 %	sequential cuts
	llll	0.5 %	sequential cuts
miss. energy	$qq\nu\nu$	28.0 %	discriminant / likelihood
	$\nu\nu\nu\nu$	4.0 %	—
	$ll\nu\nu$	2.7 %	sequential cuts
$\tau\tau$	$\tau\tau qq$	4.7 %	sequential cuts
	$\tau\tau\nu\nu$	1.3 %	—
	$\tau\tau ll$	0.5 %	sequential cuts
	$\tau\tau\tau\tau$	0.1 %	—
total		$\sim$ 95 %	

each channel. One can distinguish three categories of ZZ decays: events with four visible fermions, missing energy events or τ–events. By summing up the branching ratios of all channels investigated by DELPHI analyses, one finds that about 95% are covered (see Table 1).

2 Signal Definition

Because of the large background, a proper signal definition was necessary. Therefore the on–shell ZZ signal was defined by applying a mass–cut at generator level, asking all 4–fermion processes to fulfill the following condition[1]

$$M_Z - 10\,GeV/c^2 < M_{f\bar{f}} < M_Z + 10\,GeV/c^2 \tag{1}$$

for every fermion pairing. In ambiguous cases, when not all possible pairings fulfill Eq. 1, the event is treated as background. An exception to this rule are $f\bar{f}\nu\bar{\nu}$ processes, since here it is only useful to consider the massive fermion pairing.

Since the mass resolution depends on the channel analyzed, a correction factor R had to be applied for every channel in order to take into account the other contributing production diagrams[1].

$$R = \sigma_{NCO2} \,/\, \sigma_{4f}(window) \tag{2}$$

These factors translate the obtained cross–section values into the true Born cross–sections for ZZ–production, usually called NC02 for the two neutral current diagrams. R has been computed using EXCALIBUR[5], and WPHACT[10]. All results shown here already include the corresponding corrections.

For the year 2000 data (which was not fully analyzed when this paper was written) a more sophisticated method defining the NCO2 signal will be used, based on an event by event calculation of matrix elements for all important 4–fermion diagrams.

3 Data Samples

The results presented in this paper are based on the data taken by the DELPHI detector in the years 1997, 1998 and 1999. LEP provided a center of mass energy range of 183 GeV to 202 GeV during these two years.

Simulated events were generated using the DELPHI simulation program DELSIM[4] and were then passed through the same reconstruction and analysis chain as the data. The different 4–fermion final states were produced with EXCALIBUR relying on JETSET 7.4[6] for quark fragmentation. Two fermion background was generated using mainly PYTHIA[6], whereas two photon interactions were simulated by TWOGAM[7] and BDK[8].

4 Four jets

The $ZZ \to q\bar{q}q\bar{q}$ channel represents 49% of the ZZ final states and has a typical 4 jet topology. However WW and $q\bar{q}(\gamma)$ can lead to similar final states and since their production cross-sections are orders of magnitude larger then the doubly resonant Z pair production we are interested in, powerful discrimination techniques are required.

The method used for this channel is a probabilistic selection of hadronic ZZ events[1]. This was done by calculating for each Standard Model process leading to a four quark final state the probability of being compatible with the observed events. Three independent variables were used for this task:

- b–tagging
- reconstructed invariant mass for all possible parings
- $E_{min} * \alpha_{min}$ of all jet pairings

From the above variables one can derive a probability for every event of being compatible with the ZZ hypothesis. This leads to a distribution, where every bin directly represents the corresponding signal purity (see Figure 1).

The resulting cross-section was obtained by performing a binned maximum likelihood fit on the obtained probability distribution. Combining all energies of 1997 untill 1999 in this channel yields an averaged cross-section of $1.10 \pm 0.19 * \sigma_{SM}$ with respect to the Standard Model expectation.

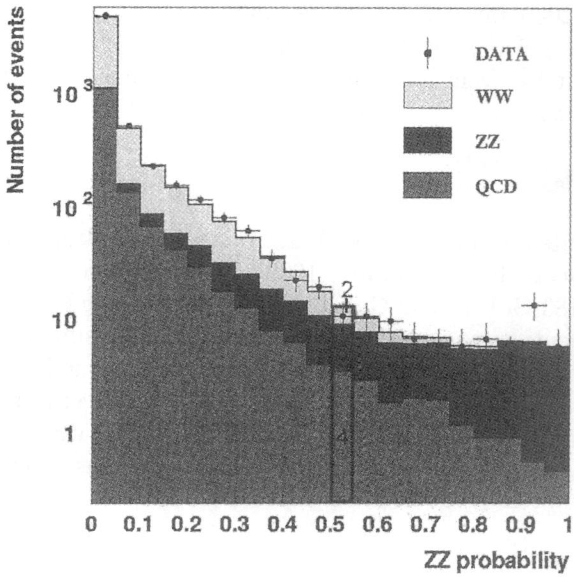

Figure 1: Distribution of ZZ probability. Every bin represents the corresponding signal purity. Detailed numbers are given as an example for 50% purity.

5 Jets and missing energy

With a branching ratio of 28% $ZZ \to q\bar{q}\nu\nu$ represents the second most important ZZ decay channel. It is characterized by a 2 jet structure with both missing and visible masses being compatible with the Z mass. Main background arises from $We\nu$ events, since the electron tends to be lost in the beam pipe. But WW and $qq(\gamma)$ events can also mimic the missing energy signal.

Therefore two different approaches are performed by two independent analyses. The first one uses a so called **I**terative **D**iscriminant **A**nalysis (IDA) technique[9]. Here a second order polynomial discrimination function is calculated from a set of variables. It maximizes the separation between signal and background by defining a multi–dimensional surface in the variable space. This function is then applied to all data and Monte Carlo events, resulting in a final discrimination variable (IDA) assigning high values for signal like events and low values for background compatible events (see Figure 2).

The second analysis uses a so called likelihood approach [3]. Here a com-

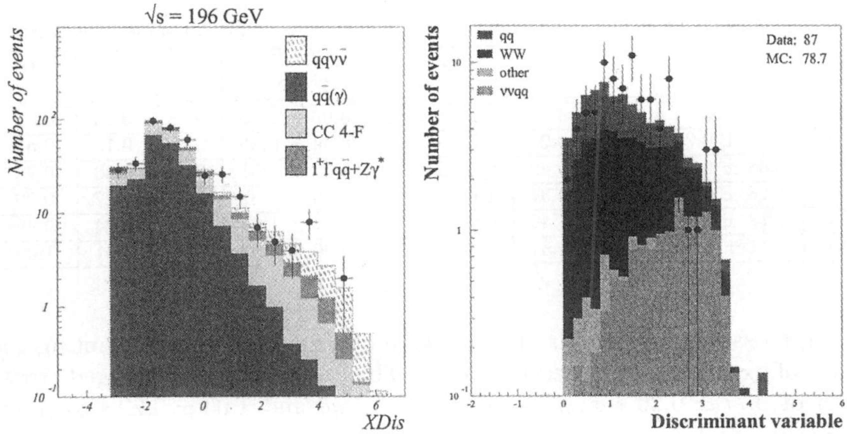

Figure 2: Both figures show the final discriminant variables at 196 GeV: the likelihood approach (left hand side) and the iterative discriminant analysis (right hand side).

bined discrimination variable (DIS) is constructed as a ratio of the product of the probability density functions defined by eight topological variables (see Figure 2). Its main purpose was to provide cross-checks for the IDA analysis.

In order to derive the NC02 cross-section, a maximum binned likelihood fit is performed on both distributions (DIS and IDA) separately. This leads to an averaged cross–section using the IDA analysis of $1.11 \pm 0.27 * \sigma_{SM}$ combining all center of mass energies since 1997 untill 1999. The corresponding results of the cross-check likelihood analysis provided $0.94 \pm 0.25 * \sigma_{SM}$. Both results are in good agreement with each other and the Standard Model expectation.

6 $l^+l^-q\bar{q}$ Channel

This channel is characterized by two hadronic jets with two isolated leptons (electrons or muons) and a branching ratio of 9.4%. Background can mainly arise from $q\bar{q}(\gamma)$ events with a converting photon or non–resonant Ze^+e^- events.

The selection procedure is based on sequential cuts on mainly two variables: the p_t of a leptonic candidate with respect to the nearest jet and the χ^2 per degree of freedom resulting from a kinematic fit constraining energy and momentum conservation. In addition all events had to have two identified leptons of opposite charges and same flavour.

Table 2: Summary of expected and obtained NC02 cross-sections. All values are in pb.

E_{cms} GeV	$q\bar{q}q\bar{q}$	$l^+l^-q\bar{q}$	$q\bar{q}\nu\nu$	$l^+l^-l^+l^-$ $l^+l^-\nu\nu$	ALL	Pred.
182.6	0.19 ± 0.24	0.70 ± 0.41	0.27 ± 0.46	0.00 ± 1.30	0.38 ± 0.18	0.25
188.6	0.59 ± 0.20	0.60 ± 0.23	0.58 ± 0.27	0.80 ± 0.66	0.60 ± 0.13	0.65
191.6	0.68 ± 0.52	0.40 ± 0.49	0.00 ± 0.77	4.67 ± 3.51	0.55 ± 0.33	0.78
195.5	1.28 ± 0.39	0.76 ± 0.37	1.66 ± 0.60	0.66 ± 1.82	1.17 ± 0.27	0.90
199.5	1.18 ± 0.38	1.02 ± 0.40	0.96 ± 0.50	1.17 ± 1.04	1.08 ± 0.24	0.99
201.6	1.11 ± 0.46	0.24 ± 0.31	1.49 ± 0.73	0.00 ± 1.92	0.87 ± 0.31	1.00

The cross-sections in this channel were obtained by simply counting the amount of expected and observed events. This resulted in an averaged cross–section of $0.77 \pm 0.26 * \sigma_{SM}$ for $ZZ \to e^+e^-q\bar{q}$ and $1.03 \pm 0.29 * \sigma_{SM}$ for $ZZ \to \mu^+\mu^-q\bar{q}$, taking into account the complete data set of all three years.

7 $l^+l^-l^+l^-$ and $l^+l^-\nu\nu$ Channels

Both channels have only few background sources but on the other hand also a tiny production rate.

In the $l^+l^-l^+l^-$ case with a branching ratio of 0.5% only roughly one event per year is expected. Since the signature is so clear, only simple cuts on variables like multiplicity, invariant mass and angle distributions were necessary to isolate signal like events. This analysis is sensitive to all combinations of leptonic flavours, except $\tau^+\tau^-\tau^+\tau^-$. The cross-section was derived by simply counting expected and observed events and one obtained an averaged cross-section of $0.79 \pm 0.61 * \sigma_{SM}$.

Events in the $l^+l^-\nu\nu$ channel typically show two charged and relatively acollinear leptons of the same flavour with large missing energy. Both invariant masses (visible and missing) have to be compatible with the Z mass. Therefore the two main background sources coming from WW and $We\nu$ were tackled by asking for events with two particles identified as electrons or muons, putting constraints on the polar angles between visible and missing momentum vectors and finally putting constraints on the accepted missing and visible masses. Determination of the cross–section was also done by event counting and yielded $1.54 \pm 0.95 * \sigma_{SM}$ averaging both years analyzed.

8 Results

Combining all different decay channels and center of mass energies one obtains an averaged NC02 cross-section of $1.04 \pm 0.12 * \sigma_{SM}$ with respect to the Standard

Model[2]. This result agrees very well within statistical uncertainty with the expectation. An overview of all results in the different channels analyzed is given in Table 2.

Acknowledgments

We are greatly indebted to our technical collaborators, to the members of the CERN–SL Division for the excellent performance of the LEP collider, and to the funding agencies for their support in building and operating DELPHI.

References

1. DELPHI Collaboration, *Measurement of the ZZ cross–section in* e^+e^- *interactions at 183-189 GeV*, Contribution to this conference, DELPHI EP-2000-089, submitted to *Physics Letters* **B**.
2. DELPHI Collaboration, *Update of ZZ production measurement in* e^+e^- *interactions using data at 192-202 GeV*, Contribution to ICHEP2000, DELPHI 2000-145.
3. DELPHI Collaboration, P. Abreu *et al.*, *Nucl. Instr. and Meth.* **A378**, 57 (1996).
4. DELSIM *Reference Manual*, DELPHI 87-97 PROG-100.
5. F.A. Berends, R. Pittau, R. Kleiss, *Comp. Phys. Comm.* **85**, 437 (1995).
6. T. Sjöstrand, *Comp. Phys. Comm.* **39**, 347 (1986); T. Sjöstrand, PYTHIA 5.6 and JETSET 7.3, CERN-TH/6488-92.
7. S. Nova, A. Olshevski, and T. Todorov, *A Monte Carlo event generator for two photon physics*, DELPHI note 90-35 PROG 152.
8. F.A. Berends, P.H. Daverveldt, R. Kleiss, *Comp. Phys. Comm.* **40**, 271-284; 285-307, 309 (1986).
9. T.G.M. Malmgren, *Comp. Phys. Comm.* **106**, 230 (1997);
   T.G.M. Malmgren and K.E. Johansson, *Nucl. Inst. Meth.* **403**, 481 (1998).
10. E. Accomando, A. Ballestrero, *Comp. Phys. Comm.* **99**, 270 (1997).

TESTS OF QED WITH MULTI-PHOTONIC FINAL STATES

KIRSTEN SACHS

Carleton University, 1125 Colonel By Drive, Ottawa, ON K15 5B6, Canada
E-mail: Kirsten.Sachs@cern.ch

In the Standard Model the process $e^+e^- \to \gamma\gamma(\gamma)$ is fully described by QED. Measurements of the differential cross-sections from the four LEP experiments are compared to the QED expectation and limits are set on parameters describing physics beyond the Standard Model. Three-photon events are used for a direct search for a photonically decaying resonance produced together with a photon.

1 Introduction

The process $e^+e^- \to \gamma\gamma(\gamma)$, called multi-photon production, is one of the few processes in high energy e^+e^- scattering which can be described by QED only. Since the only free parameter $\alpha(0)$ is precisely measured[a] the Standard Model expectation is well known. Any deviation would hint at some new physics. In general such effects can be described by cut-off parameters or in the framework of effective Lagrangian theory. Effects can for example be caused by the t-channel exchange of excited electrons or the s-channel exchange of gravitons in models with extra dimensions. Events with three photons in the final state are used to search directly for a photonically decaying resonance which is produced together with a photon. Results will be presented from the four LEP experiments based on the full LEP2 statistics, including data taken in 2000.[b]

2 Theory

The Born-level differential cross-section for the process $e^+e^- \to \gamma\gamma$ in the relativistic limit of lowest order QED is given by [1]

$$\left(\frac{d\sigma}{d\Omega}\right)_{\text{Born}} = \frac{\alpha^2}{s}\,\frac{1+\cos^2\theta}{1-\cos^2\theta}\,, \tag{1}$$

where s denotes the square of the centre-of-mass energy and θ is the scattering angle. Since the two photons are identical particles, the event angle is defined by convention such that $\cos\theta$ is positive.

[a] Since the photons are in the final state the relevant momentum transfer for the fine-structure constant is the mass of the photon which is zero.

[b] ALEPH results from data taken in 2000 are not yet available.

Possible deviations from the QED cross-section can be parametrised in terms of cut-off parameters $\Lambda_\pm$ which correspond to an additional exponential term to the Coulomb field [2] as given in Eq. 2. Alternatively, in terms of effective Lagrangian theory, [3] the cross-section depends on the mass scales (e.g. Λ') for ee$\gamma\gamma$ contact interactions or non-standard $e^+e^-\gamma$ couplings. The resulting cross-sections are of two general types, either an angular independent offset to the cross-section or similar to the form given by Eq. 2.

$$\left(\frac{d\sigma}{d\Omega}\right)_{\Lambda_\pm} = \left(\frac{d\sigma}{d\Omega}\right)_{\text{Born}} \pm \frac{\alpha^2 s}{2\Lambda_\pm^4}(1+\cos^2\theta) \tag{2}$$

Recent theories have pointed out that the graviton might propagate in a higher-dimensional space where additional dimensions are compactified while other Standard Model particles are confined to the usual 3+1 space-time dimensions. The resulting large number of Kaluza-Klein excitations could be exchanged in the s-channel of $e^+e^- \to \gamma\gamma$ scattering. This leads to a differential cross-section [4] depending on the mass scale M_S which should be of order of the electroweak scale ($\mathcal{O}(10^{2-3}\text{GeV})$) and a parameter λ which is of $\mathcal{O}(1)$. Ignoring $\mathcal{O}(M_S^{-8})$ terms $M_S^{-4} = \alpha\pi\Lambda_\pm^{-4}$ for $\lambda = \mp 1$.

The existence of an excited electron e^* with an $e^*e\gamma$ coupling would contribute to the photon production process via t-channel exchange. The resulting cross-section depends on the e^* mass M_{e^*} and the coupling constant κ of the $e^*e\gamma$ vertex relative to the eeγ vertex. [5] For large masses $M_{e^*} \approx \sqrt{\kappa}\Lambda_+$.

3 Radiative corrections

All cross-sections discussed above are calculated to $\mathcal{O}(\alpha^2)$. For higher order QED predictions an exact $\mathcal{O}(\alpha^3)$ Monte Carlo [6] and a Monte Carlo [7] for $e^+e^- \to \gamma\gamma\gamma\gamma$ in the relativistic limit are available, but no full $\mathcal{O}(\alpha^4)$ calculation. To keep theoretical uncertainties from higher orders below 1% it is essential to minimise third order corrections. This can be achieved via a proper choice of the scattering angle. Whereas in lowest order there are exactly two photons with the same scattering angle ($|\cos\theta_1| = |\cos\theta_2|$) for a measured event the two highest energy photons are in general not back-to-back ($|\cos\theta_1| \neq |\cos\theta_2|$). This leads to various possibilities for the definition of the scattering angle of the event.

The simplest quantity is the average $\cos\theta_{\text{av}} = (|\cos\theta_1| + |\cos\theta_2|)/2$ which leads to corrections of up to 30% at angles of $\cos\theta_{\text{av}} \approx 0$. These large corrections arise since the average does not change if one photon flips from one hemisphere to the other. This problem can be avoided using the difference $\cos\theta_{\text{dif}} = (|\cos\theta_1 - \cos\theta_2|)/2$, which shows otherwise the same behaviour with corrections up to 10 (15)% for large values of $\cos\theta_{\rm dif} > 0.88$ (0.95). A physics motivated definition is the angle in the centre-of-mass system of the two highest energy photons $\cos\theta^* = |\sin\frac{\theta_1-\theta_2}{2}|\,/\,(\sin\frac{\theta_1+\theta_2}{2})$. This definition leads to the smallest corrections of 3-7% within the studied angular range of $\cos\theta^* < 0.97$ and is therefore chosen for the analyses.

4 Selection

The selection of multi-photonic events relies on the photon detection in the electromagnetic calorimeter ECAL. The ECAL signature however is the same for $\gamma\gamma(\gamma)$ and $e^+e^-(\gamma)$ events. Since the Bhabha cross-section is huge this background must be suppressed by 4-5 orders of magnitude. To reject Bhabhas the tracking detectors are used to distinguish electrons from photons. This can be difficult at small scattering angles where the electrons do not travel the full extent of the tracking chamber and two particles can easily be reconstructed as one track. Also photon conversions at a small distance to the interaction point, i.e. before the first active detector layer are hard to separate. Since the conversion rate depends on the material in the detector the optimal angular range for the selection strongly depends on the experiment. The acceptance ranges given in Table 1 reflect also the different angular coverage of the ECAL. A very dangerous background is caused by low s' Bhabhas, with an invariant photon mass just above the threshold of 1 MeV. They have the same signature as a photonic event with early conversion and are badly simulated, since most Bhabha Monte Carlos impose much higher cuts on s'.

Table 1: Acceptance range and efficiency ϵ within this acceptance range of the four experiments. The efficiency might depend slightly on $\sqrt{s}$. The assumed systematic error on the efficiency $\delta\epsilon$ and the radiative corrections $\delta\rho$ are also given. Preliminary L3 results do not include systematic errors. Other systematic errors are small.

	$\cos\theta$ range	ϵ	$\delta\epsilon$	$\delta\rho$
ALEPH	$[0, 0.95]$	83%	1.3%	1.0%
DELPHI	$[0.035, 0.731]\cup$ $[0.819, 0.906]$	76%	2.5%	0.5%
L3	$[0, 0.961]$	64%	1.2%/-	-
OPAL	$[0, 0.90]$	92%	1.0%	1.0%

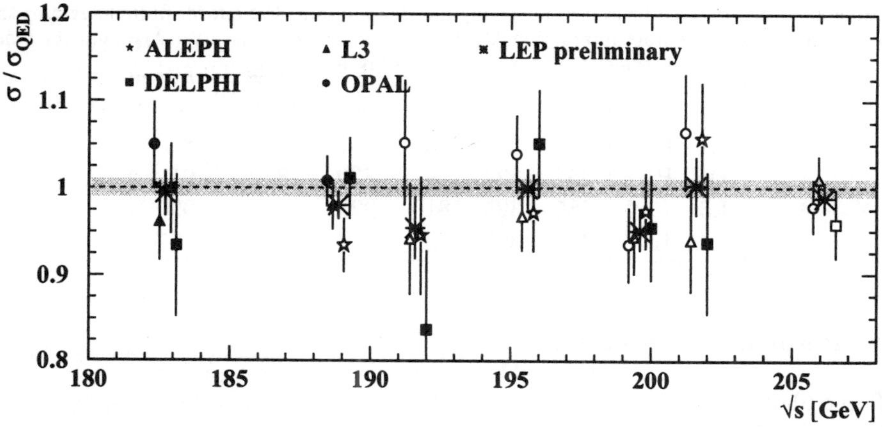

Figure 1: Measured total cross-section relative to the QED expectation. The values of single experiments are shown displaced in $\sqrt{s}$ for clarity. Filled symbols represent final results, open symbols are preliminary. The error on preliminary L3 cross-sections is statistical only. The theoretical error of 1% is shown as a shaded band and is not included in the experimental errors.

5 Cross-section results

The total and differential cross-sections are measured within the angular ranges given in Table 1. Figure 1 shows the total cross-sections normalised to the QED expectation for all four LEP experiments [8,9,10,11] and their combination. Apart from the common theoretical uncertainty the correlated systematic error between experiments is negligible. In general there is a very good agreement. The average over all energies and experiments yields $0.980 \pm 0.008 \pm 0.007$ where the first error is statistical and the second systematic. This is two standard deviations low, not accounting for the assumed theoretical error of 1% which is of the same size as the experimental error.

The angular distributions are compared to the differential cross-sections predicted by various models. No significant deviation from QED was found and limits given in Table 2 are derived. For all experiments the limit on Λ_+ is larger than the limit on Λ_-. This effect is not significant yet it implies that the observed cross-section in the central region of the detectors is smaller than expected. Small scattering angles are less sensitive to these limits though their number of events is largest.

Table 2: Limits derived from fits to the angular distribution: the cut-off parameter $\Lambda_\pm$, the mass scale for ee$\gamma\gamma$ contact interaction Λ', the mass of an excited electron M_{e^*} and the mass scale for extra dimensions M_S for $\lambda = \pm 1$.

[GeV]	Λ_+	Λ_-	Λ'	M_{e^*}	M_S λ=+1	M_S λ=-1
ALEPH	319	317	705	337	810	820
DELPHI	354	324	–	339	832	911
L3	385	325	810	325	835	990
OPAL	344	325	763	354	833	887

6 Resonance production

Three photon final states can originate from a photonically decaying resonance $X \to \gamma\gamma$, which is produced together with a photon via $e^+e^- \to X\gamma$. Figure 2 shows the invariant mass of photon pairs in events with exactly three photons. Within the small statistics the mass distribution is in good agreement with the expectation from the QED process.

If the Higgs is assumed to be this resonance X the Standard Model coupling of $H \to \gamma\gamma$ via loops of charged, massive particles is too small to lead to an observable effect. For the Standard Model Higgs the maximum of the branching ratio for $H \to \gamma\gamma$ is $2.6 \cdot 10^{-3}$ at a Higgs mass of about 125 GeV and a total Higgs width of ~ 4 MeV. For larger Higgs masses the branching ratio decreases due to the increasing $H \to W^+W^-$ contribution. However, for limits on anomalous couplings in the case of fermiophobic Higgs models the three photon final state gives information which is complementary to $e^+e^- \to HZ$ with $H \to \gamma\gamma$ or $e^+e^- \to H\gamma$ with $H \to \bar{b}b$.

Anomalous $H \to \gamma\gamma$ couplings can be described by [12]

$$\mathcal{L}_{\text{eff}}^{H\gamma\gamma} = -\frac{gM_W}{\Lambda^2}\frac{\sin^2\theta_W}{2}(f_{BB} + f_{WW} - f_{BW})HA_{\mu\nu}A^{\mu\nu}\,, \tag{3}$$

where Λ is the energy scale and the three possible couplings are f_{BB}, f_{WW} and f_{BW}. Limits on γZ interaction set strong constraints on f_{BW} [13] which is therefore set to zero. The two other parameters are in general assumed to be identical $f_{BB} = f_{WW} \equiv F$.

In Figure 3 the limit [14] on F/Λ^2 is shown from $H \to \gamma\gamma$ decay for the two processes $e^+e^- \to H\gamma$ and $e^+e^- \to HZ$, with $Z \to q\bar{q}$ and $\nu\bar{\nu}$. Although HZ production has the higher sensitivity for F/Λ^2 at low Higgs masses, $H\gamma$ production provides strong limits up to $M_H \sim 170$ GeV.

The partial width of $H \to \gamma\gamma$ is studied with the process $e^+e^- \to H\gamma$.

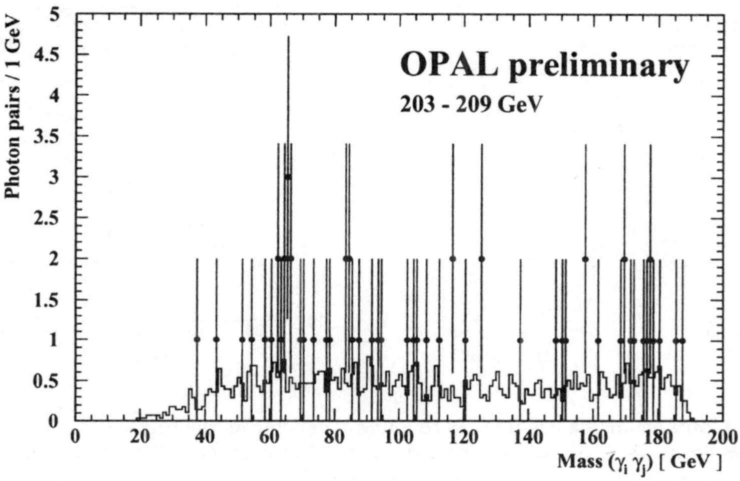

Figure 2: Invariant mass of photon pairs from events with three photons in the final state. There are three combinations per event. The points represent the data taken in 2000 by OPAL and the histogram the corresponding QED expectation. The mass resolution is about 0.5 GeV. The mass range is limited not only by the centre-of-mass energy but also by the imposed cut on the opening angle between photons.

Figure 3 shows limits [15] from three photon final states ($H \to \gamma\gamma$) which are stronger than those obtained from $b\bar{b}\gamma$ events ($H \to b\bar{b}$).

7 Conclusion

The process $e^+e^- \to \gamma\gamma(\gamma)$ provides high statistics data for the test of the Standard Model. Combining all four LEP experiments a precision of 1% for the total cross-section is reached. Since this process is dominated by QED a precise prediction is in principle possible. However, since calculations are available only up to next-to-leading order a theoretical error of about 1% has to be taken into account while searching for deviations from the Standard Model. The observed total cross-section is two standard deviations below the expectation not accounting for this theoretical error. Three photon final states are interesting for the search for photonically decaying resonances, which are produced along with a photon. The Standard Model prediction for the $H \to \gamma\gamma$ branching ratio is too small to lead to an observable cross-section at LEP hence limits on anomalous couplings are placed.

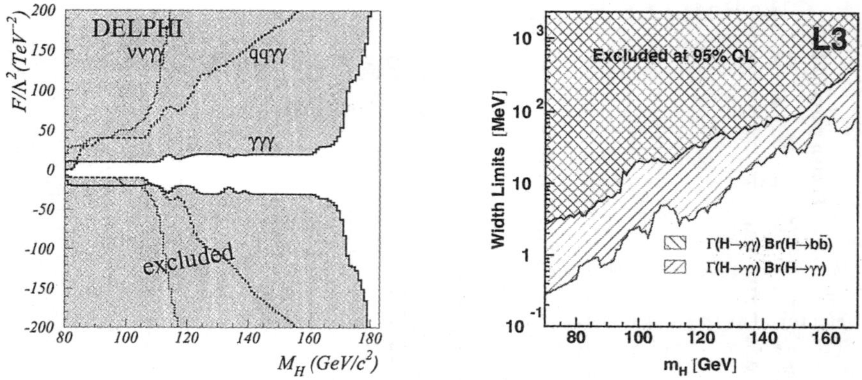

Figure 3: Limits on anomalous H $\to \gamma\gamma$ coupling depending on the Higgs mass. The DELPHI plot shows limits on F/Λ^2 derived from Hγ and HZ production. The L3 plot shows limits on the partial width H $\to \gamma\gamma$ from $e^+e^- \to$ Hγ with H $\to b\bar{b}$ and $\gamma\gamma$. Data taken until 1999 were used.

References

1. I. Harris, L.M. Brown, *Phys. Rev.* **105**, 1656 (1957);
   F.A. Berends, R. Gastmans, *Nucl. Phys.* **B61**, 414 (1973).
2. S.D. Drell, *Ann. Phys.* **4**, 75 (1958).
3. O.J.P. Éboli, A.A. Natale, S.F. Novaes, *Phys. Lett.* **B271**, 274 (1991).
4. K. Agashe, N.G. Deshpande, *Phys. Lett.* **B456**, 60 (1999).
5. A. Litke, Ph.D. Thesis, Harvard University, unpublished (1970).
6. F.A. Berends, R. Kleiss, *Nucl. Phys.* **B186**, 22 (1981).
7. F.A. Berends et al., *Nucl. Phys.* **B239**, 395 (1984).
8. ALEPH Coll., ALEPH note 2000-008 (2000).
9. DELPHI Coll., DELPHI note 2001-023 (2001).
10. L3 Coll., L3 note 2650 (2001).
11. OPAL Coll., OPAL note PN469.
12. K. Hagiwara, R. Szalapski, D. Zeppenfeld, *Phys. Lett.* **B318**, 155 (1993).
13. K. Hagiwara, S. Ishihara, R. Szalapski, D. Zeppenfeld, *Phys. Rev.* **D48**, 2182 (1993).
14. DELPHI Coll., DELPHI note 2000-082 (2000).
15. L3 Coll., L3 note 2558 (2000).

CHARMLESS HADRONIC B DECAYS AT BABAR

THOMAS SCHIETINGER
Stanford Linear Accelerator Center, Stanford, CA 94309, USA
E-mail: Thomas.Schietinger@SLAC.Stanford.edu

(for the BABAR Collaboration)

We present several searches for charmless hadronic two-body and three-body decays of B mesons from electron-positron annihilation data collected by the BABAR detector near the $\Upsilon(4S)$ resonance. We report the preliminary branching fractions $\mathcal{B}(B^0 \to \pi^+\pi^-) = (4.1 \pm 1.0 \pm 0.7) \times 10^{-6}$, $\mathcal{B}(B^0 \to K^+\pi^-) = (16.7 \pm 1.6 \pm 1.3) \times 10^{-6}$, $\mathcal{B}(B^0 \to \rho^\mp\pi^\pm) = (49 \pm 13^{+6}_{-5}) \times 10^{-6}$, $\mathcal{B}(B^+ \to \eta' K^+) = (62 \pm 18 \pm 8) \times 10^{-6}$, and present upper limits for nine other decays.

1 Introduction

The study of B meson decays into charmless hadronic final states plays an important role in the understanding of CP violation. In the Standard Model, all CP-violating phenomena are a consequence of a single complex phase in the Cabibbo-Kobayashi-Maskawa (CKM) quark-mixing matrix.[1] The Belle and BABAR collaborations have presented results[2,3] on measurements of CP-violating asymmetries in B decays into final states containing charmonium, leading to constraints on the angle β of the CKM unitarity triangle. Measurements of the rates and CP asymmetries for B decays into the charmless final states $\pi\pi$ and $K\pi$ can be used to constrain the angles α and γ of the unitarity triangle.[4]

2 Detector and data

Here we present new measurements of the branching fractions for charmless hadronic decays of B mesons in the final states[a] $\pi^+\pi^-$ and $K^+\pi^-$, and an upper limit for $B \to K^+K^-$, which are based on a data sample consisting of an integrated luminosity of 20.6 fb^{-1} taken near the $\Upsilon(4S)$ resonance ("on-resonance"), corresponding to $(22.57 \pm 0.36) \times 10^6$ $B\overline{B}$ pairs. A data sample of 2.61 fb^{-1} taken at a center-of-mass (CM) energy 40 MeV below the $\Upsilon(4S)$ resonance ("off-resonance") is used for continuum background studies.

We also report measurements of branching fractions and upper limits for B decays into non-resonant three-body modes and modes containing K^*, ρ, ω, and η' resonances, which are based on a smaller data sample (see Sec. 4).

[a] Charge conjugate states are assumed throughout, except where explicitly noted.

The data were collected with the BABAR detector at the PEP-II e^+e^- collider at the Stanford Linear Accelerator Center. The collider is operated with asymmetric beam energies, producing a boost ($\beta\gamma = 0.56$) of the $\Upsilon(4S)$ along the collision axis (z). For the analyses described in this Letter, the most significant effect of the boost relative to symmetric collider experiments is to increase the momentum range of the two-body B decay products from a narrow distribution centered at approximately 2.6 GeV/c to a broad, approximately flat distribution extending from 1.7 to 4.2 GeV/c.

At this conference, the BABAR detector has been described in detail by Jim Panetta.[5] Here we only re-emphasize that the identification of tracks as pions or kaons is based on the Cherenkov angle θ_c measured by a unique, internally reflecting Cherenkov ring imaging detector (DIRC). The K–π separation varies as a function of momentum and is better than 8 standard deviations (σ) at 1.7 GeV/c and decreases to 2.5σ at 4 GeV/c.

3 Analysis of $B^0 \to \pi^+\pi^-, K^+\pi^-, K^+K^-$

The selection of hadronic events for this analysis is based on track multiplicity and event topology. To reduce background from non-hadronic events, the ratio of Fox-Wolfram moments[6] H_2/H_0 is required to be less than 0.95 and the sphericity[7] of the event is required to be greater than 0.01.

All tracks are required to have a polar angle within the tracking fiducial region $0.41 < \theta < 2.54$ rad and a θ_c measurement from the DIRC. The latter requirement is satisfied by 91% of the tracks in the described fiducial region. We require a minimum number of Cherenkov photons associated with each θ_c measurement in order to improve the resolution. The efficiency of this requirement is 97% per track. Tracks with a θ_c within 3σ of the expected value for a proton are rejected. Electrons are rejected based on specific ionization (dE/dx) in the DCH system, shower shape in the EMC, and the ratio of shower energy to track momentum.

The kinematic constraints provided by the $\Upsilon(4S)$ initial state and relatively precise knowledge of the beam energies are exploited to efficiently identify B candidates. We define a beam-energy substituted mass $m_{\rm ES} = \sqrt{E_{\rm b}^2 - \mathbf{p}_B^2}$, where $E_{\rm b} = (s/2 + \mathbf{p}_i \cdot \mathbf{p}_B)/E_i$, and $\sqrt{s}$ and E_i are the total energies of the e^+e^- system in the CM and lab frames, respectively, and $\mathbf{p}_i$ and $\mathbf{p}_B$ are the momentum vectors in the lab frame of the e^+e^- system and the B candidate, respectively. The $m_{\rm ES}$ resolution is dominated by the beam energy spread and is approximately 2.5 MeV/c^2. Candidates are selected in the range $5.2 < m_{\rm ES} < 5.3$ GeV/c^2.

We define an additional kinematic parameter ΔE as the difference between the energy of the B candidate and half the energy of the e^+e^- system, computed in the CM system, where the pion mass is assumed for all charged B decay products. The ΔE distribution is peaked near zero for modes with no charged kaons and shifted on average -45 MeV (-91 MeV) for modes with one (two) kaons, where the exact separation depends on the laboratory kaon momentum. The resolution on ΔE is about 26 MeV. Candidates with $|\Delta E| < 0.15$ GeV are accepted.

Detailed Monte Carlo simulation, off-resonance data, and events in on-resonance $m_{\rm ES}$ and ΔE sideband regions are used to study backgrounds. The contribution due to other B-meson decays, both from $b \to c$ and charmless decays, is found to be negligible. The largest source of background is from random combinations of tracks and neutrals produced in the $e^+e^- \to q\bar{q}$ continuum (where $q = u$, d, s or c). In the CM frame this background typically exhibits a two-jet structure that can produce two high momentum, nearly back-to-back particles. In contrast, the low momentum and pseudoscalar nature of B mesons in the decay $\Upsilon(4S) \to B\overline{B}$ leads to a more spherically symmetric event. We exploit this topology difference by making use of two event-shape quantities. The variable we considered that has the greatest discriminating power is the angle θ_S between the sphericity axes evaluated in the CM frame, of the B candidate and the remaining tracks and photons in the event. The distribution of the absolute value of $\cos\theta_S$ is strongly peaked near 1 for continuum events and is approximately uniform for $B\overline{B}$ events. We require $|\cos\theta_S| < 0.9$, which rejects 66% of the background that remains at this stage of the analysis.

The second quantity used in the analysis is a linear combination of the nine scalar sums of the momenta of all tracks and photons (excluding the B candidate decay products) flowing into 10° polar angle intervals coaxial around the thrust axis of the B candidate, in the CM frame (Fisher discriminant[8] $\mathcal{F}$). Monte Carlo samples are used to obtain the values of the coefficients, which are chosen to maximize the statistical separation between signal and background events. No restrictions are placed on $\mathcal{F}$. Instead, it is used as an input variable in a maximum likelihood fit, described below.

Signal yields are determined from an unbinned maximum likelihood fit that uses the following quantities: $m_{\rm ES}$, ΔE, $\mathcal{F}$, and θ_c. The likelihood for a given candidate j is obtained by summing the product of event yield n_k and probability $\mathcal{P}_k$ over all possible signal and background hypotheses k. The n_k are determined by maximizing the extended likelihood function $\mathcal{L}$:

$$\mathcal{L} = \exp\left(-\sum_{k=1}^{M} n_k\right) \prod_{j=1}^{N} \left[\sum_{k=1}^{M} n_k \mathcal{P}_k\left(\vec{x}_j; \vec{\alpha}_k\right)\right] . \qquad (1)$$

The probabilities $\mathcal{P}_k(\vec{x}_j; \vec{\alpha}_k)$ are evaluated as the product of probability density functions (PDFs) for each of the independent variables $\vec{x}_j$, given the set of parameters $\vec{\alpha}_k$. Monte Carlo simulated data is used to validate the assumption that the fit variables are uncorrelated. The exponential factor in $\mathcal{L}$ accounts for Poisson fluctuations in the total number of observed events N. For the $K^\pm\pi^\mp$ terms the yields are rewritten in terms of the sum $n_f + n_{\bar{f}}$ and the asymmetry $\mathcal{A} = (n_{\bar{f}} - n_f)/(n_{\bar{f}} + n_f)$, where $n_f\,(n_{\bar{f}})$ is the fitted number of events in the mode $B \to f\,(\overline{B} \to \bar{f})$.

The parameters for both signal and background $m_{\rm ES}$, ΔE, and $\mathcal{F}$ PDFs are determined from data, and are cross-checked with the parameters derived from Monte Carlo simulation. The θ_c PDFs are derived from kaon and pion tracks in the momentum range of interest from approximately 42 000 $D^{*+} \to D^0\pi^+$ ($D^0 \to K^-\pi^+$) decays. This control sample is used to parameterize the θ_c resolution as a function of track polar angle.

The results of the fit are summarized in Table 1. For the decay $B^0 \to K^+K^-$ we measure the branching fraction $\mathcal{B} = (0.85^{+0.81}_{-0.66} \pm 0.37) \times 10^{-6}$. The 90% confidence level upper limit for this mode is computed as the value n_k^0 for which $\int_0^{n_k^0} \mathcal{L}_{\rm max}\, dn_k / \int_0^\infty \mathcal{L}_{\rm max}\, dn_k = 0.90$, where $\mathcal{L}_{\rm max}$ is the likelihood as a function of n_k, maximized with respect to the remaining fit parameters. The result is then increased by the total systematic error. The reconstruction efficiency is reduced by its systematic uncertainty in calculating the branching fraction upper limit. The statistical significance of a given channel is determined by fixing the yield to zero, repeating the fit, and recording the square root of the change in $-2\ln\mathcal{L}$.

Table 1: Summary of results for reconstruction efficiencies (ε), fitted signal yields (N_S), statistical significances, and measured branching fractions ($\mathcal{B}$). The total number of events entering the ML fit is 16032. For the K^+K^- mode the 90% confidence level (CL) upper limit for the branching fraction is quoted. Equal branching fractions for $\Upsilon(4S) \to B^0\overline{B}^0$ and B^+B^- are assumed.

Decay mode	ε (%)	N_S	Stat. Sig. (σ)	$\mathcal{B}(10^{-6})$
$B^0 \to \pi^+\pi^-$	45	$41 \pm 10 \pm 7$	4.7	$4.1 \pm 1.0 \pm 0.7$
$B^0 \to K^+\pi^-$	45	$169 \pm 17 \pm 13$	15.8	$16.7 \pm 1.6 \pm 1.3$
$B^0 \to K^+K^-$	43	$8.2^{+7.8}_{-6.4} \pm 3.5$	1.3	< 2.5

Event-counting analyses, based on the same variable set x_j as used in the fits, serve as cross-checks for the ML fit results. The variable ranges are generally chosen to be tighter in order to optimize the signal-to-background ratio, or upper limit, for the expected branching fractions. We count events in a rectangular signal region in the ΔE–$m_{\rm ES}$ plane, and estimate the background from a sideband area. The branching fractions measured using this technique

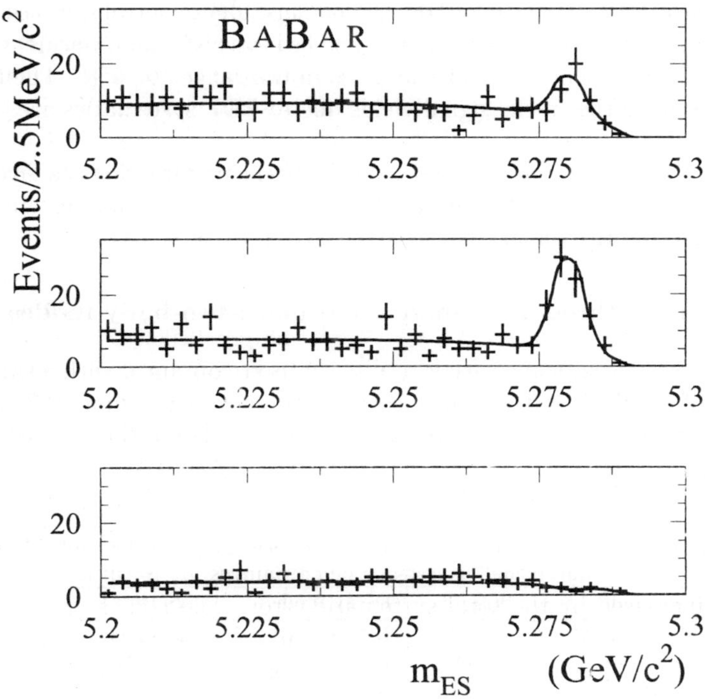

Figure 1: The $m_{\rm ES}$ distributions for candidates passing the selection criteria of the event-counting analyses for the $\pi^+\pi^-$ (top), $K^+\pi^-$ (center), and K^+K^- (bottom) mode.

are in good agreement with those arising from the ML fit analysis. Figure 1 shows the distributions in $m_{\rm ES}$ for events passing the tighter selection criteria of the event-counting analyses.

The following sources of systematic uncertainty have been considered: imperfect knowledge of the PDF shapes, which translates into systematic uncertainties in both the fit yields and asymmetries; systematic uncertainties in the detection efficiencies, which affect only the branching ratio measurements; and possible charge biases in either track reconstruction or particle identification, which affect only the asymmetries.

The PDF shapes contribute the largest source of systematic uncertainty. Systematics due to uncertainties in the PDF parameterizations are estimated either by varying the PDF parameters within 1σ of their measured uncertainties or by substituting alternative PDFs from independent control samples, and recording the variations in the fit results. The largest systematic uncertainties of this type vary across decay modes and are in the range 1%–7% . The total systematic uncertainties in the signal yields due to PDF systematics are given in Table 1.

The overall systematic uncertainties on the branching fractions as given in Table 1 are computed by adding in quadrature the PDF systematics and the systematic uncertainties on the efficiencies.

4 Non-resonant three-body modes and quasi-two-body modes

In Table 2 we summarize some earlier results that were obtained with a smaller data sample of 7.7 fb^{-1} on-resonance, corresponding to 8.8 million $B\overline{B}$ pairs, and 1.2 fb^{-1} off-resonance. For the details of these event-counting analyses we refer to Ref. 9.

Table 2: Summary of branching fraction measurements. Inequality denotes 90% CL upper limit, including systematic uncertainties.

Decay mode	$\mathcal{B}(10^{-6})$	Decay mode	$\mathcal{B}(10^{-6})$
$B^+ \to \pi^+\pi^-\pi^+$	< 22	$B^+ \to \eta' K^+$	$62 \pm 18 \pm 8$
$B^+ \to \rho^0\pi^+$	< 39	$B^0 \to \eta' K^0$	< 112
$B^0 \to \rho^{\mp}\pi^{\pm}$	$49 \pm 13^{+6}_{-5}$	$B^+ \to \omega K^+/\pi^+$	< 24
$B^+ \to K^+\pi^-\pi^+$	< 54	$B^+ \to \omega K^0$	< 14
$B^+ \to \rho^0 K^+$	< 29	$B^+ \to K^{*0}\pi^+$	< 28

5 Conclusion

In summary, we have measured branching fractions for the rare charmless decays $B^0 \to \pi^+\pi^-$, $B^0 \to K^+\pi^-$, $B^0 \to \rho^{\mp}\pi^{\pm}$, and $B^+ \to \eta' K^+$, and set upper limits on $B^+ \to \pi^+\pi^-\pi^+$, $B^+ \to \rho^0\pi^+$, $B^+ \to K^+\pi^-\pi^+$, $B^+ \to \rho^0 K^+$, $B^0 \to \eta' K^0$, $B^+ \to \omega K^+/\pi^+$, $B^+ \to \omega K^0$, and $B^+ \to K^{*0}\pi^+$. Our results are in good agreement with earlier measurements.[10]

Acknowledgments

We wish to thank our PEP-II colleagues for their outstanding efforts in providing us with excellent luminosity and machine conditions. This work is supported by the US Department of Energy and National Science Foundation, the Natural Sciences and Engineering Research Council (Canada), Institute of

High Energy Physics (China), the Commissariat à l'Energie Atomique and Institut National de Physique Nucléaire et de Physique des Particules (France), the Bundesministerium für Bildung und Forschung (Germany), the Istituto Nazionale di Fisica Nucleare (Italy), the Research Council of Norway, the Ministry of Science and Technology of the Russian Federation, and the Particle Physics and Astronomy Research Council (United Kingdom).

References

1. N. Cabbibo, *Phys. Rev. Lett.* **10**, 531 (1963); M. Kobayashi and T. Maskawa, *Prog. Theor. Phys.* **49**, 652 (1973).
2. T. Hara, these proceedings.
3. D. Lange, these proceedings.
4. M. Gronau and D. London, *Phys. Rev. Lett.* **65**, 3381 (1990); M. Gronau, J.L. Rosner and D. London, *ibid.* **73**, 21 (1994); R. Fleischer, *Phys. Lett.* **B365**, 399 (1996); R. Fleischer and T. Mannel, *Phys. Rev. Lett.* **57**, 2752 (1998); M. Neubert and J. Rosner, *Phys. Lett.* **B441**, 403 (1998); M. Neubert, *J. High Energy Phys.* **02**, 014 (1999); M. Neubert, *Nucl. Phys. Proc. Suppl.* **99**, 113 (2001).
5. J. Panetta, these proceedings.
6. G.C. Fox and S. Wolfram, *Phys. Rev. Lett.* **41**, 1581 (1978).
7. S.L. Wu, *Phys. Rep.* **107**, 59 (1984).
8. CLEO Collaboration, D.M. Asner *et al.*, *Phys. Rev.* **D53**, 1039 (1996).
9. BABAR Collaboration, B. Aubert *et al.*, *Measurements of charmless three-body and quasi-two-body B decays*, BABAR-CONF-00/15, submitted to the XXXth International Conference on High Energy Physics, Osaka, Japan.
10. CLEO Collaboration, D. Cronin-Hennessy *et al.*, *Phys. Rev. Lett.* **85**, 515 (2000); CLEO Collaboration, S.J. Richichi *et al.*, *ibid.* **85**, 520 (2000); CLEO Collaboration, C.P. Jessop *et al.*, *ibid.* **85**, 2881 (2000).

COMPUTING THE WILSON LOOP IN $\mathcal{N} = 4$ SUPER-YANG-MILLS THEORY

GORDON W. SEMENOFF
Department of Physics and Astronomy, University of British Columbia, Vancouver, British Columbia, Canada V6T 1Z1

We review computations of the Wilson loop operator in $\mathcal{N} = 4$ supersymmetric Yang-Mills theory using both the AdS/CFT correspondence and perturbation theory.

1 Gauge theory - string theory duality

One of the most interesting products of string theory during recent years is the emergence of a deep and unexpected relationship between gauge fields and gravity. It has become apparent that many gauge field theories have a dual description as a gravitational theory and vice-versa. These are typically strong to weak coupling dualities and thus are potentially very useful for understanding both gauge and gravitational phenomena in strong coupling regimes. Furthermore, the gauge and gravitational theories usually live in different dimensions. This is an example of holography, the phenomenon whereby data necessary to specify states in a higher dimensional theory can be encoded in a lower dimensional theory.

This holographic duality is at present only understood for gauge and gravitational theories which are directly related to string theory - supergravity and gauge theory with a high degree of supersymmetry. It is at present unclear whether this understanding could be expanded to include a broader array of gauge theories. This important problem is currently being intensively investigated.

It is a fascinating fact that this gauge field theory-gravity duality is not understood without string theory at an intermediate stage. The gravitational theory and the gauge theory are both obtained from the appropriate limits of string theory. These limits turn out to have an overlapping domain of validity. In that case, within the overlap, they are two different descriptions of the same dynamical system.

Another fascinating aspect is the more general duality between gauge field theory and string theory. The idea that a gauge field theory could have a string theory dual has a long history. In fact, a stringy description of strong interactions dates back to the dual resonance models which were formulated in the late 1960's. These dual resonance models were eventually supplanted by gauge theory as a fundamental theory of the strong interactions.

However, the gauge theory, quantum chromodynamics, is weakly coupled only in the ultraviolet regime, for interactions involving large momentum transfers and is strongly coupled in the infrared. It has long been speculated that some sort of string theory could give a description of QCD in the infrared regime. If the string theory had a mass gap it would explain quark confinement. Of course, it should also produce the successful aspects of Regge theory and dual model phenomenology.

The best theoretical evidence for a string theory-gauge theory duality comes from 'tHooft's large N expansion of Yang-Mills theory based on the gauge group SU(N) [1]. In that expansion, N is taken to infinity holding the 'tHooft coupling $\lambda \equiv g^2 N$ fixed. The orders of the expansion in $1/N^2$ are classified according to the Euler character of the two-dimensional surfaces on which one could draw Feynman diagrams without crossing any lines. The leading order in this expansion is called planar - it is given by all diagrams which can be drawn on a plane. The idea is that, if there is a dual string theory, it is weakly coupled in the limit where N is large and computations could well be more tractable there. In spite of this optimism, over more than a quarter of a century there has been little progress in either solving the large N limit of four dimensional Yang-Mills theory or finding a concrete model for its string theoretical dual.

2 AdS/CFT

Recently one concrete realization of a gauge theory - string theory duality has emerged. It has been *conjectured* that there is an exact mapping between $\mathcal{N} = 4$ supersymmetric Yang-Mills theory (SYM) with gauge group SU(N) on four dimensional spacetime and IIB superstring theory on background $AdS_5 \times S_5$ with N units of Ramond-Ramond 4-form flux. The radii of curvature of S_5 and AdS_5 are $R = (g_{YM}^2 N)^{1/4}$ and the string and Yang-Mills coupling constants are related by $4\pi g_s = g_{YM}^2$.

There are three levels at which the most direct form of this conjecture, due to Maldacena [2], could be correct:

- In its strongest version the correspondence asserts that there is an exact equivalence between four dimensional $\mathcal{N} = 4$ supersymmetric Yang-Mills theory and type IIB superstring theory on the $AdS_5 \times S_5$ background. This mapping would hold at all distance scales and for all values of the coupling constants.
- A weaker version asserts a duality of the 't Hooft limit of the gauge theory,

where $N \to \infty$ with the 't Hooft coupling $\lambda = g^2 N$ held fixed, and the classical $g_s \to 0$ limit of type IIA superstring theory on $AdS_5 \times S_5$. In this correspondence, corrections to classical supergravity theory from stringy effects which are of order α' would agree with corrections to the large 't Hooft coupling limit, of order $1/\sqrt{\lambda}$, but higher orders in g_s on the supergravity side and non-planar diagrams or order $1/N^2$ on the gauge theory side could disagree.

- An even weaker version is a duality between the 't Hooft limit where one also takes the strong coupling limit $\lambda \to \infty$ and the low energy, supergravity limit of type IIB superstring theory on $AdS_5 \times S_5$. In this case, there would be order α' and g_s corrections to supergravity which might not agree with order $1/N^2$ and $1/\sqrt{\lambda}$ corrections to $\mathcal{N} = 4$ supersymmetric Yang-Mills theory.

Even the last, weakest version of the correspondence has profound consequences. Previous to it, the only successful quantitative tool for supersymmetric Yang-Mills theory was perturbation theory in g^2, the Yang-Mills coupling constant. This is limited to the regime where g^2 and λ are small. Also, although some qualitative features of the large N limit are known, it is not possible to sum planar diagrams explicitly. The AdS/CFT correspondence enables one to compute correlation functions in the large N, large λ limit. This limit contains the highly nontrivial sum of all planar Feynman diagrams, and emphasizes those diagrams which have infinitely many vertices.

It should be emphasized that, in its simplest form, this gauge theory - string theory mapping is not the hoped for one between an asymptotically free and infrared strongly coupled gauge theory to a string theory, but instead a mapping between two conformally invariant theories. (For attempts at more realistic theories, see [3,4]). $\mathcal{N}$=4 supersymmetric Yang-Mills theory has vanishing beta function and its coupling constant doesn't run. If the coupling is small, it has an asymptotic expansion at weak coupling which is valid at all distance scales.

The best evidence in support of the AdS/CFT correspondence comes from symmetry arguments. The global symmetries of both $\mathcal{N} = 4$ supersymmetric Yang-Mills theory and type IIB string theory on $AdS_5 \times S_5$ are identical. They have the same global super-conformal group $SU(2,2 \mid 4)$ (whose bosonic subgroup is $SO(4,2) \times SU(4)$). Not only are the global symmetries the same, but some of those objects which carry the representations of the symmetry group—the spectrum of chiral operators in the field theory and the fields in supergravity theory—can, to some degree, be matched [5]. Furthermore, both theories are conjectured to have an Montonen-Olive $SL(2, Z)$ duality acting on their coupling constants. Once a correspondence between the maximally symmetric theories is established, one could hope for a similar mapping between a broader class of theories which are obtained by perturbing the models on either side of the correspondence with relevant operators.

3 Perturbative checks of AdS/CFT

3.1 $\mathcal{N}=4$ supersymmetric Yang-Mills theory

The field content and other conventions of four dimensional $\mathcal{N}=4$ supersymmetric Yang-Mills theory in four dimensions can be read off from the Euclidean space action

$$S=\int d^4x \frac{1}{2g^2}\left(\frac{1}{2}\left(F^a_{\mu\nu}\right)^2+\left(\partial_\mu\phi^a_i+f^{abc}A^b_\mu\phi^c_i\right)^2+\bar{\psi}^a i\gamma^\mu\left(\partial_\mu\psi^a+f^{abc}A^b_\mu\psi^c\right)+\right.$$

$$\left.+if^{abc}\bar{\psi}^a\Gamma^i\phi^b_i\psi^c-\sum_{i<j}f^{abc}f^{ade}\phi^b_i\phi^c_j\phi^d_i\phi^e_j+\partial_\mu\bar{c}^a\left(\partial_\mu c^a+f^{abc}A^b_\mu c^c\right)+\xi(\partial_\mu A^a_\mu)^2\right)$$

where $F^a_{\mu\nu}=\partial_\mu A^a_\nu-\partial_\nu A^a_\mu+f^{abc}A^b_\mu A^c_\nu$. f^{abc} are the structure constants of $SU(N)$, $\left[T^a,T^b\right]=if^{abc}T^c$, and generators are normalized so that $\mathrm{Tr}T^aT^b=\frac{1}{2}\delta^{ab}$,. ψ^a is a sixteen component 10-D Majorana spinor. $\Gamma^A=(\gamma^\mu,\Gamma^i)$, for $\mu=1,...,4$ and $i=5,...,10$ are ten real 16×16 Dirac matrices (in the 10-dimensional Majorana representation with the Weyl constraint). All fields are $N\times N$ matrices and transform in the adjoint representation of the gauge group.

The Wilson loop operator is a phase factor associated with the trajectory of a heavy charged particle in the fundamental representation of the gauge group, which in our case is $SU(N)$. The loop operator of most interest in the Ads/CFT correspondence is the one which couples to classical quantities in the superstring theory in the simplest way provides a source for a classical string. It is

$$W(C)=\frac{1}{N}\,\mathrm{Tr}\exp\oint_C d\tau\;\left(iA_\mu(x)\dot{x}_\mu+\Phi_i(x)y_i\right), \tag{1}$$

where $x_\mu(\tau)$ is a parameterization of the loop, $y_i=\sqrt{\dot{x}^2}\theta_i$ and θ_i is a point on the five dimensional unit sphere ($\theta^2=1$). This operator measures the holonomy of a heavy W-boson whose mass results from spontaneous breaking of $SU(N)$ gauge symmetry to $SU(N-1)\times U(1)$ with $<\Phi_I>/|<\Phi>|=\theta_I$. In the AdS/CFT correspondence, it is computed by finding the area of the world-sheet of the classical string in $AdS_5\times S_5$ whose boundary is the loop C, which in turn lies on the boundary of AdS_5 [6,7,8].

Though it has been used for many computations of the strong coupling limits of gauge theory quantities, it is difficult to obtain an explicit check of the AdS/CFT correspondence. The reason for this is the fact that the correspondence with supergravity computes gauge theory in the large λ limit, with corrections from tree level string effects being suppressed by powers of $1/\sqrt{\lambda}$ and sometimes computable to the next order. On the other hand, the only other analytical tool which can be used systematically in the gauge theory is perturbation theory which is an asymptotic expansion in small λ. Generally, the only quantities for which these expansions have an overlapping range of validity is for quantities which are so protected by supersymmetry that they do not depend on the coupling constant.

An example is the entropy density that the gauge theory has when it is heated to a temperature T. Because it is a conformal field theory, the entropy density should get its dimensions from powers of the temperature. It should thus be given by an expression like

$$S/V = -f(g_{YM}^2, N)\pi^2 N^2 T^4/6 \tag{2}$$

where f is a function of the dimensionless Yang-Mills coupling constant and N. In the large N limit, it should be a function of λ. It is thought to interpolate smoothly between its weak coupling limit which can be obtained by (re-)summing Feynman diagrams,

$$f = 1 - \frac{32\lambda}{2\pi^2} + \frac{(3+\sqrt{2})\lambda^{3/2}}{\pi^3} + \ldots$$

and its strong coupling limit which is found using the AdS/CFT correspondence. There it is the Beckenstein-Hawking entropy of a black hole with Hawking temperature T,

$$f = \frac{3}{4} + \frac{45}{32}\zeta(3)\lambda^{-3/2} + \ldots.$$

Though these asymptotic expansions are known, it is indeed not possible to check that this function indeed interpolates across the gap between weak and strong coupling.

There are now a few known examples of quantities which are non-trivial functions of the coupling constant and whose large N limit is computable and is thought to be known to all orders in perturbation theory in planar diagrams. These functions can be extrapolated to their large λ limit and compared with the predictions of the AdS/CFT correspondence. The first example is the expectation value of the circular Wilson loop, which was computed by Erickson *et.al.*[9]. The contribution of a subset of all Feynman graphs, the planar rainbow diagrams, was found at each order in λ and the sum of all orders was taken to obtain the result

$$\langle W[\text{circle}]\rangle = \frac{2}{\sqrt{\lambda}} I_1(\sqrt{\lambda}) \quad \approx \quad \sqrt{\frac{2}{\pi}} \frac{e^{\sqrt{\lambda}}}{\lambda^{3/4}} \text{ as } \lambda \to \infty \tag{3}$$

where $I_1(x)$ is a modified Bessel function. It was also shown explicitly that the leading corrections to the sum of rainbow diagrams cancels identically.

It was conjectured that this cancellation would also occur at higher orders and the result (3) was thus the exact sum of all planar diagrams. Some support for this conjecture was developed in the literature[10]. They also observed that the sum over Feynman diagrams could be obtained for all orders in the $1/N$ expansion and had a beautiful argument that, in the large λ limit, these higher orders produced the expected higher genus string corrections.

The other quantities that are also probably exactly known are correlators of the circular Wilson loop operator with so-called chiral primary operators. The chiral primary is

$$\mathcal{O}_k^I = \frac{(8\pi^2)^{k/2}}{\sqrt{k}\lambda^{k/2}} C^I_{i_1\ldots i_k} \mathrm{Tr}\Phi^{i_1}\ldots\Phi^{i_k}, \tag{4}$$

The sum of all ladder and rainbow-like planar Feynman diagrams was found[11]. The result is

$$\frac{\langle W[\text{circle}]\,\mathcal{O}_k^I\rangle}{\langle W[\text{circle}]\rangle} = 2^{k/2-1}\sqrt{k\lambda}\,\frac{I_k\left(\sqrt{\lambda}\right)}{I_1\left(\sqrt{\lambda}\right)}\,\frac{R^k}{L^{2k}}\,Y^I(\theta) \tag{5}$$

where L is the distance of the operator insertion from the center of the loop and R is the radius of the circular loop. It was also argued[11] that radiative corrections to this expression cancel exactly.

The operator product expansion coefficients are highly non-trivial functions of λ. Furthermore, in the large λ limit, this correlator agrees with a computation of it using the AdS/CFT correspondence which was originally performed by Berenstein *et al.*[12].

References

1. G. 't Hooft, "A Planar Diagram Theory For Strong Interactions," *Nucl. Phys.* **B72**, 461 (1974).
2. J. Maldacena, "The large N limit of super-conformal field theories and supergravity," *Adv. Theor. Math. Phys.* **2**, 231 (1998); *Int. J. Theor. Phys.* **38**, 1113 (1998)] [hep-th/9711200].

3. J. Polchinski and M.J. Strassler, "The string dual of a confining four-dimensional gauge theory," hep-th/0003136.
4. I.R. Klebanov and M.J. Strassler, "Supergravity and a confining gauge theory: Duality cascades and chiSB-resolution of naked singularities," *JHEP* **0008**, 052 (2000) [hep-th/0007191].
5. E. Witten, "Anti-de Sitter space and holography," *Adv. Theor. Math. Phys.* **2**, 253 (1998) [hep-th/9802150].
6. J. Maldacena, "Wilson loops in large N field theories," *Phys. Rev. Lett.* **80**, 4859 (1998) [hep-th/9803002].
7. S. Rey and J. Yee, "Macroscopic strings as heavy quarks in large N gauge theory and anti-de Sitter supergravity," hep-th/9803001.
8. N. Drukker, D.J. Gross and H. Ooguri, "Wilson loops and minimal surfaces," *Phys. Rev.* **D60**, 125006 (1999) [hep-th/9904191].
9. J.K. Erickson, G.W. Semenoff and K. Zarembo, "Wilson loops in N = 4 supersymmetric Yang-Mills theory," *Nucl. Phys.* **B582**, 155 (2000) [hep-th/0003055].
10. N. Drukker and D.J. Gross, "An exact prediction of N = 4 SUSYM theory for string theory," hep-th/0010274.
11. G.W. Semenoff and K. Zarembo, "More exact predictions of SUSYM for string theory," hep-th/0106015.
12. D. Berenstein, R. Corrado, W. Fischler and J. Maldacena, "The operator product expansion for Wilson loops and surfaces in the large N limit," *Phys. Rev.* **D59**, 105023 (1999) [hep-th/9809188].
13. S. Lee, S. Minwalla, M. Rangamani and N. Seiberg, "Three-point functions of chiral operators in D = 4, N = 4 SYM at large N," *Adv. Theor. Math. Phys.* **2**, 697 (1998) [hep-th/9806074].
14. J.K. Erickson, G.W. Semenoff, R.J. Szabo and K. Zarembo, "Static potential in N = 4 supersymmetric Yang-Mills theory," *Phys. Rev.* **D61**, 105006 (2000) [hep-th/9911088].
15. J. Erickson, G.W. Semenoff and K. Zarembo "BPS vs. non-BPS Wilson loops in N = 4 supersymmetric Yang-Mills theory," *Phys. Lett.* **B466**, 239 (1999) [hep-th/9906211].

NON-EQUILIBRIUM PHASE TRANSITION DYNAMICS BEYOND THE GAUSSIAN APPROXIMATION

SUPRATIM SENGUPTA, S.P. KIM and F.C. KHANNA
Department of Physics, University of Alberta,
Edmonton - T6E 5P1, Canada.
E-mail: sengupta@phys.ualberta.ca, spkim@phys.ualberta.ca,
khanna@phys.ualberta.ca

We discuss aspects of domain formation and growth during a quenched second order phase transition. The canonical Louiville-von Neumann (L-vN) formalism is introduced and used to study a quantum- mechanical toy model of two coupled quartic oscillators. The resulting coupled non-linear equations are solved numerically to $\mathcal{O}(\lambda)$ in the small coupling parameter λ using the Multiple Scale Perturbation Theory. This toy model provides important insight into the effects of mode-coupling between hard and soft modes and gives an indication of how the non-gaussian effects can terminate domain growth.

1 Introduction

Spontaneous symmetry breaking (SSB) phase transitions are ubiquitous in many systems both in particle physics and in condensed matter physics and is also a crucial component of the Standard Model of Particle Physics. Recently, a lot of work has been addressed towards understanding the dynamics of SSB phase transition. In particular, investigations of quenched second order phase transitions have provided important insights into phenomenon as diverse as domain formation and growth in generic second order SSB phase transitions[1], formation of Disoriented Chiral Condensates (DCC)[2] during a chiral phase transition, and formation of topological defects such as domain walls, monopoles and cosmic strings[3].

During a quenched second order transition, the time-scale in which the effective potential changes is much smaller than the relaxation time-scale of the scalar (order-parameter) field because of the phenomenon of critical slowing down. This destabilizes the field and leads to its subsequent out-of-equilibrium evolution as it tries to relax to a new vacuum state having a non-vanishing vacuum expectation value. Formation of uncorrelated domains occurs because the field in different regions of space, which are causally disconnected, evolve independently during the out-of-equilibrium roll down process of the field. During this process, the long wavelength modes of the scalar field start growing exponentially and this leads to the establishment of large-scale correlations which are responsible for increase in the domain size[1].

Our aim in this article is to address the issues related to the non-Gaussian effects on domain growth. The canonical L-vN formalism[4] is introduced to study the growth of domains in a linearized spontaneously broken scalar field theory, after neglecting the effect of fluctuations. As a precursor to studying the effect of non-linear mode coupling on domain formation and growth in a quantum field theory, we investigate the simpler and more tractable quantum mechanical toy model of two coupled quartic oscillators where one of the oscillators has negative frequency squared. This model exactly mimics the effect of interaction between a hard mode and a soft mode in the corresponding quantum field theory problem.

We solve the resulting non-linear equations perturbatively, by using an improved perturbation scheme called the multiple scale perturbation theory (MSPT) [5]. It is found that the non-linear effects are able to terminate the growth of the unstable oscillator only at the first order ($\mathcal{O}(\lambda)$) in perturbation theory.

In the next section, we will briefly discuss the recently introduced L-vN formalism [4] and subsequently use it to make an estimate for the domain size in the Gaussian approx. The section provides an overview of the results obtained by Kim and Lee [4]. In section 3 we will make use of this canonical formalism to study a quantum mechanical toy model of two coupled quartic oscillators. In section 4, we conclude with a summary of our main results.

2 The Liouville - Von-Neumann Formalism

The L-vN formalism is based on the assumption that a quantum state of a non-equilibrium system evolves according to the time-dependent Schrodinger equation i.e.

$$i\hbar \frac{\partial |\Psi(t)>}{\partial t} = \hat{H}(t)|\Psi(t)> \tag{1}$$

where $\hat{H}(t)$ is the time-dependent Hamiltonian of the system. Also, from a statistical point-of view, the density matrix of the non-equilibrium system is assumed to obey the quantum Liouville - von Neumann equation.

$$i\hbar \frac{\partial \hat{\rho}(t)}{\partial t} = [\hat{H}(t), \hat{\rho}(t)] \tag{2}$$

In the quantum-mechanical framework, Lewis and Reisenfeld have shown that for a time-dependent Hamiltonian system; if $\hat{\mathcal{O}}(t)$ is any operator satisfying the quantum L-vN equation i.e.

$$i\hbar \frac{\partial \hat{O}(t)}{\partial t} = [\hat{H}(t), \hat{O}(t)] \tag{3}$$

then the exact quantum states of the system can be obtained in terms of the eigenfunctions of the operator $\hat{O}(t)$. The role of the operator $\hat{O}(t)$ will be played by the number operator $\hat{N}(t)$ in our case.

Consider a non-interacting scalar field theory where the Hamiltonian of the system when expressed in terms of field momentum modes is given by

$$\hat{H}(t) = \frac{1}{2\pi^3} \int d^3k[\pi_k(t)^2 + (k^2 + m^2(t))\phi_k^2(t)] \tag{4}$$

where $\pi_k(t) = \frac{d\phi_k}{dt}$ and $[\phi_k(t), \pi_{k'}(t)] = i\hbar\delta(\vec{k} - \vec{k'})$. The onset of phase transition is modelled by changing the sign of the mass squared term from positive to negative at $t = 0$. This amounts to inverting the oscillator potential, thereby destabilizing the field which was initially fluctuating around the vacuum expectation value $< \phi >= 0$. As we shall see below, the growth of long wavelength modes occurs during the out-of-equilibrium evolution of the destabilized field.

In the L-vN approach, annihilation and creation operators are defined as

$$\hat{a}_k(t) = i[f_k^*(t)\hat{\pi}_k - \dot{f}_k^*(t)\hat{\phi}_k] \tag{5}$$

$$\hat{a}_k^\dagger(t) = -i[f_k(t)\hat{\pi}_k - \dot{f}_k(t)\hat{\phi}_k] \tag{6}$$

The commutation relations $[\hat{a}_k, \hat{a}_{k'}^\dagger] = \delta_{k,k'}$ lead to the Wronskian condition $\hbar(\dot{f}_k^*(t)f_k(t) - \dot{f}_k(t)f_k^*(t)) = i$.

The entire Fock space can be constructed by repeated action of the creation operator on the Gaussian vacuum state which is defined by $\hat{a}_k(t)|0, t >= 0$. Requiring $\hat{a}_k(t)$ and $\hat{a}_k^\dagger(t)$ to satisfy the quantum L-vN equations leads to the following classical equations for the auxilliary field modes

$$\frac{d^2 f_k(t)}{dt^2} + \omega_k^2(t) f_k(t) = 0 \tag{7}$$

$$\frac{d^2 f_k^*(t)}{dt^2} + \omega_k^2(t) f_k^*(t) = 0 \tag{8}$$

The equations admit solutions of the form

$$f_k(t) = \frac{1}{\sqrt{2\hbar\omega_{i,k}}} e^{-i\omega_{i,k}t}; \qquad \omega_{i,k} = \sqrt{(k^2 + m^2)} \tag{9}$$

before the phase transition ($t < 0$) and

$$f_{k,h}(t) = \frac{1}{\sqrt{2\hbar\omega_{f_h,k}}}[cos(\omega_{f_h,k}t) - i\frac{\omega_{i,k}}{\omega_{f_h,k}}sin(\omega_{f_h,k}t)] \tag{10}$$

where $\omega_{f_h,k} = \sqrt{(k^2 - m^2)}$ with $k^2 > m^2$ and

$$f_{k,s}(t) = \frac{1}{\sqrt{2\hbar\omega_{f_s,k}}}[cosh(\omega_{f_s,k}t) - i\frac{\omega_{i,k}}{\omega_{f_s,k}}sinh(\omega_{f_s,k}t)] \tag{11}$$

where $\omega_{f_s,k} = \sqrt{(m^2 - k^2)}$ with $k^2 < m^2$ for $t > 0$ i.e. after the phase transition.

The solutions and their derivatives, *before* and *after* the phase transition, must match at $t = 0$. The 2-point correlation function with respect to the vacuum and thermal states are respectively given by

$$G_v(\vec{x},\vec{y};t) =< 0,t|\Phi(\vec{x},t)\Phi(\vec{y},t)|0,t >= \frac{1}{2\pi^3}\int d^3k[\hbar^2 f_k^*(t)f_k(t)]e^{i\vec{k}.(\vec{x}-\vec{y})} \tag{12}$$

$$G_T(\vec{x},\vec{y};t) = \frac{1}{Z_{N,k}}\sum_{n_k=0}^{\infty} < n_k,t|e^{-\beta\hbar(\hat{N}_k+1/2)\omega_k}\Phi(\vec{x},t)\Phi(\vec{y},t)|n_k,t > \tag{13}$$

where $Z_{N,k}$ is the partition function and $\omega_{i,k} = \sqrt{(k^2 + m^2)}$. After some algebra, the contribution to the 2-point correlation function from the soft modes is found to be

$$G_{T,f_s}(\vec{x},\vec{y};t) = G_{T,f_s}(\vec{0},T)\frac{sin(\sqrt{(m/2t)r})}{\sqrt{(m/2t)r}}e^{-\frac{mr^2}{8t}} \tag{14}$$

Therefore, this gives the standard Gaussian result for the domain size $\xi_D = \sqrt{\frac{8t}{m}}$.

3 The Quartic Coupled Oscillators Model

The estimate of the domain size in the previous section was obtained in the Gaussian approximation. The effect of non-linear coupling was neglected as a result of which, domain growth is unbounded, as is clear from Eq. (14). While the linear approximation may be justified during the early stages of the out-of-equilibrium evolution of the field, when fluctuations are small, the non-linear effects cannot be ignored if a correct estimate of the domain size, taking into account the non-Gaussian effects, is desired. Moreover, as we shall see below, non-linear effects are responsible for terminating the unbounded growth of domains evidenced from the Gaussian result. In this section we will make an attempt to go beyond the Gaussian approximation by investigating a quantum mechanical toy model of two coupled quartic oscillators. The more complicated quantum field theory problem will be addressed in a subsequent work [6].

We consider the model Hamiltonian

$$\hat{H}(t) = \frac{\hat{p}_1^2}{2m} + \frac{1}{2}\omega_1^2\hat{q}_1^2 + \frac{\hat{p}_2^2}{2m} + \frac{1}{2}\omega_2^2\hat{q}_2^2 + \frac{\lambda}{4!}\hat{q}_1^4 + \frac{\lambda}{4!}\hat{q}_2^4 + \frac{6\lambda}{4!}\hat{q}_1^2\hat{q}_2^2 \tag{15}$$

where $\omega_j^2 = k_j^2 + m^2(t)$; with $\omega_1^2 > 0$ and $\omega_2^2 < 0$.

The couplings have been chosen such that they mimic the interaction between a *hard mode* and a *soft mode* in a quantum field theory system. As before, the Gaussian number states in Fock space can be constructed $|n_j, n_k; t >= \frac{(\hat{A}_j^\dagger)^{n_j}}{\sqrt{n_j!}}|0_j; t > \frac{(\hat{A}_k^\dagger)^{n_k}}{\sqrt{n_k!}}|0_k; t >$.

The Hamiltonian, expressed in terms of the annihilation and creation operators can be separated into a Gaussian(quadratic part) and a quartic part which can be thought of as a perturbation $\hat{H}(t) = \hat{H}_g(t) + \lambda\hat{H}_p(t)$. The Gaussian part involves operators which are *quadratic* in $\hat{A}_j$ and/or $\hat{A}_j^\dagger$ while the non-gaussian (quartic) part contains operators which are *quartic* in $\hat{A}_j$ and/or $\hat{A}_j^\dagger$. The vacuum state for the full Hamiltonian is expressed as a perturbation series in powers of λ.

$$|\Psi_0; t >= |0_1, 0_2; t > + \sum_{l=0}^{\infty} \lambda^{(l)} \sum_{n1,n2} c_{n_1,n_2}^{(l)}(t)|n_1, n_2; t > \tag{16}$$

The non-gaussian contributions to the gaussian vacuum state appear in the form of particle excitations of the Gaussian Hamiltonian. The time-dependent perturbation expansion coefficients can be obtained from the fact that $|\Psi_0; t >$ satisfies the time-dependent Schrödinger equation for the full Hamiltonian. The only *non-vanishing* contributions to $\mathcal{O}(\lambda)$ in the perturbation expansion arise from the following coefficients

$$c_{4_1,0_2}^{(1)}(t) = -\frac{i\hbar^3\sqrt{6}}{12}\int dt[v_1^{*4}(t)] \tag{17}$$

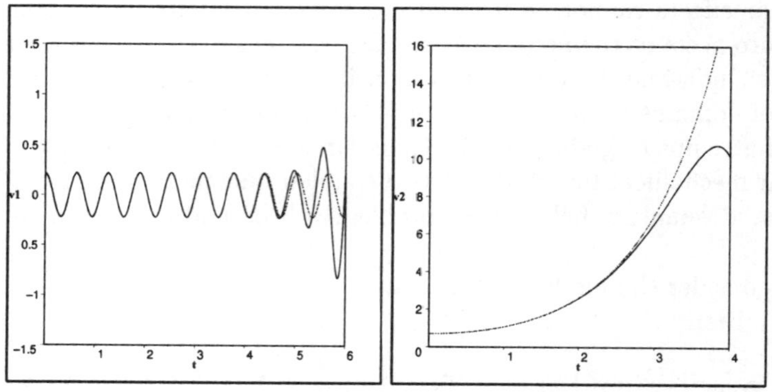

Figure 1: (a)Real part of the solution for $v_1(t)$ to $\mathcal{O}(\lambda)$. (b)Real part of the solution for $v_2(t)$ to $\mathcal{O}(\lambda)$. Parameters used are $\omega_1 = 10, \omega_2 = 1, \lambda = 0.01, \hbar = 1$.

$$c^{(1)}_{0_1,4_2}(t) = -\frac{i\hbar^3\sqrt{6}}{12}\int dt[v_2^{*4}(t)] \tag{18}$$

$$c^{(1)}_{2_1,2_2}(t) = -\frac{i\hbar^3}{2}\int dt[v_1^{*2}(t)v_2^{*2}(t)]. \tag{19}$$

Therefore to $\mathcal{O}(\lambda)$, the corrected non-gaussian vacuum state is

$$|\Psi_0;t>=|0_1,0_2;t> +\lambda[c^{(1)}_{4_1,0_2}(t)|4_1,0_2;t> +c^{(1)}_{0_1,4_2}(t)|0_1,4_2;t> +c^{(1)}_{2_1,2_2}(t)|2_1,2_2;t>] \tag{20}$$

By expressing the annihilation and creation operators in terms of auxilliary variables $v_j(t)$ and $v_j^*(t)$ and requiring that $\hat{A}_j$ and $\hat{A}_j^\dagger$ satisfy the quantum L-vN equation, the following equations for the auxilliary variables are obtained

$$\ddot{v}_j(t) + \omega_j^2(t)v_j(t) + \frac{\lambda\hbar^2}{2}(\sum_{k=1}^{2} v_k^* v_k)v_j(t) = 0 \tag{21}$$

$$\ddot{v}_j^*(t) + \omega_j^2(t)v_j^*(t) + \frac{\lambda\hbar^2}{2}(\sum_{k=1}^{2} v_k^* v_k)v_j^*(t) = 0. \tag{22}$$

The above equations are a set of coupled non-linear equations that have to be solved perturbatively. We have used the MSPT method [5] to solve the above set of eqs. to $\mathcal{O}(\lambda^\infty)$

$$v_1(t) = v_1^{(0)}(t) + \lambda v_1^{(1)}(t) \quad (23)$$

$$v_1(t) = v_1^{(0)}(t) + \lambda v_1^{(1)}(t). \quad (24)$$

It is possible to obtain an analytical solution to $\mathcal{O}(\lambda^0)$. However, the role of non-linear coupling in terminating the exponential growth of the unstable oscillator is manifest only at $\mathcal{O}(\lambda)$ as evident from Fig. 1, where a numerical solution of the above set of Eqs. to $\mathcal{O}(\lambda)$ is plotted.

4 Conclusions

In this article, we have discussed the formation and growth of domains during a quenched second order phase transition. After a brief introduction to the Louiville - von Neumann formalism, we use it to investigate a toy model of two coupled quartic oscillators in which one of the oscillators is unstable. The non-linear field equations for the auxilliary variables are solved using the MSPT scheme. We find that the non-Gaussian effects are capable of terminating the exponential growth of the unstable mode only at $\mathcal{O}(\lambda)$. We plan to use this scheme for investigating the non-gaussian effects on domain growth in model quantum field theory in a future work.

References

1. D. Boyanovsky, D.S. Lee and A. Singh, *Phys. Rev.* **D48**, 800, (1993).
2. K. Rajagopal and F. Wilczek, *Nucl. Phys.* **B404**, 577, (1993). See also D. Boyanovsky, H.J. de Vega and R. Holman, *Phys. Rev.* **D51**, 734, (1995); F. Cooper, Y. Kluger and E. Mottola and J.P. Paz, *Phys. Rev.* **D51**, 2377, (1995).
3. P. Laguna and W.H. Zurek, *Phys. Rev. Lett.* **78**, 2519 (1997); A. Yates and W.H. Zurek, *ibid.* **80**, 5477 (1998); N.D. Antunes, L.M.A. Bettencourt, and W.H. Zurek, *ibid.* **82**, 2824 (1999).
4. S.P. Kim and C.H. Lee, *Phys. Rev.* **D62**, 057901, (2000).
5. C.M. Bender and L.M.A. Bettencourt, *Phys. Rev.* **D54**, (1996).
6. S.P. Kim, S. Sengupta and F.C. Khanna, work in progress.

ATLAS PHYSICS POTENTIAL

J. SJÖLIN
Stockholm University, Fysikum, Box 6730, S-113 85 Stockholm, Sweden
E-mail: sjolin@physto.se

A highlighted overview of the expected performance and physics potential of the ATLAS detector is given. ATLAS is one of two general purpose detectors that will measure the output from collisions at the CERN Large Hadron Collider (LHC). Both the detectors and the collider has entered a machine production phase and expect to start data taking in 2006.

1 The LHC

The Large Hadron Collider (LHC) is a proton-proton collider currently under construction at CERN. It will replace the LEP electron-positron collider and utilize the same 4.3 Km radius tunnel. Due to the design center-of-mass energy of 14 TeV, the superconducting magnets have to produce a magnetic field of 8.4 T (the field is slightly higher then the naive 5.4 T since the magnets are segmented). To do this in a stable and safe manner has been one of the major technological challenges.

The first real data taking run is planned to start in 2006 with a $10^{33}\text{cm}^{-2}\text{s}^{-1}$ (10 fb^{-1}/year) luminosity per experiment. Later on the plan is to operate the LHC at the design luminosity of $10^{34}\text{cm}^{-2}\text{s}^{-1}$ (100 fb^{-1}/year). However, even for the initial low luminosity option the physics program is impressive.

2 ATLAS physics input

The high luminosity combined with the large proton-proton cross-section put very large demands on the rate handling abilities of the detector. Since the bunches will cross every 25 ns and the expected rate at design luminosity is

$$\text{Rate} = \sigma \times L = 80 \text{ mb} \times 10^{34}\text{cm}^{-2}\text{s}^{-1} = 1 \text{ GHz},$$

there will on average be 25 interactions per bunch crossing, mostly dominated by soft interactions. The high-p_T spectrum is expected to be dominated by gluon and quark QCD events. Both these types of events will be a background for any new physics and an advanced trigger is needed in order reduce the data stream. Complete events are expected to be stored at a rate of 100 Hz. How the signal and background situation compares to other experiments is shown in Table 1.

Table 1: Events Produced at low luminosity.

Process	Events/s	Events/year	Previous colliders
$W \to e\nu$	15	10^8	10^4 LEP, 10^7 Tevatron
$Z \to ee$	1.5	10^7	10^7 LEP
$t\bar{t}$	0.8	10^7	10^5 Tevatron
$b\bar{b}$	10^5	10^{12}	10^8 BaBar
$\tilde{g}\tilde{g}$ (1 TeV)	0.001	10^4	-
H (800 GeV)	0.001	10^4	-
QCD ($p_T > 200$ GeV)	10^2	10^9	10^7 Tevatron

3 ATLAS design

ATLAS is a general purpose detector. However, during the design the detector was optimized according to numerous constraints, among the more important were

- Excellent lepton and photon identification and measurement in a range from GeV to TeV in p_T. This is required by soft B-hadrons and searches for heavy particles. This is achieved by
  - High resolution LAr EM calorimeter, both in energy and position (e.g. $H \to \gamma\gamma$).
  - Huge stand alone muon spectrometer (up to 23m from the I.P.).
  - Powerful tracker for accurate momentum measurement and tag of secondary vertices.
- Large acceptance, including full-coverage ($|\eta| < 5$) calorimetry for accurate jet and missing E_T measurements. Needed e.g. for SUSY searches.
- Trigger with low p_T thresholds ($p_T > 6$ GeV).

4 ATLAS performance

The detector subsystems of ATLAS are now mainly in the phase of production. The performance of the design has been verified in test-beams and in detailed full simulations. This enormous effort has been summarized in the ATLAS Detector and Physics Performance Technical Design Report [1]. A simple estimation of the performance is given by the following parameterizations

- Charged particles

- $\sigma/p_T \sim 5 \times 10^{-4} p_T(\text{GeV}) \oplus 0.01$ (I.D.).
- $\sigma/p_T \sim 0.07$ @ 1 TeV (Muon spectrometer).

- Electromagnetic calorimetry
  - $\sigma/E \sim 0.1/\sqrt{E(\text{GeV})} \oplus 0.01$.
- Hadronic calorimetry
  - $\sigma/E \sim 0.5/\sqrt{E(\text{GeV})} \oplus 0.03$.
- Absolute lepton energy scale
  - $\leq 0.1\%$, $Z \to ll$ (goal $\sim 0.02\%$).
- Absolute jet energy scale
  - $\sim 1\%$, $t\bar{t} \to bW^+\bar{b}W^-, W \to jj$.
- Particle identification and rejection
  - $e/jet \sim 10^{-5}$ (70% eff.), $\gamma/jet \sim 10^{-3}$ (80% eff.), $b_{jet}/u_{jet} \sim 0.01$ (50% eff.).

5 Physics menu

5.1 Searches for SM Higgs

One of the main goals of ATLAS is of course to reveal the source for the generation of mass. In the Standard Model (SM) the Higgs particle is the only missing ingredient and its mass could be anywhere between the current LEP limit of 114 GeV up to 1 TeV. The production mechanisms for SM Higgs at LHC ordered in cross-section are shown in Figure 1. The detection turns out to be most difficult in the low mass range (110-140 GeV).

5.2 Low mass Higgs

Due to the large QCD background the first two diagrams are not very useful for the hadronic Higgs decays which have the largest branching-ratios, see Figure 1. Hence the discovery in the low mass region will rely on the clean $H \to \gamma\gamma$ mode plus the $H \to b\bar{b}$ in association with $t\bar{t}$, see Figure 2. The mass resolution of the electromagnetic calorimeter plays a crucial role in the background rejection of $H \to \gamma\gamma$ since the Higgs width is expected to be small, see lower right plot in Figure 1.

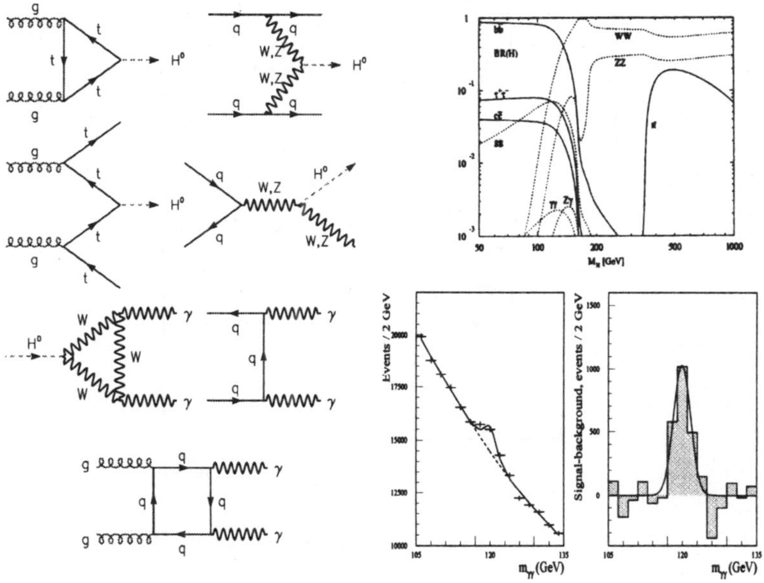

Figure 1: Production diagrams (upper left) and decay branching ratios (upper right) for SM Higgs. The main diagram responsible for the $H \to \gamma\gamma$ and its dominating background are shown in the lower left plots. The lower right plots shows the $\gamma\gamma$ spectrum with a Higgs mass at 120 GeV.

5.3 Searches for Supersymmetry

Probably the most favorable class of new physics is low energy supersymmetry (SUSY). It passes the SM precision tests in an impressive way and also makes some new solid predictions, e.g. GUT running of the coupling constants and a low mass Higgs.

In the minimal version (MSSM), with an imposed so called R-parity, the supersymmetric particles will be pair produced by strong interactions with large cross-sections.

The actual mechanism for SUSY breaking (SSB) is maybe the most interesting one and in order to do predictions some assumptions must be made. Two plausible scenarios have been studied in great detail: Gravity Mediated SSB (SUGRA) and Gauge Mediated SSB (GMSB). If low energy SUSY exist, it will manifest itself at LHC very early on as deviations from the SM in the inclusive signals, e.g. see Table 2. In the next step, an interesting handle to estimate the order of the SSB scale at least in the SUGRA case, turns out to

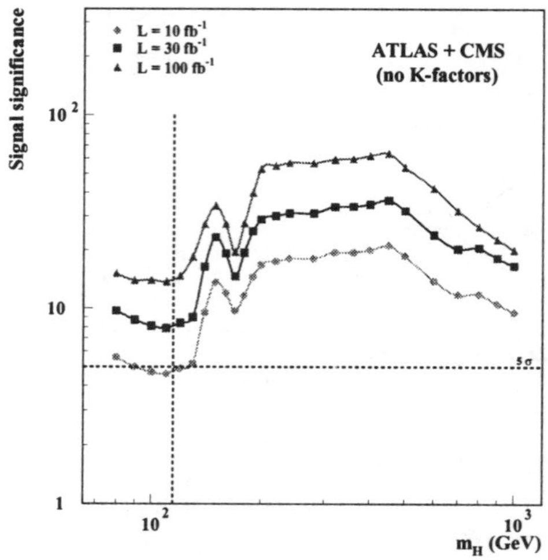

Figure 2: Expected LHC discovery potential for SM Higgs.

Table 2: Squark and gluino discovery limits within SUGRA.

Scenario	14 TeV (100 fb^{-1})	14 TeV (1000 fb^{-1})
$m_{\tilde{q}} \sim m_{\tilde{g}}$	2.1	2.7
$m_{\tilde{q}} \sim 2m_{\tilde{g}}$	2.0	2.3
$2m_{\tilde{q}} \sim m_{\tilde{g}}$	1.9	2.2

be the peak of the distribution of the inclusive variable M_{eff} defined as the scalar sum of the transverse momentum of the four leading jets plus missing energy in the detector, see Figure 3.

5.4 SUSY precision measurements

Hints about the SSB mechanism can be found in the supersymmetric mass spectrum. A big effort has been made to reconstruct the decay chains and estimate the mass resolutions. In this process the selection of a particular SSB model is crucial since the dominating background usually is SUSY itself. A robust tool for the estimation of the mass resolution are edges in an effective mass spectrum which is insensitive to cross-sections and efficiencies. Figure 3 shows an example of the effective mass for a minimal GMSB model with the

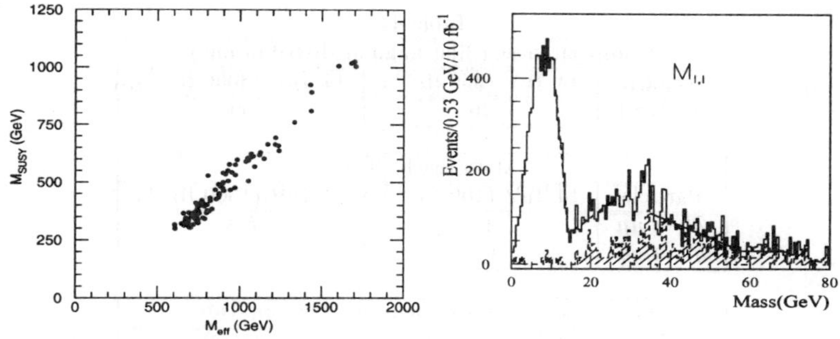

Figure 3: The effective mass M_{eff} vs. the effective SSB scale (left). The right plot shows an example of the same-flavor-opposite-signed leptons effective mass for the minimal GMSB model: $M_m = 250$ TeV, $\Lambda_m = 30$ TeV, $N_5 = 3, \tan\beta = 12, \mu =$ positive. The hatched area shows the SM background.

Table 3:

Standard model parameters		
Parameter	Error	Luminosity (fb^{-1})
m_h (< 500 GeV)	0.1 %	300
m_W	< 25 MeV	10
m_t	< 2 GeV	10
$\sin^2\theta_{\text{eff}}$	$\sim 10^{-4}$	100

decay chain

$$\tilde{q}_R \to \tilde{\chi}^0_{1,2}q \to \tilde{l}^-_R l^+ q \to \tilde{\tau}_1(\tau)l^- l^+ q \to \tilde{G}\tau(\tau)l^- l^+ q.$$

In many of the minimal SSB models the model parameters can be fully inferred with 10-100 GeV precision from the measured mass constraints.

5.5 SM precision measurements

One should not forget that the large statistics of a hadron collider in many cases mean very precise measurements. ATLAS can in many cases improve on the current SM limits set from e.g. the LEP experiments, see Table 3. One interesting note is that the limit on the m_W depends very much on reach in the absolute lepton energy scale.

5.6 Exotics

Of course anything can happen as soon as one enters a new energy domain. The sensitivity to exotic physics has been studied in many different scenarios,

Table 4:

Compositeness (dijet angular distributions)		
Scenario	14 TeV (300 fb^{-1})	14 TeV (3000 fb^{-1})
Λ (TeV)	40	60

Mass reach (TeV)		
Particle	14 TeV (100 fb^{-1})	14 TeV (1000 fb^{-1})
$Z' \to \mu\mu$	4.5	5.4
$q^* \to gq$	6.5	7.5

Factorisable extra dimensions (M_D in TeV)		
δ	14 TeV (100 fb^{-1})	14 TeV (1000 fb^{-1})
2	9	12
3	6.8	8.3
4	5.8	6.9

Non-factorisable (R-S) extra dimensions	
Resonance mass reach (TeV)	Luminosity (fb^{-1})
1.7	100

e.g. the different plausible theories of hidden dimensions, see Table 4.

Acknowledgments

The author would like to thank all the other numerous collaboration members that contributed to the results and for asking me to speak on their behalf.

References

1. ATLAS Performance TDR, CERN/LHCC/99-14,15.

DEEP INELASTIC SCATTERING AT HIGH Q^2

ALEX TAPPER
for the H1 and ZEUS collaborations
Blackett Laboratory, Imperial College of Science and Technology,
Prince Consort Road, London, SW7 2BW, United Kingdom.
E-mail: a.tapper@ic.ac.uk

Measurements of the cross sections for charged and neutral current deep inelastic scattering at HERA are presented. Comparing data from electron-proton and positron-proton collisions highlights the effect of the exchange of Z^0 and W bosons at high Q^2. The structure function xF_3^{NC} is extracted from the measured neutral current cross sections and found to be in agreement with the Standard Model expectation.

1 Introduction

Deep inelastic scattering (DIS) of leptons off nucleons probes the structure of matter at small distance scales. Two types of DIS interactions are possible at HERA: the neutral current (NC) reactions $e^-p \to e^-X$ and $e^+p \to e^+X$, where a photon or Z^0 boson is exchanged and the charged current (CC) interactions $e^-p \to \nu X$ and $e^+p \to \bar{\nu}X$, where a $W^\pm$ boson is exchanged.

The HERA accelerator collides electrons or positrons with protons. In the years 1994 to 1997 HERA collided positrons of energy 27.5 GeV with protons of energy 820 GeV , giving collisions at a centre of mass energy of 300 GeV. In this period the ZEUS and H1 detectors collected data samples with integrated luminosities of 47.7 pb^{-1} and 35.6 pb^{-1}, respectively. In 1998 and early 1999 HERA collided electrons of energy 27.5 GeV with protons of energy 920 GeV, giving collisions at a centre of mass energy of 320 GeV. In this period the ZEUS and H1 detectors collected data samples with integrated luminosities of 16.4 pb^{-1}. In addition measurements from the H1 collaboration based on 45.9 pb^{-1} of positron-proton collision data collected in 1999 and 2000 are presented.

The kinematics of charged current and neutral current deep inelastic scattering processes are defined by the four-momenta of the incoming lepton (k), the incoming proton (P), the outgoing lepton (k') and the hadronic final state (P'). The four-momentum transfer between the electron and the proton is given by $q = k - k' = P' - P$. The square of the energy in the centre of mass frame is given by $s = (k + P)^2$. The description of DIS is usually given in terms of three Lorentz invariant quantities, which may be defined in terms of the four-momenta k, P and q:

- $Q^2 = -q^2$, the negative square of the four-momentum transfer,
- $x = \frac{Q^2}{2p \cdot q}$, the Bjorken scaling variable,
- $y = \frac{q \cdot P}{k \cdot P}$, the fraction of the energy transferred to the proton in its rest frame.

These variables are related by $Q^2 = xys$, when the masses of the particles can be neglected.

Measurements of the neutral and charged current deep inelastic scattering cross sections are presented as functions of x and Q^2, in the kinematic region $Q^2 > 200\text{GeV}^2$.

2 Experimental setup

The ZEUS and H1 detectors are described in detail elsewhere[1,2]. The measurements presented make use primarily of the calorimeters, tracking detectors and luminosity measurement detectors. Selection of neutral current DIS events is based on the identification of a scattered electron or positron. The primary signature of charged current DIS events is missing transverse momentum from the final-state neutrino, which escapes undetected.

3 Cross sections

The double-differential Born-level cross section for the neutral current deep inelastic scattering processes $e^-p \to e^-X$ and $e^+p \to e^+X$, with unpolarised beams, is given by:

$$\frac{d^2\sigma^{NC}_{Born}(e^{\pm}p)}{dxdQ^2} = \frac{2\pi\alpha^2}{xQ^4}[Y_+F_2^{NC}(x,Q^2) - y^2F_L^{NC}(x,Q^2) \mp Y_- xF_3^{NC}(x,Q^2)] \quad (1)$$

where $Y_{\pm} = 1 \pm (1-y)^2$, and α is the QED coupling constant. The neutral current structure functions in the quark parton model, where $F_L^{NC}(x,Q^2) = 0$, are given by:

$$F_2^{NC}(x,Q^2) = \frac{1}{2}\sum_q [(V_q^L)^2 + (V_q^R)^2 + (A_q^L)^2 + (A_q^R)^2][xq(x,Q^2) + x\bar{q}(x,Q^2)] \quad (2)$$

$$xF_3^{NC}(x,Q^2) = \sum_q [V_q^L A_q^L - V_q^R A_q^R][xq(x,Q^2) - x\bar{q}(x,Q^2)] \quad (3)$$

where the sums run over all quarks, q, in the proton. The structure function xF_3^{NC} has contributions from the interference between the photon and Z^0 exchange amplitudes, and pure Z^0 exchange, and violates parity. The functions V_q and A_q contain the couplings of the electron to the photon and Z^0. The NC reduced cross section, $\tilde{\sigma}_{NC}$, is defined to be:

$$\tilde{\sigma}_{NC}(x,Q^2) = \frac{1}{Y_+}\frac{xQ^4}{2\pi\alpha^2}\frac{d^2\sigma_{Born}^{NC}}{dxdQ^2} \quad (4)$$

The double-differential Born-level cross section for the unpolarised charged current deep inelastic scattering processes $e^-p \to \nu X$ and $e^+p \to \bar{\nu}X$ are given by:

$$\frac{d^2\sigma_{Born}^{CC}(e^-p)}{dxdQ^2} = \frac{G_F^2}{2\pi}\frac{M_W^4}{(Q^2+M_W^2)^2}[(u+c)+(1-y)^2(\bar{d}+\bar{s})] \quad (5)$$

$$\frac{d^2\sigma_{Born}^{CC}(e^+p)}{dxdQ^2} = \frac{G_F^2}{2\pi}\frac{M_W^4}{(Q^2+M_W^2)^2}[(\bar{u}+\bar{c})+(1-y)^2(d+s)] \quad (6)$$

where M_W is the mass of the W boson and G_F is the Fermi constant. The CC reduced cross section, $\tilde{\sigma}_{CC}$, is defined to be:

$$\tilde{\sigma}_{CC} = \frac{2\pi}{G_F^2}\left(\frac{Q^2+M_W^2}{M_W^2}\right)^2\frac{d^2\sigma_{Born}^{CC}}{dxdQ^2} \quad (7)$$

4 Results and interpretation

Figure 1 (left) shows the single-differential cross sections $d\sigma/dQ^2$ for NC and CC DIS in both e^+p and e^-p scattering, measured by the H1 collaboration[10,8,9]. The measurements are well described by the Standard Model evaluated using the NLO QCD fit from the H1 collaboration[10] to their e^+p data only. The NC cross sections fall approximately six orders of magnitude over the Q^2 range 200 GeV^2 to 30000 GeV^2. At the higher end of the Q^2 range it can be seen that the cross section for e^-p scattering is higher than that for e^+p. The Standard Model predicts a higher cross section for e^-p interactions, due to constructive interference between the photon and Z^0 exchange amplitudes, compared to e^+p interactions where destructive interference is expected. The cross sections for CC DIS are lower than those for NC DIS and fall less steeply due to the propagator mass term. It can be seen that the CC DIS cross section for e^-p scattering is significantly higher than that for e^+p interactions. This is expected from the cross section formulae in which e^-p scattering is dominated by the u quark density and e^+p interactions probe the smaller d quark density and are suppressed by a factor of $(1-y)^2$. Shown in Figure 1 (right) are the reduced NC cross sections for e^+p and e^-p scattering, measured by the ZEUS collaboration[6]. The cross sections are plotted at fixed values of x as functions of Q^2, and are well described by the Standard Model evaluated with either CTEQ4D[3] or MRST(99)[4] parton density functions (PDFs). The effects of interference between photon and Z^0 exchange amplitudes can clearly be seen at higher values of Q^2 where the cross section for e^-p scattering is higher than that for e^+p interactions.

The NC cross sections for e^-p and e^+p scattering can be combined to extract the structure function xF_3^{NC}. Figure 2 shows the structure function xF_3^{NC} extracted by the ZEUS[6] and H1[8] collaborations. Both are shown at fixed values of Q^2 as functions of x, and are found to be well described by the Standard Model prediction.

Figure 3 (left) shows the reduced cross section for charged current e^+p interactions at fixed values of Q^2 as a function of x measured by the ZEUS collaboration[7]. It can be seen that the data are well described by the Standard Model. At low Q^2 and low x it can be seen from the dashed and dotted lines, which show the contributions from the valence and sea quarks, that the measurements are sensitive to the sea quark distribution. Figure 3 (right) shows the extracted values of the u and d valence quark distributions from the H1 collaboration[9], as a function of Q^2 at fixed values of x. The extracted quark densities are compared with the Standard Model prediction evaluated using the CTEQ5M, MRST and H1 NLO QCD fit PDFs.

5 Summary and future prospects

Cross sections for neutral and charged current deep inelastic scattering interactions have been presented by the H1 and ZEUS collaborations. In all cases the Standard Model gives a good description of the data. In the neutral current case the data at high Q^2 are consistent with the effects of interference between processes where a photon is exchanged and a Z^0 is exchanged. The structure function xF_3^{NC} has been extracted for the first time at HERA.

The HERA upgrade programme is scheduled to provide a factor of five increase in luminosity and longitudinally polarised lepton beams towards the end of 2001. The precision of NC and CC DIS cross section measurements will benefit from the large volume of data. In particular determination of the u and

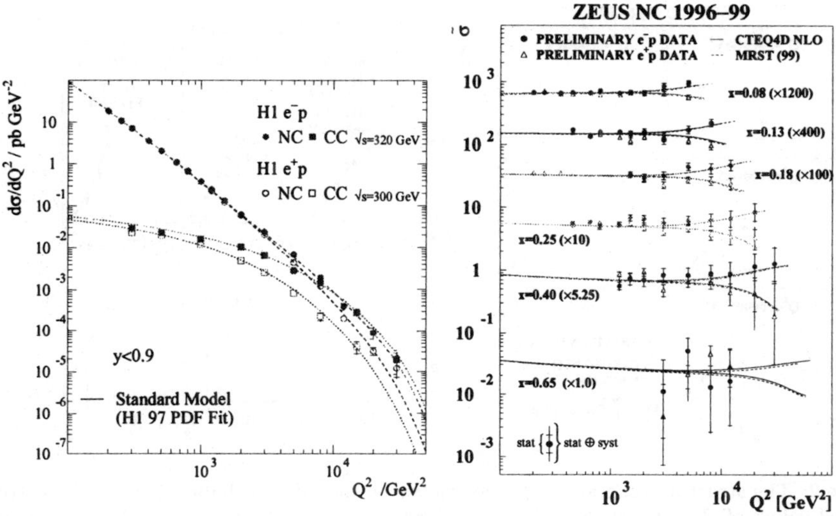

Figure 1: The cross section $d\sigma/dQ^2$ is shown (left) for NC (circles) and CC (squares) DIS interactions for both e^+p (open symbols) and e^-p (filled symbols) scattering from the H1 collaboration. ZEUS collaboration measurements of the NC reduced cross section for e^+p (open triangles) and e^-p (filled circles) are shown (right), at values of fixed x as functions of Q^2. Also shown are the Standard Model predictions evaluated using the CTEQ4D and MRST(99) parameterisations of the parton density functions.

d quark densities at high Q^2 and at high x from the CC DIS cross sections will be possible with much improved accuracy and in the case of NC DIS, fits to determine the gluon density distribution and the value of the strong coupling constant, α_S, will achieve a higher precision than currently possible. The introduction of longitudinally polarised lepton beams for the H1 and ZEUS experiments will allow the investigation of the chiral nature of the Standard Model. Searches for right-handed charged currents and the determination of the vector and axial-vector couplings to the u and d quarks will be among the measurements possible at high Q^2 with high luminosity and longitudinally polarised lepton beams.

Acknowledgments

I would like to thank Dr. K. Long for his advice on this subject and the Imperial College London HEP group for providing the financial assistance which allowed me to attend the Lake Louise Winter Institute.

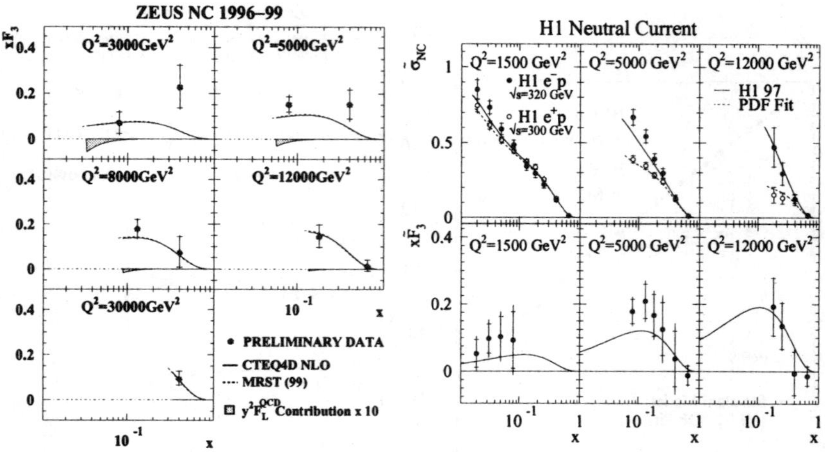

Figure 2: The structure function xF_3^{NC} extracted by the ZEUS collaboration (left), plotted at fixed values of Q^2 between 3000 GeV^2 and 30000 GeV^2 as a function of x. Also shown are the Standard Model predictions evaluated using the CTEQ4D and MRST(99) parameterisations of the parton density functions. The H1 collaboration measurements (right) illustrate the extraction of xF_3^{NC}. The upper half of the diagram shows the reduced NC cross sections for e^+p (open circles) and e^-p (filled circles) at fixed values of Q^2 from 1500 GeV^2 to 12000 GeV^2 as functions of x. The lower half shows the xF_3^{NC} extracted from the cross section measurements. The measurements are compared to the Standard Model predictions evaluated using PDFs from the H1 QCD fit.

References

1. ZEUS Collab., U. Holm (ed.), *The ZEUS Detector*, Status Report (unpublished), DESY, 1993.
2. H1 Collab., I. Abt *et al.*, *Nucl. Instrum. Methods* **A36**, 310 (1997).
3. H.L. Lai *et al.*, *Phys. Rev.* **D55**, 1280 (1997).
4. A.D. Martin *et al.*, *Eur. Phys. J.* **C14**, 133 (2000).
5. M. Botje, *Eur. Phys. J.* **C14**, 285 (2000).
6. ZEUS Collab., Abstracts 1048 and 1049, XXX International Conference of High-Energy Physics, Osaka, July 27 - August 2 2000.
7. ZEUS Collab., J. Breitweg *et al.*, *Eur. Phys. J.* **C12**, 3, 411 (2000).
8. H1 Collab., C. Adloff *et al.*, accepted by *Eur. Phys. J.* **C12**, (2000).
9. H1 Collab., Abstract 975, XXX International Conference of High-Energy Physics, Osaka, July 27 - August 2 2000.
10. H1 Collab., C. Adloff *et al.*, *Eur. Phys. J.* **C13**, 609 (2000).

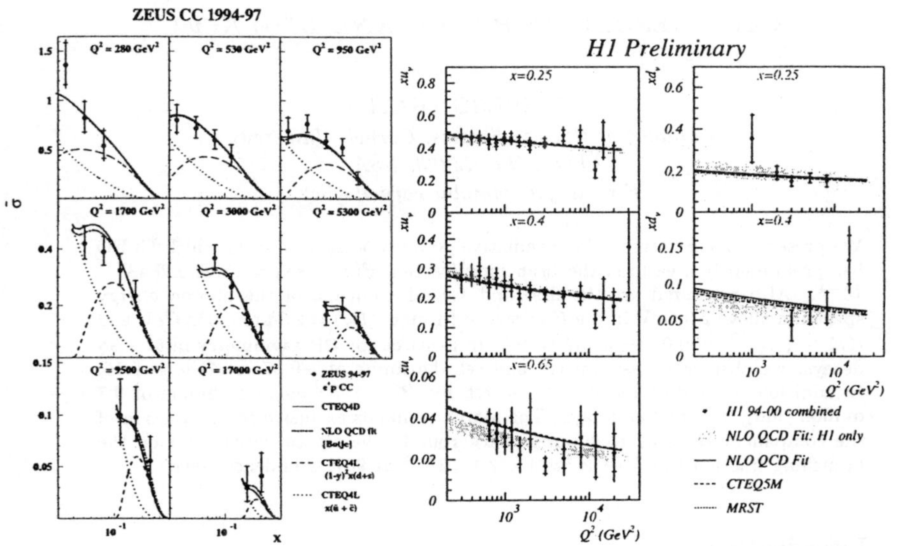

Figure 3: The reduced CC cross section, $\tilde{\sigma}_{CC}$, for e^+p scattering measured by the ZEUS collaboration (left), plotted at fixed values of Q^2 as a function of x. The ZEUS data are confronted with the Standard Model prediction evaluated using the CTEQ4D PDFs and ZEUS NLO QCD fit predictions. The values of the u and d valence quark densities extracted by the H1 collaboration are shown (right) for fixed values of x as a function of Q^2. The extracted quark densities are compared with the Standard Model prediction evaluated using the CTEQ5M, MRST and H1 NLO QCD fit PDFs.

CLEO RESULTS FOR $b \to s\gamma$ AND $B^{\pm} \to K^{\pm}\nu\bar{\nu}$

JOHN GREGG THAYER
Laboratory of Nuclear Studies, Cornell University,
Ithaca, NY, 14853, USA
E-mail: jgt@mail.lns.cornell.edu

We present a recent study of the radiative decay process $b \to s\gamma$ in which CLEO has preliminarily measured the branching fraction $\mathcal{B}(b \to s\gamma) = (3.03 \pm 0.48) \times 10^{-4}$. Also measured are the first and second moments of the photon energy spectrum above 2.0 GeV in the B meson rest frame, $\langle E_\gamma \rangle = 2.345 \pm 0.032$ GeV and $\langle (E_\gamma - \langle E_\gamma \rangle)^2 \rangle = 0.021 \pm 0.007$ GeV^2. In a search for CP asymmetry in $b \to s\gamma$ decays, we observe no asymmetry and set the limits at $-0.27 < \mathcal{A}_{CP} < 0.10$. In addition, we report results of a search for $B^{\pm} \to K^{\pm}\nu\bar{\nu}$ in a sample of 9.7 million charged B meson decays. The search demands exclusive reconstruction of the companion $\bar{B}$ decay to suppress background. We set an upper limit on the branching fraction $\mathcal{B}(B^{\pm} \to K^{\pm}\nu\bar{\nu}) < 2.4 \times 10^{-4}$ at 90% confidence level.

1 Introduction

This talk presents results from two analyses by the CLEO Collaboration using data collected by the CLEO II detector which is described in detail elsewhere[1]. The B mesons are produced via $e^+e^- \to \Upsilon(4S) \to B\bar{B}$ in CESR, a symmetric e^+e^- storage ring. The first analysis describes measurements made of the inclusive $b \to s\gamma$ decay. The second details a search for the rare process $B \to K\nu\bar{\nu}$.

2 $b \to s\gamma$

The characteristic signature of $B \to X_s\gamma$ is a peak in the photon energy spectrum near 2.4 GeV. Unfortunately there are two to three orders of magnitude more photons with these energies produced in non-B meson decays (continuum) than in our signal. These come from initial state radiation ($e^+e^- \to \gamma q\bar{q}$) and photons from the decay of hadrons ($\pi^0, \eta, \omega, \eta', etc.$) produced in $e^+e^- \to q\bar{q}$. CLEO takes some data just below the $B\bar{B}$ threshold (OFF Resonance) in addition to data at the $\Upsilon(4S)$ (ON Resonance.) This allows us to understand these continuum backgrounds. Unfortunately, the size of these backgrounds are so large that a simple "ON-OFF" subtraction of the photon yield would be dominated by statistical errors. What this means is that we must first suppress these backgrounds before performing the subtraction.

Previous analyses have chosen the low photon energy cutoff to be 2.2 GeV

Ref.[2] and 2.1 GeV Ref.[3]. These were chosen to remove background from other B decays. The photon energy spectrum from other B meson decays falls rapidly with increasing energy and is essentially gone by 2.2 GeV, but since theoretical predictions of the $b \to s\gamma$ photon energy cutoff vary, these introduce a systematic error that is lessened by our choice in this analysis of 2.0 GeV for the low energy cutoff.

The analysis begins by selecting a candidate photon, rejecting any which can be combined with another photon in the event to reconstruct a π^0 or η. This affects a reduction on both $B\bar{B}$ and continuum backgrounds. We then use a combination of three types of information to further reduce the continuum backgrounds: event shape variables, pseudoreconstruction information, and information about any lepton that might be present. Shape variables include ratios of Fox-Wolfram[4] parameters (event geometry,) and energy distributions about the direction of the photon candidate. The variables are discussed in more detail in Ref. 2. Pseudoreconstruction tries to reconstruct the B meson from the candidate photon, a kaon, and $1-4$ pions. If an acceptable combination is found, the χ^2 of the fit and the angle between the candidate B and the thrust axis of the rest of the event are added to the cadre of discriminating variables. If the event contains a lepton, we add the the lepton energy and the angle between the lepton and candidate photon. These variables are then combined using neural nets that were trained with Monte Carlo. This results in each event getting a weight between 0.0 and 1.0 based on the neural net output. This weight represents how "signal-like" or "continuum-like" each event is. The use of these weights reduces the amount of continuum background to the level of ~ 3 times the amount of $b \to s\gamma$ signal.

After suppressing the continuum background in this way, we subtract the off resonance data from the on resonance data. This subtraction makes it unnecessary to understand the continuum background or how the neural net works to suppress it. With the continuum background removed, we are still left with $B\bar{B}$ backgrounds as seen in the top of Figure 1. This background is dominated by photons from π^0 and η decays that have evaded our veto. Background from other B decay processes are determined by using Monte Carlo that has been tuned to match the data yields of π^0, η, η', and (approximately) ω. This Monte Carlo includes charmless hadronic B decay and $b \to u\ell\nu$. We also allow for radiative ψ decay, $a_1 \to \pi\gamma$, $\rho \to \pi\gamma$, and final state radiation. We determine the background from neutral hadrons using the lateral distribution of electromagnetic calorimeter showers. We then subtract the $B\bar{B}$ background to obtain the photon energy spectrum shown in the bottom of Figure 1.

Since our signal region is above 2.0 GeV, the region $1.5-2.0$ GeV indicates how well we have accounted for the $B\bar{B}$ backgrounds. The yield in this region

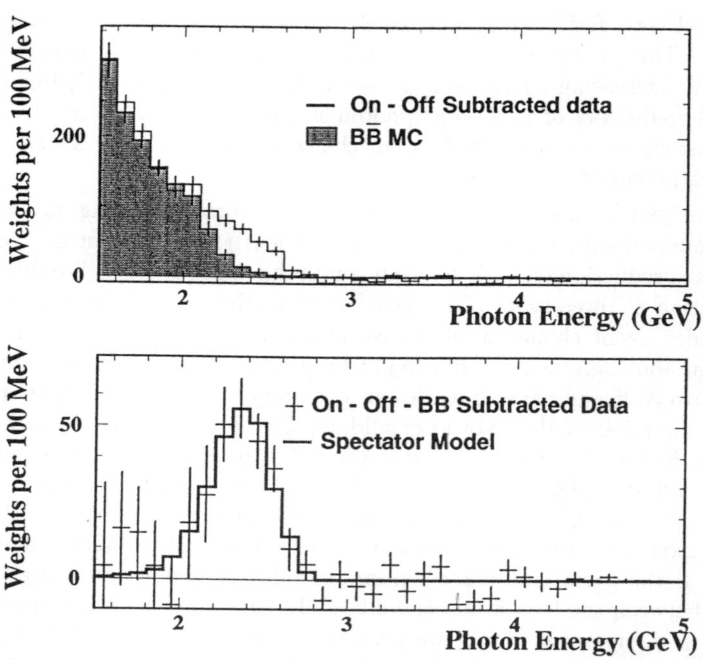

Figure 1: Laboratory photon energy spectrum which has not been corrected for experimental resolution after continuum subtraction and before (top) and after (bottom) subtraction of the $B\bar{B}$ backgrounds. Included is a fit using the Spectator Model.

is consistent with zero. To obtain the $b \to s\gamma$ branching fraction, we sum the weights between 2.0 and 2.7 GeV from the $B\bar{B}$ and continuum subtracted set. The yield is $240.7 \pm 31.7 \pm 13.5$ *weights*, where the first error is statistical and the second is systematic. For photons above 2.0 GeV the signal efficiency is $(4.26 \pm 0.25) \times 10^{-2} \frac{weights}{event}$. The error in this efficiency accounts for model dependence and other inaccuracies in the Monte Carlo samples. Our data sample contains 9.70×10^6 $B\bar{B}$ pairs. We take 91.5% as the fraction of $b \to s\gamma$ decays with a photon energy above 2.0 GeV Ref.[5]. After correcting for $b \to d\gamma$, we obtain

$$\mathcal{B}(b \to s\gamma) = (3.03 \pm 0.48) \times 10^{-4}.$$

This result is in good agreement with the Standard Model prediction of $\mathcal{B}_{SM}(b \to s\gamma) = (3.29 \pm 0.33) \times 10^{-4}$.

We also extract the first and second moments from the photon energy spectrum. To do this we correct for the effects of detector resolution and the motion of the B meson in the lab. As a result we obtain the moments in the

B rest frame for $E_\gamma > 2.0\ GeV$. These are

$$\begin{aligned} \langle E_\gamma \rangle &= 2.345 \pm 0.030 \pm 0.011\ GeV \\ \langle (E_\gamma - \langle E_\gamma \rangle)^2 \rangle &= 0.021 \pm 0.007 \pm 0.002\ GeV^2. \end{aligned}$$

In events that are pseudoreconstructed or have a lepton, we attempt to use available information to tag the flavor of the b. We can then measure the rates of $b \to s\gamma$ and $\bar{b} \to \bar{s}\gamma$ separately. We then construct the CP Asymmetry as

$$\mathcal{A}_{CP} \equiv \frac{\Gamma(b \to s\gamma) - \Gamma(\bar{b} \to \bar{s}\gamma)}{\Gamma(b \to s\gamma) + \Gamma(\bar{b} \to \bar{s}\gamma)}.$$

This procedure, described in detail in Ref. 6, results in the 90% confidence level limits

$$-0.27 < \mathcal{A}_{CP} < 0.10.$$

3 $B^\pm \to K^\pm \nu\bar{\nu}$

To search for $B^\pm \to K^\pm \nu\bar{\nu}$ decays we fully reconstruct each $\Upsilon(4S) \to B^+B^-$ event in the simultaneous decay modes $B^+ \to K^+\nu\bar{\nu}$ ("signal B") and $B^- \to D^{(*)0}(n\pi)^-$ ("companion B"). For the signal B we accept any single track which passes track quality requirements and fails lepton identification.

For the companion B, we take advantage of the large (46%) $b \to cu\bar{d}$ branching fraction and seek to reconstruct $B^- \to D^{(*)0}(n\pi)^-$, accepting either D^0 or $D^{*0} \to D^0(\gamma, \pi^0)$ and reconstructing the D^0 in the following eight modes, $K^-\pi^+$, $K^-\pi^+\pi^0$, $K^-\pi^+\pi^-\pi^+$, $K^0\pi^+\pi^-$, $K^-\pi^+\pi^0\pi^0$, $K^-\pi^+\pi^-\pi^+\pi^0$, $K^0\pi^+\pi^-\pi^0$, and $K^0\pi^0$. Based on the reconstructed D^0 mass, the π^0 mass, and the kaon and pion particle identification information, we compute a χ^2 quality factor and use it to reject poor D^0 candidates. The $(n\pi)^-$ system may be any of the following: π^-, $\pi^-\pi^+\pi^-$, $\pi^-\pi^+\pi^-\pi^+\pi^-$, $\pi^-\pi^0$, $\pi^-\pi^+\pi^-\pi^0$, or $\pi^-\pi^0\pi^0$.

Backgrounds arising from $e^+e^- \to q\bar{q}$ events (continuum) are suppressed by computing the direction of the thrust axis of the companion B candidate and measuring the angle θ to the direction of the K^+ candidate. Additional backgrounds from $e^+e^- \to \tau^+\tau^-$ are suppressed by requiring the Fox-Wolfram[4] moments ratio H2/H0 to be less than 0.5, which favors spherical topologies. Contributions from two-photon events ($e^+e^- \to \gamma^{(*)}e^+e^-$) are negligible.

The identification of acceptable candidates for the K^+, D^0, and $n\pi$ system, together with the absence of extra tracks or significant extra neutral energy, marks the appearance of a signal candidate. We now characterize these candidates by the kinematic properties of the *companion* B. In particular we use the total momentum $\vec{P}_B$ and energy E_B of the companion B, computed from momenta and energies of its daughter products. These raw quantities are then recast as the more useful beam-constrained mass and energy difference variables

$$\begin{aligned} M(B) &\equiv (E_{\rm beam}^2 - \vec{P}_B)^{1/2} \\ \Delta E &\equiv E_B - E_{\rm beam}. \end{aligned}$$

We select events whose $M(B)$ falls within 2.5 standard deviations of the true B mass, and extract the signal yield by fitting the resulting ΔE distribution. The net signal efficiency for the analysis is $\varepsilon = 1.8 \times 10^{-3}$. The efficiency is calculated using the form factor model of Ref. 7, but it changes only negligibly if instead we use 3-body phase space and a constant matrix element. The signal fit shape is determined by Monte Carlo and residual backgrounds are modeled by a linear distribution. We fit the ΔE distribution by an extended unbinned maximum likelihood method[8] to obtain the total yield of signal and background. Three events remain after all selection criteria are applied and the central value of the fitted yield is 0.81 events (Figure 2.) The background level is consistent with Monte Carlo expectations given the selection criteria and the size of the data sample.

The branching ratio is related to the signal yield $N_{\rm sig}$ by $\mathcal{B}(B^\pm \to K^\pm \nu\bar{\nu}) = N_{\rm sig}/N_{B\bar{B}}\varepsilon$ where $N_{B\bar{B}} = 9.66 \times 10^6$ is the number of charged B mesons in the data sample and ε is the efficiency as given above. We cross check the efficiency by conducting a separate analysis identical to this one in all key respects except that the $K^\pm\nu\bar{\nu}$ target signal is replaced by $D^{*0}\ell^-\nu$ whose branching fraction is large and well-measured. To ensure as much topological similarity to the $K^\pm\nu\bar{\nu}$ case as possible, we restrict this ancillary analysis to the low-multiplicity sub-mode, $D^0 \to K^-\pi^+$. The discrepancy between data and Monte Carlo yields is 1.3σ.

In total, the relative systematic error on efficiency is 24.4%. We integrate the systematics-convolved likelihood function shown in Figure 2c to obtain a 90% confidence upper limit $\mathcal{B}_{90}$ on $\mathcal{B}(B^\pm \to K^\pm\nu\bar{\nu})$. Using

$$0.90 = \frac{\int_0^{\mathcal{B}_{90}} \mathcal{L}(\mathcal{B})d\mathcal{B}}{\int_0^1 \mathcal{L}(\mathcal{B})d\mathcal{B}},$$

we find

$$\mathcal{B}_{90}(B^\pm \to K^\pm\nu\bar{\nu}) < 2.4 \times 10^{-4}.$$

4 Conclusion

We have presented a recent study of the radiative decay process $b \to s\gamma$ in which CLEO has preliminarily measured the branching fraction $\mathcal{B}(b \to s\gamma) =$

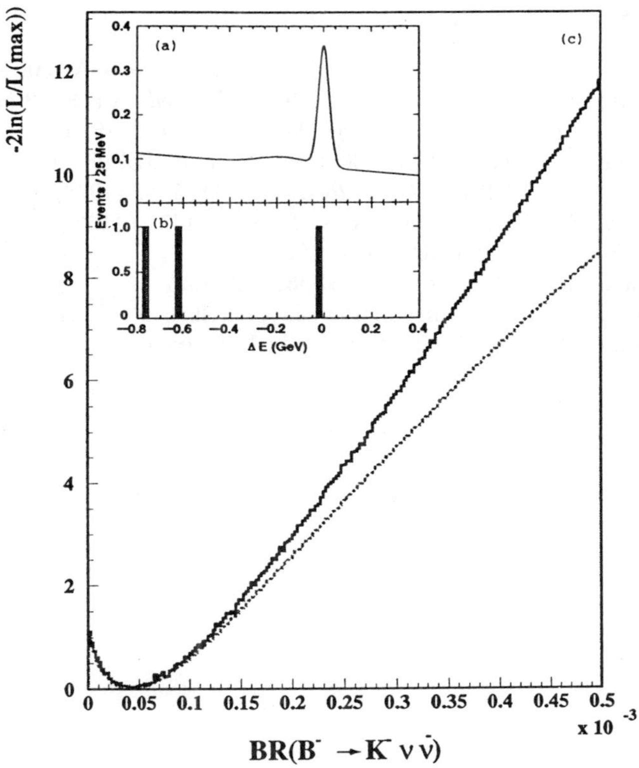

Figure 2: Final Results. (a) The combined fit of signal and background of (b) the three events that remain after all cuts. (c) The solid (dotted) curves are the negative log-likelihood curves from the fit in units of $\mathcal{B}(K^- \to \nu\bar{\nu})$ excluding (including) the 24% systematic error.

$(3.03 \pm 0.48) \times 10^{-4}$ and the first and second moments of the B rest frame photon energy spectrum above 2.0 GeV, $\langle E_\gamma \rangle = 2.345 \pm 0.032\ GeV$ and $\langle (E_\gamma - \langle E_\gamma \rangle)^2 \rangle = 0.021 \pm 0.007\, GeV^2$. In a search for CP asymmetry in $b \to s\gamma$ decays, we observed no asymmetry and set the limits at $-0.27 < \mathcal{A}_{CP} < 0.10$. We have also reported the results of a search for $B^\pm \to K^\pm \nu\bar{\nu}$ in a sample of 9.7 million charged B meson decays. We have set an upper limit on the branching fraction $\mathcal{B}(B^\pm \to K^\pm \nu\bar{\nu}) < 2.4 \times 10^{-4}$ at 90% confidence level.

Acknowledgments

We gratefully acknowledge the effort of the CESR staff in providing us with excellent luminosity and running conditions. This work was supported by the National Science Foundation, the U.S. Department of Energy, and the Natural Sciences and Engineering Research Council of Canada.

References

1. Y. Kubota *et. al.* (CLEO), *Nucl. Instrum. Methods* **A320**, 66 (1992); T.S. Hill *et. al.* (CLEO), *Nucl. Instrum. Methods* **A418**, 32 (1998).
2. M.S. Alam *et. al.* (CLEO), *Phys. Rev. Lett.* **74**, 2885 (1995).
3. S. Glenn *et. al.* (CLEO), CLEO Conf 98-17, ICHEP98, 1011.
4. G. Fox and S. Wolfram, *Phys. Rev. Lett.* **41**, 1581 (1978).
5. A.L. Kagan and M. Neubert, *Eur. Phys. J.*, **C7**, 5 (1999).
6. T.E. Coan *et. al.* (CLEO), hep-ex/0010075.
7. P. Colangelo *et al*, *Phys. Lett.* **B395**, 339 (1997).
8. L. Lyons, W. Allison and J. Panella Comellas, *Nucl. Instrum. Methods* **A245**, 530 (1986); R. Barlow, *Nucl. Instrum. Methods* **A297**, 496 (1990).

AN IMPROVED MEASUREMENT OF THE ANOMALOUS MAGNETIC MOMENT OF THE POSITIVE MUON

ALEXEI TROFIMOV

Department of Physics, Boston University, Boston, MA 02215, USA

E-mail: trofimov@buphy.bu.edu

For the (g-2)μ Collaboration [1]

Abstract. A new precise measurement of the magnetic anomaly of the muon, $a_\mu = (g-2)/2$, has been made at the Brookhaven Alternating Gradient Synchrotron, based on the data taken in 1999. The new result, $a_\mu = 11\ 659\ 202(14)(6) \times 10^{-10}$ (1.3 ppm) is in good agreement with previous measurements. The difference between the current theoretical value from the Standard Model and the average experimental value is $a_\mu^{exp} - a_\mu^{SM} = 43(16) \times 10^{-10}$.

1 Introduction

The anomalous magnetic moment of the muon, $a_\mu = (g-2)/2$, describes the relative deviation of the muon gyromagnetic ratio from the value predicted in Dirac theory. The sensitivity of the muon magnetic anomaly to heavier particles makes a precise measurement of a_μ a powerful probe of physics beyond the Standard Model. The "new physics" could manifest itself in a disagreement between the experimentally measured value and the theoretical prediction.

In the Standard Model, the anomalous magnetic moment of the muon is dominated by the QED radiative correction, and also includes contributions from hadronic and electroweak loops:

$$a_\mu^{SM} = a_\mu^{QED} + a_\mu^{Had} + a_\mu^{Weak}, \tag{1}$$

with the theoretical prediction of the sizes of each contribution[2] given in Table 1. The uncertainty in the theoretical value comes mainly from the first-order hadronic contribution, which cannot be calculated from first principles in QCD, but is determined using dispersion theory and experimental data from electron-positron annihilation into hadrons, and hadronic τ-decays. A detailed discussion of the theory of a_μ is provided in another contribution to these proceedings [3].

2 Experimental Technique

Experiment E821 at Brookhaven Alternating Gradient Synchrotron (AGS) is designed to measure a_μ to 0.35 ppm. The general method is similar to that

Table 1: Standard Model prediction for a_μ.

Contribution	Value, $\times 10^{-10}$	Relative Size, ppm
a_μ^{QED}	11 658 470.56(0.29)	$\pm$ 0.025
a_μ^{Had}	673.9(6.7)	57.80 $\pm$ 0.57
a_μ^{Weak}	15.1(0.4)	1.30 $\pm$ 0.03
Total a_μ^{SM}	11 659 159.6(6.7)	$\pm$ 0.57

of the last CERN experiment [4], with a number of improvements. Polarized muons from pion decay, $\pi^+ \to \mu^+ \nu_\mu$, are stored in a highly homogeneous dipole magnetic field, with the necessary vertical focusing provided by electrostatic quadrupoles. The difference between the spin precession frequency ω_s, and the cyclotron frequency ω_c is given by:

$$\vec{\omega_a} = \vec{\omega_s} - \vec{\omega_c} = -\frac{e}{mc}[a_\mu \vec{B} - (a_\mu - \frac{1}{\gamma^2 - 1})\vec{\beta} \times \vec{E}], \tag{2}$$

The dependence of ω_a on the electric field can be eliminated by storing muons with the "magic" $\gamma = 29.30$, corresponding to a muon momentum $p_\mu = 3.094$ GeV/c. The anomalous magnetic moment is then determined as:

$$a_\mu = \frac{\omega_a/\omega_p}{\lambda - \omega_a/\omega_p}, \tag{3}$$

where ω_p is the free proton precession frequency in the same magnetic field seen by the muons, and the ratio of muon to proton magnetic moments $\lambda = \mu_\mu/\mu_p = 3.183\ 345\ 39(10)$ [5].

The pions are produced in the collision of the proton beam from the AGS, delivered in 6 separate 50-ns-wide bunches over the 2.5 s cycle, with a nickel target. Most of the pions decay in the secondary beam line, and the highly polarized decay muons with the momentum close to the "magic" value are selected and injected into the storage ring. On the average, of the order of 5×10^4 muons are stored per AGS cycle.

The storage ring is a superferric magnet [6], 14 m in diameter, providing a 1.45-T magnetic field. The muons are injected into the storage through a superconducting inflector [7], and are placed onto stable orbits by means of a magnetic kick. The magnetic field is monitored continuously by some 380 NMR probes [8] embedded in the top and bottom walls of the storage ring. At

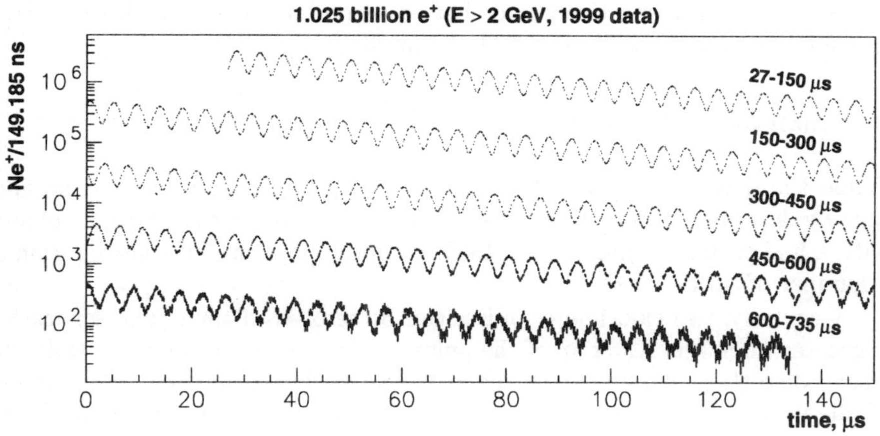

Figure 1: Time distribution of decay positrons (number of decays per cyclotron period). The muon lifetime in the laboratory is close to 64.4 μs due to time dilation. The events from the time range 32-600 μs after the injection were used in the analysis.

regular time intervals, typically twice a week, the field throughout the storage region is mapped using 17 NMR probes mounted on a trolley. The probes in the trolley are calibrated against a standard probe [9]. The magnetic field is then averaged over the azimuth and weighted with the muon distribution, to give the measurement of the field (ω_p) seen by the stored muons.

Due to parity violation in the weak decay, $\mu^+ \to e^+\overline{\nu_\mu}\nu_e$, in the muon's rest frame, the positrons are emitted preferentially along the muon spin direction. In the laboratory frame, this translates into a time-dependent variation of the number of high-energy positrons, observed in 24 calorimeter detectors [10] placed symmetrically around the inside of the ring (Fig. 1). The calorimeter signals are sampled by waveform digitizers, and then fitted to an average pulse shape to obtain the arrival time and the energy of the positrons. The data are collected for about 750 μs following the beam injection. The time spectrum of decay positrons is:

$$n(E,t) = N_0(E) \cdot e^{-t/\gamma\tau} \cdot [1 + A(E) \cdot \cos(\omega_a t + \phi(E))], \tag{4}$$

where the normalization constant N_0, decay asymmetry A, and the phase ϕ depend on the energy threshold used in the selection of positrons; and τ is the muon lifetime.

The experiment started taking data in 1997. The first two results [11,12], based on the data taken during the runs in 1997 and 1998, had precisions of 13 and 5 ppm respectively and showed a good agreement with the CERN measurements and the Standard Model prediction.

3 Analysis of the 1999 Data

While the decay positron data collected during our 1998 run could be adequately described by Eq. 4, the 15-fold increase in statistical power, combined with a higher data taking rate in 1999, required more careful consideration of a number of subtle effects [13].

Generally, two calorimeter pulses can be resolved if their time separation is greater than 3 ns. Overlap of the pulses within the resolution time (pile-up) causes distortion in the reconstructed time and energy spectra. The number of pile-up pulses is proportional to the instantaneous decay rate squared and to the minimum separation time of the reconstruction algorithm. Since both the phase and the asymmetry of pulses depend on the pulse energy, the pile-up phase ϕ_p and asymmetry A_p are different from ϕ and A in Eq. 4. Pile-up can be described by a function of the form:

$$p(t) = N_0 e^{-2t/\gamma\tau} \cdot [n_p + A_p \cdot \cos(\omega_a t + \phi_p)] \cdot [1 + a_p e^{-\frac{1}{2}(t/\tau_p)^2}], \qquad (5)$$

where N_0, τ and ω_a are same as in Eq. 4, n_p is the fraction of pile-up pulses with respect to the normalization constant N_0; a_p and τ_p are respectively the amplitude and the characteristic decay time of the pile-up enhancement at early times due to the bunched structure of the beam.

A new technique, "pile-up subtraction" [13], allowed for extensive studies of this phenomenon. In the "subtraction", the pile-up pulses are constructed from well-resolved pairs of pulses separated by an offset time of about 10 ns. The time spectrum obtained in such manner can be fitted to get a better estimate of the pile-up parameters; or it can be subtracted from the data set altogether, to give an approximation of a "pile-up-free" sample. The pile-up phase ϕ_p is highly correlated to the frequency ω_a, and was fixed in the fit, after studies (as well as the pile-up enhancement parameters, a_p and τ_p).

A combination of a non-ideal kick, and the storage region phase space not being filled out uniformly, causes betatron oscillation of the average radial position of the stored muon distribution. This was observed in the data from scintillator fiber detectors, which can be moved into the storage region to measure the beam profile. As the radial acceptances of the calorimeter detectors are not flat, such coherent betatron oscillation (CBO) affects the observed time distribution of decay positrons. This could be easily seen in the residuals of the fit with the 5-parameter function given by Eq. 4. After extensive studies we arrived at the following form of the CBO correction to the decay rate:

$$b(t) = 1 + A_B \cdot \cos(\omega_B t + \phi_B) \cdot e^{-(t/\tau_B)^2}. \tag{6}$$

In Eq. 6, A_B, ω_B, ϕ_B and τ_B refer to the amplitude, frequency, phase and characteristic decay time of CBO, generally different from the parameters in Eq. 4.

The muon losses cause the decay rate to decrease with time. In the experiment, we can study lost muons using coincident signals in two or more adjacent scintillator detectors situated in front of the calorimeters. We found the necessary correction to the decay rate of the form:

$$l(t) = 1 + a_{\mu L} \cdot e^{-\frac{1}{2}(t/\tau_{\mu L})^2}, \tag{7}$$

with $a_{\mu L}$ and $\tau_{\mu L}$, respectively the muon loss amplitude and decay time.

The data from 22 detectors were fitted, separately and combined, to

$$f(t) = [n(t) + p(t)] \cdot b(t) \cdot l(t), \tag{8}$$

where $n(t)$, $p(t)$, $b(t)$, $l(t)$ are described respectively by Eqs. 4 - 7. Three out of 16 parameters in Eq. 8 were fixed in the fit: the pile-up phase ϕ_p, and the parameters describing pile-up enhancement due to beam bunching, a_p and τ_p. The results obtained by using this function are shown in Fig. 2.

Four independent analyses of ω_a, slightly differing in the data selection and fitting methods, and the form of fit function, were performed [13]. The results were found to agree within 0.3 ppm, well within the expected statistical variation of 0.4 ppm, due to the differences in the analyzed data sets. Combining the results of the four analyses, and including the correction for the electric field and the betatron pitching motion (+0.81 ppm), we found $\omega_a/2\pi$ = 229 072.8 ± 0.3 Hz (1.3 ppm), where the error is dominately statistical.

Two independent analyses of the magnetic field [13], using different selections of about 150 fixed NMR probes each, agreed to within 0.03 ppm. The value of the field $<B>$ averaged over the muon distribution, in terms of the free NMR frequency of the proton in water, was found to be $\omega_p/2\pi$ = 61 791 256 ± 25 Hz (0.4 ppm). The leading contributions to the associated systematic error came from uncertainties in the calibration of the trolley NMR probes, and inhomogeneity due to the inflector fringe field.

Combining the results for ω_a and ω_p, using Eq. 3, brings the new experimental value:

$$a_\mu = 11\ 659\ 202(16) \times 10^{-10}\ (1.3\ \text{ppm}). \tag{9}$$

The leading systematic errors for the analysis of the 1999 data are listed in Table 2. The new result is in good agreement with previous measurements.

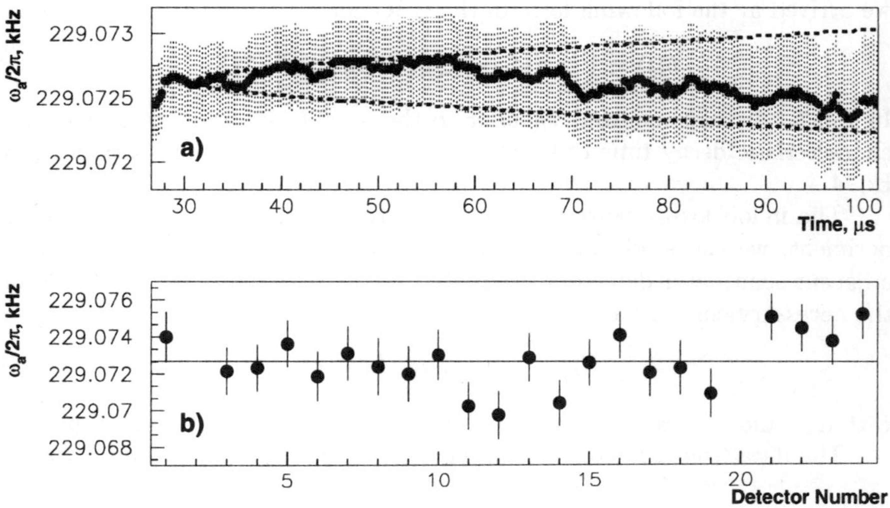

Figure 2: a) The fitted frequency $\omega_a/2\pi$ vs. start time of the fit. The shaded area represents a one standard deviation statistical error band. The broken line shows the size of expected statistical fluctuations with respect to the result at 32 μs. b) Results of the fit to the spectra from individual detectors. Detectors 2 and 20 were not used in the analysis due to hardware problems.

The difference between the new experimental average, $a_\mu^{exp} = 11\ 659\ 203(16)$ $\times 10^{-10}$, and the theoretical prediction [2] (Table 1) is

$$a_\mu^{exp} - a_\mu^{SM} = 43(16) \times 10^{-10} \ (1.3 \text{ ppm}), \tag{10}$$

where the uncertainties are added in quadrature.

4 Future Plans

During our run in 2000, approximately 4.7 billion high-energy decay positrons ($E > 2$ GeV, after 32 μs) were collected, almost 5 times as many as before. The analysis of these data has begun. In addition to increased statistical power, the systematic errors on both $<B>$ and ω_a measurements are expected to decrease. The magnetic field uniformity has improved significantly after a new inflector was installed. AGS background and muon losses were substantially lower than in 1999 due to improvements made in the beam line and storage technique respectively.

Table 2: Leading systematic errors for the a_μ measurement.

Source of errors	Size, ppm	
Calibration of trolley probes	0.20	
Inflector fringe field	0.20	
Interpolation with fixed probes	0.15	
Uncertainty from muon distribution	0.12	
Trolley measurements	0.10	
Absolute calibration of standard probe	0.05	
Total systematic error on ω_p		0.4
Pile-up	0.13	
AGS background	0.10	
Muon losses	0.10	
Timing shifts	0.10	
E field and pitch correction	0.08	
Binning and fitting procedure	0.07	
Coherent betatron oscillation	0.05	
Total systematic error on ω_a		0.3
Total		0.5

In February-April 2001 we collected data with negative muons. The data from this run, combined with that from another one planned for 2002, will bring the statistical error on a_{μ^-} to the level reached for a_{μ^+}. Measuring the magnetic anomaly for both the positive and the negative muon will provide a sensitive test of CPT-invariance.

At the same time we expect a reduction in the uncertainty on the theoretical prediction, after new experimental results [14] are included in the calculation of the hadronic contribution.

Acknowledgments

I would like to thank everyone in the $(g-2)_\mu$ collaboration whose hard work made this result possible. I am very grateful to the organizers of the Winter Institute for giving me a chance to participate in the meeting, and for their generous financial support. Support from the National Science Foundation grant is also very much appreciated.

References

1. The (g-2)$_\mu$ Collaboration for 1999: H.N. Brown[b], G. Bunce[b], R.M. Carey[a], P. Cushman[i], G.T. Danby[b], P.T. Debevec[g], M. Deile[l], H. Deng[l], S.K. Dhawan[l], V.P. Druzhinin[c], L. Duong[i], E. Efstathiadis[a], F.J.M. Farley[l], G.V. Fedotovich[c], S. Giron[i], F. Gray[g], D. Grigoriev[c], M. Grosse-Perdekamp[j], A. Grossmann[f], M.F. Hare[a], D.W. Hertzog[g], V.W. Hughes[l], M. Iwasaki[k], K. Jungmann[f], D. Kawall[l], M. Kawamura[k], B.I. Khazin[c], J. Kindem[i], F. Krienen[a], I. Kronkvist[i], R. Larsen[b], Y.Y. Lee[b], I. Logashenko[a,c], R. McNabb[i], W. Meng[b], J. Mi[b], J.P. Miller[a], W.M. Morse[b], D. Nikas[b], C.J.G. Onderwater[g], Y. Orlov[d], C. Özben[b], J.M. Paley[a], C. Polly[g], J. Pretz[l], R. Prigl[b], G. zu Putlitz[f], S.I. Redin[l], O. Rind[a], B.L. Roberts[a], N. Ryskulov[c], Y.K. Semertzidis[b], Yu.M. Shatunov[c], E. Sichtermann[l], E. Solodov[c], M. Sossong[g], A. Steinmetz[l], L.R. Sulak[a], C. Timmermans[i], A. Trofimov[a], D. Urner[g], P. von Walter[f], D. Warburton[b], D. Winn[e], A. Yamamoto[h], D. Zimmerman[i]
   [a] *Boston U.*, [b] *BNL*, [c] *Budker Institute of Nuclear Physics (Russia)*, [d] *Cornell U.*, [e] *Fairfield U.*, [f] *Universität Heidelberg (Germany)*, [g] *U. of Illinois*, [h] *KEK (Japan)*, [i] *U. of Minnesota*, [j] *RIKEN-BNL Research Center*, [k] *Tokyo Institute of Technology (Japan)*, [l] *Yale U.*
2. A. Czarnecki, W.J. Marciano, *Nucl. Phys. (Proc. Suppl.)* **B76**, 245 (1999).
3. W.J. Marciano, these proceedings.
4. J. Bailey *et al.*, *Nucl. Phys.* **B150**, 1 (1979).
5. W. Liu *et al.*, *Phys. Rev. Lett.* **82**, 711 (1999).
6. G.D. Danby *et al.*, *Nucl. Instrum. Meth.* **A457**, 151 (2001).
7. F. Krienen *et al.*, *Nucl. Instrum. Meth.* **A283**, 5 (1989).
   A. Yamamoto *et al.*, Proc. of 15th Int. Conf. on Magnet Technology, Science Press Beijing, 246 (1998).
8. R. Prigl *et al.*, *Nucl. Instrum. Meth.* **A374**, 118 (1996).
9. X. Fei *et al.*, *Nucl. Instrum. Meth.* **A394**, 3499 (1997).
10. S. Sedykh *et al.*, *Nucl. Instrum. Meth.* **A455**, 346 (2000).
11. R.M. Carey *et al.*, *Phys. Rev. Lett.* **82**, 1632 (1999).
12. H.N. Brown *et al.*, *Phys. Rev.* **D62**, 091101 (2000).
13. H.N. Brown *et al.*, *Phys. Rev. Lett.* **86**, 2227 (2001).
14. R.R. Akhmetshin *et al.*, *Phys. Lett.* **B475**, 190 (2000).
    S. Anderson *et al.*, *Phys. Rev.* **D61**, 112002 (2000).
    J.Z. Bai *et al.*, *Phys. Rev. Lett.* **84**, 594 (2000).
    Z.G. Zhao, *Int. J. Mod. Phys.* **A15**, 3739 (2000).

PHYSICAL INTERPRETATION OF STRING THEORIES AND THE ORIGIN OF CHARGE

F. WINTERBERG

University of Nevada, Reno
Department of Physics, MS / 220
Reno, NV, 89557, USA
E-mail: winterbe@physics.unr.edu

Assuming that nature works like a computer with SU2 as the fundamental group, position space should be three dimensional with SU2 isomorph to SO3. With all of physics reduced to Plancks units then suggests that the vacuum is a positive-negative Planck mass plasma, in which each Planck length volume is occupied by one positive or negative Planck mass interacting with other Planck masses over a Planck length by the Planck force. With equal sign Planck mass particles repelling and unequal sign Planck mass particles attracting each other, the vacuum is stable, with quantum mechanics resulting from the Zitterbewegung of interacting positive with negative Planck mass particles. Charge is explained by the quantum mechanical zero point (i.e. Zitterbewegung) fluctuations of Planck mass particles bound in quantized vortices of the superfluid Planck mass plasma, whereby strings are explained as line vortices. The introduction of negative masses is possible in an exactly nonrelativistic theory where the particle number operator commutes with the Hamilton operator, but where Lorentz invariance must be explained as a dynamic symmetry as in the pre-Einstein theory of relativity by Lorentz and Poincar.

1 Introduction

The many similarities between condensed matter and high energy physics suggest the existence of an aether. In analogy to charge-neutral condensed matter it should be mass-neutral, which means that it should be composed of both positive and negative masses. A comparison of Newtons point particle with Helmholtz line vortex dynamics, and of Einsteins and Yang Mills field theories (see Table 1), shows that in Newtons mechanics one has mass times acceleration = force, whereas in Helmholtz line vortex dynamics one has mass times velocity = force. The same kind of hierarchical displacement is seen by comparing Einsteins gravitational field theory with Yang Mills theories. In Einsteins theory the curvature tensor is made up from Christoffel symbols which are derivatives of the potentials, whereas in Yang Mills theories the curvature tensor (in color space for example) is made up from the potentials. This comparison suggests that QCD is nothing more than some kind of vortex dynamics, of vortices in strongly condensed regions of a superfluid aether[1].

2 The Planck Aether Hypothesis

Because it was Planck, who in 1911 came to the conclusion that quantum mechanics implied the existence of a zero point vacuum energy, somehow replacing the 19th century aether models, and because he had already back in 1899 emphasized that all of physics should be reduced to the units named after him, I propose what I have called the Planck aether hypothesis[2]:

- The ultimate building blocks of matter are Planck mass particles which obey the laws of classical Newtonian mechanics and that there are also negative Planck mass particles.

- A positive Planck mass particle exerts a short range repulsive and a negative Planck mass particle a likewise attractive force, with the magnitude of the force equal to the Planck force and the range equal a Planck length.

- Space is filled with an equal number of positive and negative Planck mass particles whereby each Planck length volume is in the average occupied by one Planck mass particle.

Making this hypothesis I can derive:

- Nonrelativistic quantum mechanics as an approximation with departures from this approximation suppressed by the Planck length.

- Lorentz invariance as a dynamic symmetry for energies small compared to the Planck energy.

- A spectrum of quasiparticles like the particles of the standard model. The Planck mass m_p, length r_p and time $t_p = r_p/c$ are derived from $m_p r_p c = \hbar$ and $G m_p^2 = \hbar c$, where G is Newtons constant. The hypothesis permits a finitistic (non-Archimedean) formulation[3], which can be seen as an attempt to make discrete number theory part of physics, as Einstein made geometry part of physics. The two signs of the Planck mass particles are suggested by the assumption that SU2 is the fundamental symmetry of nature, with the finitistic (non-Archimedean) formulation of the theory indicating that nature works like a computer with discrete elements, excluding the noncompact Lorentz group as the fundamental kinematic symmetry.

3 Derivation of Quantum Mechanics

From the force law of the Planck aether hypothesis, Planck mass particles of equal sign repel and those of unequal sign attract each other, with a negative Planck mass particle acting like a hole of depth $m_p c^2$ on a positive Planck mass particle and vice versa. During the collision of a positive with a negative Planck mass particle the kinetic energy remains constant, but the momentum fluctuates. The momentum fluctuation is equal to $\Delta p = F_p t_p = m_p c$, where $F_p = m_p c^2 / r_p$ is the Planck force and $t_p = r_p / c$ the collision time for a Planck mass particle reaching the velocity of light by falling into the potential $m_p c^2$. But since $m_p r_p c = \hbar$ it follows that $\Delta p = \hbar / r_p$ as it would be expected from Heisenbergs uncertainty relation. The basic quantum mechanical uncertainty is thus explained to result from the interaction with a sea of hidden negative masses, with the recoil momentum absorbed by the Planck aether. It is therefore of no surprise that the Schrdinger equation for the Planck mass particles can be derived solving the Boltzmann equation for the Planck aether. By taking the first two moments of this equation, one obtains the law of mass conservation (continuity) and Eulers equation with a quantum potential, which by the Madelung back transformation lead to the Schrdinger equation of a positive and negative Planck mass particle. Approximating the short range (over r_p) interaction between the Planck mass particles by a delta function one can write down a nonrelativistic operator field equation for the positive and negative mass component of the Planck aether:

$$i\hbar \frac{\partial \Psi_\pm}{\partial t} = \mp \frac{\hbar^2}{2m_p} \nabla^2 \Psi_{pm} \pm (2\hbar c r_p^2 (\Psi_{pm}^\dagger \Psi_\pm - \Psi_\mp^\dagger \Psi_\mp) \Psi_\pm \tag{1}$$

where the operators $\Psi_\pm, \Psi_\pm^\dagger$ have to obey the canonical commutation relations

$$[\Psi_\pm(\underline{r}) \Psi_\pm^\dagger(\underline{r\prime})] = \delta(\underline{r} - \underline{r\prime}), [\Psi_\pm(\underline{r}) \Psi_\pm(\underline{r\prime})] = [\psi_\pm^\dagger(\underline{r}) \Psi_\pm^\dagger(\underline{r\prime})] = 0 \tag{2}$$

Equations (1) and (2) can be used as a starting point to make calculations, in particular with the Hartree and Hartree Fock approximation. Equation (1) has the form of a nonlinear nonrelativistic Heisenberg type unified field theory equation, with the important difference that it contains a mass, the absence of which in Heisenbergs nonlinear relativistic spinor equation led to insurmountable problems.

4 Lorentz Invariance

Negative masses are possible only in an exactly nonrelativistic theory where the Hamilton operator commutes with the particle number operator, keeping the number of particles constant. There, then, Lorentz invariance must be explained as a dynamic symmetry as in the pre-Einstein theory of relativity by Lorentz and Poincar, with true deformations of rods and slower going clocks in absolute motion against an aether.

Lorentz invariance is in the present theory derived as follows:

- With the Hartree-Fock approximation of (1) one finds that for wave lengths large to the Planck length the Planck aether has compressional waves propagating with the velocity of light mediating attractive forces like the phonons in the theory of superconductivity.
- In its groundstate the superfluid Planck aether has an energy spectrum of the form
$$f(\omega) = const.\omega^3 \tag{3}$$
as for the experimentally established phonon energy spectrum of superfluid helium.
- The energy spectrum (3) is the only one which is Lorentz invariant, but it also has the form of the quantum mechanical zero point energy spectrum which causes the repulsive quantum force.
- With both the wave equation for the compressional waves and the zero point energy Lorentz invariant, Lorentz invariance is established as a dynamic symmetry for energies small compared to the Planck energy.
- The same reasoning applies to other wave modes propagating with the velocity of light.

5 Vortex Solutions

With the Hartree-Fock approximation of (1) one obtains two kinds of vortex solutions, one where the positive and negative Planck mass particles co-rotate, and one where they counter-rotate. The Planck aether hypothesis therefore leads on a very fundamental level to the doubling of states as it was emphasized by Heisenberg as a necessary condition for any theory of this kind, to explain the two types of fermions (electrons and neutrinos). As in a superfluid, the vortices are quantized with the vortex core radius equal the Planck length, where the velocity reaches the velocity of light. It is these vortex filaments which give a physical explanation of strings. Like one dimensional mathematical lines strings can have no connection to physical reality.

6 Origin of Charge

If bound in vortex filaments of radius r_p, the Planck mass particles execute quantum mechanical fluctuations with a kinetic energy density

$$\epsilon_k \sim m_p c^2 / r_p^3 \tag{4}$$

We compare this energy density with the magnitude of the gravitational energy density

$$\epsilon_g = g^2, g = \sqrt{G} m_p / r_p^2 \tag{5}$$

a Planck mass particle would have as the source of a Newtonian gravitational field at the distance r_p. With $Gm_p^2 = \hbar c$ and $m_p r_p c^2 = h$ one finds that $\epsilon_k \approx \epsilon_g$. With the zero point fluctuations acting as the source of a virtual phonon field mediating an inverse square law attractive force, gravitational charge is explained by the zero point fluctuations of Planck mass particles bound in vortex filaments. With all interactions becoming equal at the Planck length, all other charges are explained likewise.

7 Vortex lattice waves

Because there are equal numbers of positive and negative Planck mass particles, the Planck aether can by spontaneous symmetry breaking without expenditure of energy be transformed into a lattice of closed vortex rings. Hydrodynamic stability, as in the Karman vortex street, suggests that the ring radius and separation of rings must be $\sim 10^3$ times larger than the vortex core radius. Such a vortex lattice can propagate two kinds of waves, one antisymmetric (already discovered by Kelvin in 1887) and one symmetric. The antisymmetric mode can describe electromagnetic and the symmetric one gravitational waves. The vortex lattice therefore unifies Maxwells and Einsteins field equations.

8 Dirac Spinors

Because there are negative besides positive masses the Planck aether hypothesis can explain Dirac spinors. In the simplest way a Dirac spinor can be understood as a pole-dipole particle, that is a mass dipole with a superimposed mass monopole. In the Planck aether the mass dipole is explained as the $\pm 10^{12}$ GeV resonance energy of the vortex rings, with the mass pole coming from the small positive gravitational interaction energy of a positive with a negative mass. The electron masses come from the co-rotating and the neutrino masses from the counter-rotating vortex solution. The much smaller gravitational interaction in the counter-rotating solution explains the smallness of the neutrino masses.

9 Concluding Remark

No theory is very credible without some kind of empirical evidence. I believe that there is at least some evidence: It is what has been named quintessence. As it now appears, about 70% of the matter in the universe is dark energy, 26.5% nonbaryonic dark matter, with a mere 3.5% baryonic. Compare this with the phonon-roton spectrum of superfluid helium where the same division in parts exists between the roton energy gap and the roton kinetic energy[4]. Since rotons are a kind of cavitons, the roton fluid has a negative pressure, as for the cosmic quintessence. This could be, of course, just a coincidence.

Table 1 Hierarchical Displacement of Einstein and Yang-Mills Field Theories as Related to the Hierarchical Displacement of Newton Point-Particle and Helmholtz Line-Vortex Dynamics.

Newton point-particle dynamics and Einstein's gravitational field theory		Kinematic quantities	Helmholtz line-vortex dynamics and Yang-Mills field theories	
		r	ψ	Velocity potential
			f	Gauge functions
Newtonian potential	φ	$\dot{r}$	$-\nabla\psi$	Force on line vortex
Metric tensor	g_{ik}		A	Gauge potentials
Force on point particle Gravitational force field expressed by Christoffel symbols	$-\nabla\varphi$ Γ	$\ddot{r}$	$W = \text{Curl} A - g^{-1} A \otimes A$	Yang-Mills force field expressed by charge-space curvature tensor
Einstein's filed equations expressed by metric-space curvature tensor	$R = \text{Curl}\Gamma + \Gamma\otimes\Gamma$			

References

1. F. Winterberg, *International Journal of Theoretical Physics* **34**, 265 (1995).
2. F. Winterberg, *Z.f. Naturforsch* **52a**, 183 (1997).
3. F. Winterberg, *International Journal of Theoretical Physics* **32**, 261 (1993).
4. F. Winterberg, *International Journal of Theoretical Physics* **34**, 399 (1995).

Lake Louise Winter Institute - 2001

LIST OF PARTICIPANTS

S. Adler	Brookhaven National Laboratory
J.P. Archambault	University of Alberta
D. Ashery	TRIUMF
P. Azzi	INFN Padova
G. Blair	University of London
I. Blokland	University of Alberta
J. Brooke	University of Bristol
N. Buchanan	University of Alberta
B. Campbell	University of Alberta
G. Collazuol	INFN-Firenze
S. Cucciarelli	Perugia, Italy
A. Czarnecki	University of Alberta
W. Decking	DESY - TESLA
N. De Filippis	INFN Bari
P. Deglon	University of Geneva
N. Delerue	Marseille, France
E. Elfgren	Universite de Montreal
F. Ellinghaus	DESY - Zeuthen
L. Ewell	Brookhaven National Laboratory
A. Frolov	CITA, University of Toronto
M. Fujiwara	University of Tokyo/CERN
E. Gallas	FERMILAB
D. Haas	University of Basel
R. Hall-Wilton	DESY - York Group
J. Hansen	University of Oslo
T. Hara	Osaka University
J. Hardy	Carleton University
R. Hemingway	Carleton University
P. Jackson	TRIUMF
E. Jankowski	University of Alberta
K. Kaminsky	University of Alberta
D. Karlen	Carleton University
J. Kennedy	The University of Glasgow

F.C. Khanna	University of Alberta
K. Klein	Heidelberg, Germany
M. Kos	Queen's University
M. Kovash	University of Kentucky
T. Kuhl	University of Bonn
D. Lange	SLAC
D. Leahy	University of Calgary
M. Lefebvre	University of Victoria
W. Marciano	Brookhaven National Laboratory
D. Maybury	University of Alberta
T. Meyer	Cornell University
G. Moloney	University of Melbourne
T. Nakaya	Kyoto University
J. Panetta	University of Pennsylvania
N. Parashar	FERMILAB/ Northeastern University
V. Pavlunin	Purdue University
V. Poireau	CEA Saclay, France
J.M. Poutissou	TRIUMF
J. Rehn	University of Karlsruhe
H. Robertson	University of Washington
N. Rodning	University of Alberta
G. Ross	Oxford University
K. Sachs	CERN
T. Schietinger	SLAC
G. Semenoff	University of British Columbia
S. Sengupta	University of Alberta
D. Shaw	University of Alberta
J. Sjolin	Stockholm Universitet, Sweden
G. Sprouse	State University of New York
T. Stanescu	University of Alberta
A. Tapper	Imperial College, London
G. Thayer	Cornell University
B. Trocme	LAPP, France
A. Trofimov	Boston University
M. Vincter	University of Alberta
D. Waller	Carleton University

R. White	Imperial College of Science, London
F. Winterberg	University of Nevada
M. Wobisch	DESY